TP245/c5/m6

AF713245

PLEASE
RETURN TO
EXXON
LINDEN
INFORMATION
CENTER

MODERN CHLOR-ALKALI TECHNOLOGY
Volume 2

Editor:
C. JACKSON
Electrochemical Technology Group of
the Society of Chemical Industry
and I.C.I. Limited, Mond Division
Northwich, Cheshire

Published for the
SOCIETY OF CHEMICAL INDUSTRY
London by

ELLIS HORWOOD LIMITED
Publishers · Chichester

First published in 1983 by

ELLIS HORWOOD LIMITED
Market Cross House, Cooper Street, Chichester, West Sussex, PO19 1EB, England

The publisher's colophon is reproduced from James Gillison's drawing of the ancient Market Cross, Chichester.

Distributors:

Australia, New Zealand, South-east Asia:
Jacaranda-Wiley Ltd., Jacaranda Press,
JOHN WILEY & SONS INC.,
G.P.O. Box 859, Brisbane, Queensland 40001, Australia

Canada:
JOHN WILEY & SONS CANADA LIMITED
22 Worcester Road, Rexdale, Ontario, Canada.

Europe, Africa:
JOHN WILEY & SONS LIMITED
Baffins Lane, Chichester, West Sussex, England.

North and South America and the rest of the world:
Halsted Press: a division of
JOHN WILEY & SONS
605 Third Avenue, New York, N.Y. 10016, U.S.A.

© **1983 Society of Chemical Industry/Ellis Horwood Limited**

British Library Cataloguing in Publication Data
Modern chlor-alkali technology
Vol. 2
1. Sodium hydroxide
I. Jackson, C.
661'.322 TP245.S/

Library of Congress Card No. 81-131882

ISBN 0-85312-525-2 (Ellis Horwood Ltd., Publishers)
ISBN 0-470-27471-9 (Halsted Press)

Typeset in Press Roman by Ellis Horwood Ltd.
Printed in Great Britain by R. J. Acford, Chichester.

COPYRIGHT NOTICE –

All Rights Reserved. No part of this publication may be reproduced, stored in a retrieval system, or transmitted, in any form or by any means, electronic, mechanical, photocopying, recording or otherwise, without the permission of Ellis Horwood Limited, Market Cross House, Cooper Street, Chichester, West Sussex, England.

Table of Contents

Introduction – *C. Jackson, ICI (Mond Division) plc* 15

SECTION A – THE DEVELOPMENT OF DIAPHRAGMS AND DIAPHRAGM ELECTROLYZERS

Chapter 1 – CHARACTERIZATION OF ASBESTOS DIAPHRAGMS FOR CHLOR-ALKALI ELECTROLYSIS
K. A. Poush and D. L. Caldwell, Dow Chemical;
J. W. Van Zee and R. E. White, Texas A & M University

1.1 Introduction 21
1.2 Diaphragm Characterization 22
1.3 Experimental Determination of MacMullin Number and Thickness 22
1.4 Diaphragm Experimental Study 24
1.4.1 Results 26
1.4.2 Comparison with theoretical model 26
1.5 Discussion and Conclusions 30
1.6 References 30
1.7 Open Discussion 30

Chapter 2 – PPG BIPOLAR DIAPHRAGM ELECTROLYZER – NEW DESIGN FOR LOW POWER CONSUMPTION
R. J. Scott, PPG Industries Inc.

2.1 Introduction 33
2.2 Model V-1144 Bipolar Diaphragm Electrolyzer 33
2.3 Improved Design – Prototype Electrolyzer 34
2.4 Model V-1161 Electrolyzer Commercialization 35
2.5 Design and Operating Characteristics – Model V-1161 Versus Model V-1144 36
2.6 Discussion 37

Chapter 3 – SYNTHETIC DIAPHRAGMS FOR CHLORINE CELLS
J. J. Bowser, W. L. Gore & Associates Inc.

3.1 Introduction . . . 38
3.2 The Advantages of a PTFE Diaphragm . . . 38
3.2.1 Reduced cell voltage . . . 39
3.2.2 Good physical uniformity . . . 39
3.2.3 Improved chemical stability . . . 39
3.2.4 Extended lifetime . . . 39
3.2.5 Asbestos free . . . 39
3.3 Alternative PTFE Synthetic Diaphragm . . . 39
3.3.1 Deposited fibrous fluoropolymers . . . 40
3.3.2 Calendered PTFE . . . 40
3.3.3 PTFE fabrics and felts . . . 40
3.3.4 Expanded PTFE . . . 40
3.4 The Gore-Tex® Diaphragm Manufacturing Process . . . 40
3.5 Properties of the Gore-Tex® Diaphragm . . . 42
3.6 Gore-Tex® Diaphragm Cell Performance . . . 44
3.7 Cladding of Cathode Boxes with Gore-Tex® Diaphragm . . . 47
3.8 The Present Development Position . . . 49
3.9 Acknowledgement . . . 49

SECTION B – THE DEVELOPMENT OF MEMBRANES AND MEMBRANE ELECTROLYZERS

Chapter 4 – HIGH PERFORMANCE MEMBRANES FOR THE CHLOR-ALKALI INDUSTRY
S. M. Ibrahim, E. H. Price, R. A. Smith, E. I. DuPont de Nemours & Co.

4.1 Introduction . . . 53
4.2 Nafion® Membrane Products . . . 53
4.3 Nafion® 901 . . . 54
4.3.1 Current efficiency . . . 54
4.3.2 Voltage . . . 55
4.3.3 Power consumption . . . 55
4.3.4 Performance stability . . . 55
4.3.5 Membrane life . . . 56
4.3.6 Durability . . . 56
4.3.7 Current density . . . 58
4.3.8 Temperature . . . 58
4.3.9 Caustic and chlorine quality . . . 58
4.3.10 Convenience . . . 59
4.3.11 Availability . . . 59
4.3.12 Supply dependability . . . 59
4.4 Summary . . . 59
4.5 Discussion . . . 60

Chapter 5 – AZEC SYSTEM – NEW ION-EXCHANGE MEMBRANE CHLOR-ALKALI PROCESS
M. Nagamura, H. Ukihashi and O. Shiragami, Asahi Glass Co. Ltd.

5.1 Introduction . . . 61
5.2 The General Aspect of Flemion . . . 62
5.3 AZEC System . . . 64
5.3.1 Flemion DX . . . 64
5.3.2 Electrode system . . . 68
5.3.3 Electrolyzer . . . 70
5.4 Conclusion . . . 74
5.5 References . . . 75
5.6 Discussion . . . 75

Chapter 6 – ASAHI CHEMICAL'S PERFLUOROCARBOXYLIC ACID MEMBRANE
M. Seko, H. Miyauchi, J. Ohmura and K. Kimoto, Asahi Chemical Industry Co. Ltd.

6.1 Introduction . . . 76
6.2 Membrane Containing Carboxylic Acid . . . 77
6.3 Ion-exchange Groups and Membrane Structure . . . 81
6.4 Multilayer Membrane Prepared by Chemical Treatment . . . 82
6.5 Thickness of Carboxylic Acid Layer, Current Efficiency and Membrane Resistance . . . 83
6.6 Carboxylic Acid Formation . . . 84
6.7 Monomer Structure . . . 85
6.8 Hydrogen Gas Adhesion on the Surface of the Membrane . . . 89
6.9 Typical Asahi Chemical Membrane . . . 90
6.10 Power Consumption of the Membrane Chlor-Alkali Process . . . 92
6.11 New Plant for Manufacture of Ion-exchange Membranes for Chlor-Alkali Production . . . 92
6.12 References . . . 94

Chapter 7 – DEVELOPMENT OF ASAHI CHEMICAL CHLOR-ALKALI TECHNOLOGY
M. Seko, A. Yomiyama, S. Ogawa and H. Ono, Asahi Chemical Industry Co. Ltd.

7.1 Introduction . . . 97
7.2 Characteristics of the Membrane Process . . . 99
7.2.1 Limiting currect density . . . 99
7.2.2 Dimensional stability of membrane, electrode and cell . . . 101
7.2.3 Neutralization of brine by hydrochloric acid . . . 101

7.2.4 Optimum concentration of caustic soda . . . 103
7.2.5 Operating pressure . . . 103
7.3 Basic Design Concept of Asahi Chemical's Electrolyzer . . . 104
7.3.1 High uniform current density . . . 104
7.3.2 Low energy consumption at high current density . . . 105
7.3.3 High capacity individual cell . . . 105
7.3.4 Bipolar and monopolar cells . . . 105
7.3.5 Corrosion resistant cell materials . . . 110
7.3.6 Dimensional stability . . . 111
7.3.7 Long anode service life . . . 111
7.4 Experience with Other Types of Electrolyzers . . . 111
7.4.1 Rubber cell bodies . . . 111
7.4.2 Plastic cell materials . . . 111
7.4.3 Plate-type cells . . . 112
7.5 Recent Improvements in Commercial Electrolyzers . . . 112
7.5.1 Membranes . . . 112
7.5.2 Rough surface membranes . . . 112
7.5.3 Electrodes . . . 113
7.5.4 Electrolyzers . . . 116
7.5.5 Power consumption . . . 118
7.6 References . . . 118
7.7 Discussion . . . 120

Chapter 8 – OLIN MEMBRANE CELL TECHNOLOGY
P. A. Danna and K. E. Woodard, Olin Corporation

8.1 Program Scope . . . 121
8.2 Component Development and Evaluation . . . 122
8.2.1 Anode coatings . . . 122
8.2.2 Membranes . . . 123
8.2.3 Cathode coatings . . . 126
8.3 Brine Quality . . . 128
8.4 Cell Startup . . . 128
8.5 Engineering Packages and Development . . . 128
8.6 Reference . . . 131

Chapter 9 – UHDE'S MEMBRANE CELL TECHNOLOGY
H. Isfort, Uhde GmbH

9.1 Introduction . . . 132
9.2 First Membrane-cell Generation (1975 to 1979) . . . 133
9.3 Second Membrane-cell Generation (1979 to 1981) . . . 139
9.4 Third Membrane-cell Generation (Since 1981) . . . 151

Chapter 10 – ENERGY SAVING ADVANCES IN THE CHLOR-ALKALI INDUSTRY
V. H. Thomas and E. J. Rudd, Eltech Systems Corporation

10.1 Introduction 159
10.2 Energy Requirements in Chlor-alkali Processes 160
10.3 Cell Voltage 164
10.3.1 Diaphragm cell process 166
10.3.2 Membrane cell process 166
10.3.3 Oxygen cathodes 168
10.4 Future Goals 170
10.5 Discussion 170

SECTION C – APPLICATION OF MEMBRANE CELL TECHNOLOGY

Chapter 11 – CRITERIA FOR THE SELECTION OF MEMBRANE CELL TECHNOLOGY
E. Pearson, Robert Spence Associates

11.1 Summary 177
11.2 Background 177
11.3 System Definition 178
11.4 Technical Criteria 180
11.4.1 Membranes 180
11.4.2 Raw materials 180
11.4.3 Products and byproducts 182
11.4.4 Operating characteristics 183
11.4.5 Maintenance 183
11.4.6 Technology trajectory 184
11.5 Commercial Considerations 184
11.5.1 Engineering capability 184
11.5.2 Licence terms and conditions 184
11.5.3 Operational support capability 185
11.5.4 Licensor background 185
11.6 Evaluation 185
11.7 References 187
11.8 Discussion 188

Chapter 12 – CONSIDERATIONS IN CONVERSION OF EXISTING CHLOR-ALKALI PLANTS TO MEMBRANE CELL OPERATION
T. F. O'Brien, Catalytic Incorporated

12.1 Introduction 190
12.2 Overview 191

12.3 Brine Processing 192
12.3.1 Brine treatment and resaturation 192
12.3.2 Secondary brine treatment 194
12.3.3 Secondary dechlorination 195
12.4 Chlorine Processing 197
12.4.1 Chlorine cooling 198
12.4.2 Chlorine drying 199
12.4.3 Acid handling 199
12.4.4 Compression 200
12.4.5 Liquefaction 200
12.4.6 Tail gas recovery 200
12.4.7 Other systems 201
12.5 Caustic Processing 201
12.5.1 Caustic recycle 201
12.5.2 Caustic evaporation 202
12.5.3 Product storage and handling 203
12.6 Cell Room 203
12.7 Partial Conversions 204
12.8 Economics 206
12.9 Summary 210
12.10 Acknowledgement 211
12.11 Discussion 212

SECTION D – ELECTRODE COATINGS

Chapter 13 – DEVELOPMENTS IN COATINGS FOR METAL ANODES
D. A. Denton, J. T. DeSouza, J. H. Entwisle, D. Lee and H. G. Wilson, ICI (Mond Division) plc

13.1 Introduction 215
13.2 Development of Coatings for Mercury Cells 215
13.3 Anode Coatings for Diaphragm Cells 222
13.4 Anode Coatings for Membrane Cells 228
13.5 Conclusions 232

Chapter 14 – THE DEVELOPMENT OF LOW OVERVOLTAGE CATHODES
D. E. Brown, P. O. Fogarty, M. N. Mahmood and A. K. Turner, BP Research Centre

14.1 Introduction 233
14.2 Experimental Techniques 233
14.3 Performance of the BP Electrocatalyst 234
14.3.1 Hydrogen evolution in NaOH–NaCl electrolyte 234
14.3.2 The effect of hypochlorite on coated electrodes 235

14.3.3 Simulated diaphragm application 236
14.3.4 Hydrogen evolution in NaOH electrolyte 239
14.3.5 Electrocatalyst morphology 242
14.4 Conclusions 243
14.5 Acknowledgements 244
14.6 References 244
14.7 Discussion 244

Chapter 15 – ELECTROCATALYSIS AND THE CHLORINE EVOLUTION REACTION

J. A. Harrison, Department of Physical Chemistry

15.1 Introduction 246
15.2 The Computer Method 246
15.3 Data Handling 247
15.4 Chlorine Formation 251
15.4.1 Manual results 251
15.4.2 Computer results 258
15.5 Chlorine Reduction 261
15.6 Discussion 262
15.7 Acknowledgement 263
15.8 Appendix 263
15.8.1 Consider as a one step reaction 263
15.8.2 Consider as a two step reaction 265
15.9 References 266
15.10 Open Discussion 268

Chapter 16 – KINETICS AND MECHANISM OF CHLORINE EVOLUTION ON METALOXIDE ELECTRODES

L. J. J. Janssen, Eindhoven University of Technology on behalf of the Royal Society of Chemistry

16.1 Introduction 271
16.2 Theory 272
16.3 Experimental Results 272
16.3.1 Electrochemical RuO_2 electrode 274
16.3.2 Chemical RuO_2 electrode 276
16.3.3 RTO electrode 278
16.3.4 The MO-electrode 283
16.4 Discussion and Conclusions 283
16.5 References 284
16.6 Open Discussion 285

SECTION E – CHLORATE MANUFACTURE

Chapter 17 – ENERGY SAVING IN CHLORATE PRODUCTION WITH THE USE OF A FUEL CELL
I. H. Warren, Chemetics International Company

17.1 Introduction 289
17.2 The Fuel Cell and its Operating Characteristics 290
17.3 The Chlorate Cell and its Operating Characteristics. 293
17.4 Integration of a Fuel Cell with the Chlorate System 295
17.5 The Economics of Integration of a Fuel Cell with the Chlorate System 299
17.6 Discussion and Conclusions 301
17.7 Acknowledgements 303
17.8 References 303
17.9 Open Discussion 303

Chapter 18 – LCD MK II – A NEW SODIUM CHLORATE CELL DESIGN
C. G. Rader, E. M. Spore and J. C. DeLong, Occidental Chemical Corporation

18.1 Overall Configuration 305
18.2 The Anode Assembly 307
18.3 Cathode Assembly 308
18.4 Reactor Top 308
18.5 Internal Design 309
18.6 Prototype Cell. 311
18.7 Cell Operation. 313
18.8 Conclusion 314

SECTION F – ENERGY AND THE CHLOR-ALKALI BUSINESS SCENE

Chapter 19 – FACTORS AFFECTING THE FUTURE AND THE CHLORINE INDUSTRY
D. S. Davies, Duncan Davies Associates/Tosco Corporation

19.1 Introduction 317
19.2 Policymaking – Forecasting – Discontinuity – Timescale – The Headlines 318
19.3 Controlling Factors 319
 19.3.1 The product pattern of the chlorine industry: substitutes 319
 19.3.2 Toxicology and environment 321

19.3.3 Demand – overall price elasticity – market saturation and overcapacity 321
19.3.4 Geography and politics 323
19.3.5 Timing 324
19.4 Countries and Products 324
19.4.1 Products 324
19.4.2 Places 325
19.5 Conclusions 326
19.6 Reference 326
19.7 Discussion 326

Chapter 20 – IMPACT OF RISING ENERGY COSTS ON THE WORLDWIDE STRUCTURE OF THE CHLOR-ALKALI INDUSTRY

G. L. Innes, Charles H. Kline & Co. Ltd.

20.1 Introduction 328
20.2 Industrial Countries 330
20.3 Latin America 332
20.4 Middle East and Mediterranean 333
20.5 Southeast Asia 334
20.6 Australia 335
20.7 Canada 335
20.8 Risks 335
20.9 Distribution of New Production 336
20.10 Surviving the Global Shakeout 336
20.10.1 Reducing costs 337
20.10.2 Differentiating the product 337
20.10.3 Selling knowledge and services 338
20.10.4 Redeploying assets 338
20.11 Discussion 339

Chapter 21 – PAST AND FUTURE TRENDS IN THE EUROPEAN CHLOR-ALKALI INDUSTRY AND THEIR IMPACT ON CHLORINE TECHNOLOGY

S. M. Andrew and J. P. H. Shaw, ICI (Mond Division) plc

21.1 Introduction 340
21.2 Industrial Beginnings 340
21.3 Early Mercury Cell Development 348
21.4 Organic Chlorine Derivatives 349
21.5 Future Trends 350
21.6 Summary 353
21.7 Acknowledgements 354

Chapter 22 – THE AMERICAN BUSINESS SCENE FROM THE PERSPECTIVE OF THE CHLOR-ALKALI INDUSTRY

C. A. Hansen,
The Chlorine Institute, LCP Chemicals & Plastics Inc.

22.1 Introduction 355
22.2 The Past Decade 355
22.2.1 Energy effects 356
22.2.2 New technology 356
22.2.3 End-use patterns and process developments 357
22.3 Future Trends 359
22.3.1 Production and capacity 359
22.3.2 Exports 360
22.3.3 New technology 361
22.3.4 Profitability 361
22.3.5 Role of government 362
22.4 Conclusions 362

SECTION G – ENVIRONMENTAL/SAFETY

Chapter 23 – ENVIRONMENTAL TRENDS AND NORTH AMERICAN CHLOR-ALKALI INDUSTRY CHALLENGES: PAST AND PROJECTED

E. J. Laubusch, The Chlorine Institute Inc.

23.1 Introduction 367
23.2 The US EPA 369
23.3 Water Pollution Control 370
23.4 Air Pollution Control 372
23.5 Drinking Water 373
23.6 Waste Management 373
23.7 Conclusions 375

Chapter 24 – CHLORINE HANDLING AND SAFETY IN THE EUROPEAN SITUATION

R. Papp,
Produits Chimiques Ugine Kuhlmann on behalf of the BITC

24.1 Introduction 376
24.2 The Work of BITC 380
24.3 Legislation – European Directives 382
24.4 Conclusions 384
24.5 Discussion 384

Index 387

Introduction

C. JACKSON

The papers published in this book were all presented at the International Chlorine Symposium held in London in June 1982. This symposium, the third of a series, was jointly sponsored by the Electrochemical Technology Group of the Society of Chemical Industry and the Electrochemistry Group of the Faraday Division of the Royal Society of Chemistry. Although only three years have passed since the last Chlorine Symposium held in London [1] and the Industry now finds itself in the depths of recession, the pace of development and the interest generated in the theme of the current symposium – the effect of high energy costs on the chlorine industry – can be judged by the fact that 24 papers were presented to an audience of over 260 delegates from 28 countries.

The book opens with a section devoted to the diaphragm electrolysis process starting with an examination of a method for the better characterization of asbestos diaphragms. Next follows a description of a new cell design for reduced power consumption and the final chapter of this section offers the tantalizing possibility of the eclipse of asbestos by synthetic diaphragms.

Over the past few years the chlorine industry has faced rapidly escalating energy costs, increasing environmental pressure on the use of mercury and asbestos and on the transporting of liquid chlorine, so it is not altogether surprising that the pace of development of membrane cell technology from laboratory curiosity to commercial reality has been rapid. This is reflected in the next two sections of the book which contain in all some nine chapters.

In the first of these the leading US and Japanese membrane manufacturers describe their products and in the case of the Japanese, the parallel development of electrolyzer technology. A chapter from a major chlorine producer on electrolyzer design plus a fascinating description of the advancement of membrane electrolyzers from the first commercial unit to today's highly efficient designs combine to illustrate the divergent thought which has been applied to solving the many common problems. The exciting possibility of making even greater power savings by combining membrane technology and oxygen depolarized cathodes is explored in the next chapter and rounds off this section of the book.

Having covered electrolyzer design the book moves on in the next section to discuss the application of membrane electrolyzer technology. Any potential customer needs to know the answers to many different questions before he makes a decision and one method for assessing competing technologies is presented. The second paper describes in detail the necessary considerations for the conversion of an existing chlorine plant to membranes and discusses the suitability of the existing plant equipment.

Whatever the technology, no electrolyzer is complete without electrodes and electrode coating development continues. The opening chapter of the next section describes the developments that are being pursued for mercury, diaphragm and latterly membrane electrolyzers in one major company. In keeping with the theme of the symposium, ever increasing energy costs have stimulated work into the development of low hydrogen overvoltage cathode coatings and one such coating is described in the next chapter.

Then follow two papers sponsored by the Royal Society of Chemistry. Both deal with the chlorine evolution reaction and offer different interpretations of the observed phenomena. One of them vividly illustrates how the use of computers in the laboratory is revolutionizing electrochemical research.

Developments in chlorine production are described next with contributions from two companies active in the field. One chapter describes some advances in electrolyzer hardware whilst the other puts forward the economic attractions of linking fuel cells with the chlorate process.

It is wholly appropriate in today's economic climate to consider the chlorine business scene and this is done in the context of increasing energy costs in the penultimate section of this book.

The opening chapter considers the chlorine industry as a mature industry and in considering the implications of this draws some disturbing parallels with steel and spells out the need for the industry to shake off complacency and lead change in order to survive.

In subsequent papers views on the worldwide, US and European markets are covered and attention is drawn to opportunities for new plants in the Gulf, Latin America and Australia and for the replacement of existing capacity in other developed countries – all indicating the future widespread use of membrane technology.

The concluding section deals with the environmental trends in the US and safety considerations of chlorine handling and transport in Europe and indicates possible future scenarios.

The Electrochemical Technology Group of the Society of Chemical Industry offer thanks to the several people and organizations whose help ensured the success of this symposium. In particular to:

1. The contributors of papers.

2. The session Chairmen: Dr A. T. Churchman (Electricity Council Research Centre)
M. P. DeCuyper (Solvay et Cie. Brussels)
Dr B. S. Gilliatt (ICI Mond Division plc)
Dr M. N. Dudley (IMI Marston Ltd)
Prof. M. Fleischmann (University of Southampton)
3. The Electrochemistry Group of the Faraday Division of the Royal Society of Chemistry for active participation and joint sponsorship.
4. The US Chlorine Institute for aid with printing costs and active participation.
5. The BITC for active participation.
6. ICI (Mond Division) plc for aid with printing costs and active participation.
7. The Associated Octel Co. Ltd for generous hospitality.
8. The organizing committee: Mr J. H. Collins (Chairman)
Mr C. Jackson (Secretary) and
Mr I. J. M. Girvan (all of ICI Mond Division plc)
Mr B. Lott (The Associated Octel Co. Ltd)
Dr R. D. Giles (International Nickel Co. Ltd)
Mr D. Bell (BP Chemicals Ltd)

REFERENCE

[1] Coulter, M. O., (ed.), *Modern Chlor-Alkali Technology*, Ellis Horwood Ltd, 1980.

SECTION A

The Development of Diaphragms and Diaphragm Electrolyzers

CHAPTER 1

Characterization of Asbestos Diaphragms for Chlor-alkali Electrolysis

K. A. POUSH, D. L. CALDWELL, Texas Division, Dow Chemical USA, Chlor-Alkali Research, Freeport, Texas, USA 77541

J.W. VAN ZEE, R. E. WHITE,† Department of Chemical Engineering, Texas A & M University, College Station, Texas, USA 77843

1.1 INTRODUCTION

The diaphragm is the key to the efficient operation of a diaphragm chlorine caustic cell. Although the development of novel electrode materials has resulted in marked improvements in cell voltage and operating life during the past decade, diaphragm technology has not advanced materially since the invention of the deposited asbestos diaphragm 50 years ago. Additional major increases in cell energy efficiency, which depends both on cell voltage and current efficiency, must await significant diaphragm improvements.

The preparation of commercial chlorine cell diaphragms remains more art than science. Although recipes that yield good results have been developed over the years, improvements have normally been obtained only as the result of tedious trial-and-error experimentation. One reason is the absence of suitable diaphragm characterizing parameters. Recently, Koh [1] presented a parameter for characterizing diaphragms; unfortunately, the utility of his approach is unclear because he presented values that cannot be checked for proprietary reasons. In a recent paper by the present authors [2] (see also Hine [3]), a theoretical basis is presented for the use of the MacMullin number, or resistance ratio, to predict cell current efficiencies and separator voltage drop. The present work provides a laboratory-scale procedure for the determination of MacMullin number values and a laboratory-scale method of determining the utility of the MacMullin number concept.

† To whom correspondence should be addressed.

1.2 DIAPHRAGM CHARACTERIZATION

Two measurable quantities are used here to characterize the diaphragm: the MacMullin number (N_m) and the thickness (t). The MacMullin number can be expressed in terms of bulk diaphragm parameters [4] (i.e. porosity, tortuosity, hydraulic radius and permeability). Here the MacMullin number is defined in readily measurable quantities as follows:

$$N_m = r/r_0 \tag{1}$$

where r/r_0 is the ratio of the electrical resistance of an electrolyte-saturated porous medium to the resistance of an equivalent volume of electrolyte. The MacMullin number is a measure of the ability of the porous media to prevent the movement of the dissolved species in the electrolyte; consequently, it is used explicitly in a one-dimensional dilute solution flux equation which can be used to describe the movement of ionic species within a diaphragm [2].

The diaphragm thickness is also a characterizing parameter. The diaphragm thickness has practical utility for characterization of both laboratory-scale and production-scale cell operation because it can be measured easily.

1.3 EXPERIMENTAL DETERMINATION OF MACMULLIN NUMBER AND THICKNESS

The procedure used in this work to obtain N_m and an estimate of the accuracy of the measurement consisted of the following steps. First, an 8-inch square diaphragm was drawn using standard techniques. Next, this diaphragm was quartered to make four 4-inch square diaphragms. Then, the MacMullin number was determined for each section using the apparatus shown in Fig. 1.1 according to the following procedure:

1. Soak the diaphragm in saturated brine 16–24 hours.
2. Measure the standard resistance ($R_{S,\text{before}}$) of the test cell filled with saturated brine without the diaphragm but with the window holder in place.
3. Assemble the diaphragm into the holder and place in the tank.
4. Measure the resistance with the diaphragm in place (R).
5. Repeat step 2 and obtain $R_{S,\text{after}}$. Then calculate $R_S = (R_{S,\text{before}} + R_{S,\text{after}})/2$.
6. Calculate N_m from the following equation:

$$N_m = \frac{R - R_S}{r_0} + 1 \tag{2}$$

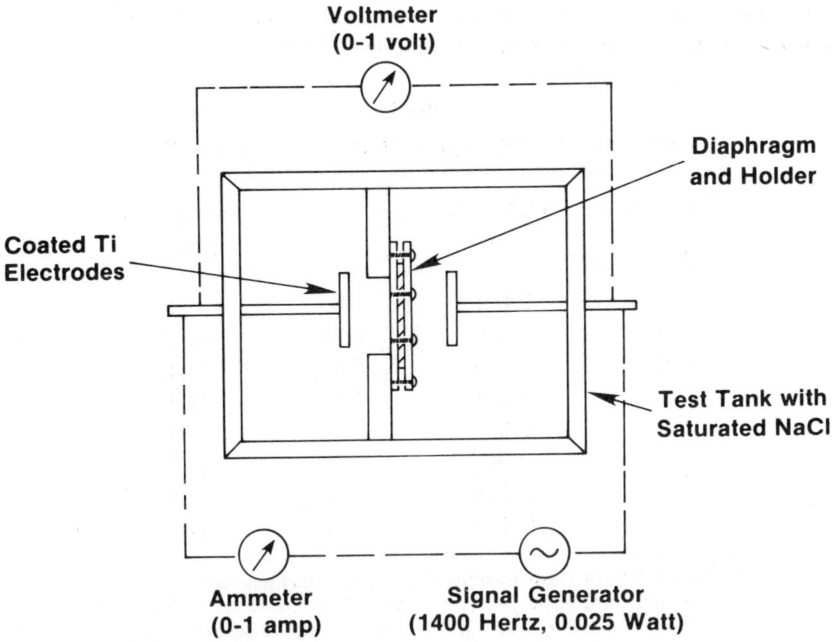

Fig. 1.1 – A schematic of the experimental apparatus used to measure N_m.

where r_0 is obtained from the resistivity of the electrolyte (ρ), diaphragm thickness (t), and the cross-sectional area (A) of the diaphragm:

$$r_0 = \rho \frac{t}{A} \tag{3}$$

The accuracy of N_m obtained in this manner for each of the four 4-inch square diaphragms was approximately ±5%.

The thickness of a diaphragm was determined by measuring the thickness of the dry 4-inch square diaphragm at nine points (corners, in between the corners, and in the center) and then by averaging these measurements. The accuracy of this method was approximately ±5%. The assumption here is that the thickness of the dry diaphragm adequately represents the thickness during cell operation. This assumption was checked by measuring the thickness of the wet diaphragm before and after the laboratory cell experiments (described below). These before and after measurements did not vary from the dry measurement average by more than ±5%.

Table 1.1 presents the values obtained for N_m and t for the diaphragms used in this study together with a high (H), low (L) classification scheme† used in the factorial design of the experiments, as described below.

Table 1.1

Experimental diaphragm characterization properties

Diaphragm	Thickness (in)	N_m	Classification‡ (N_m, t)
A	0.066	3.45	L, L
B	0.109	4.63	L, H
C	0.061	5.15	H, L
D	0.128	7.45	H, H
E	0.071	5.30	H, L
F	0.067	3.30	L, L
G	0.110	4.85	L, H
H	0.119	7.50	H, H

‡ The diaphragm classifications H and L for thickness correspond to thick and thin, respectively, in Table 1.2. The arbitrary division of high (H) and low (L) at $N_m = 5$ does not affect the experimental design of Table 1.2 because regression techniques were used to analyze the data.

1.4 DIAPHRAGM EXPERIMENTAL STUDY

The performance of a chlorine–caustic cell depends upon the diaphragm characterization parameters (N_m and t) and the independent operating variables of current density, head differential, and HCl feed concentration. Here, the experimental approach for studying the diaphragm cell performance was as follows:

1. Characterize several chrysotile asbestos diaphragms by determining N_m and thickness for each.
2. Obtain the results for a factorial experiment (see Table 1.2) for these two characterizing parameters and the three operating variables.
3. Statistically analyse the experimental results for relationships between the diaphragm characterization parameters and cell performance as indicated by the current efficiency.
4. Compare the experimental results with the theoretical model of Reference [2].

The eight diaphragms whose N_m and t values are shown in Table 1.1 were mounted in laboratory chlorine cells and their N_m and t values checked at the

† The arbitrary division of high (H) and low (L) at $N_m = 5$ in Table 1.1 does not affect the experiment because regression techniques were used to analyze the data.

Table 1.2

Random assignment of diaphragm characterizing parameters and operating variables for diaphragm model verification

Diaphragm N_m	High (H)				Low (L)				
Diaphragm thickness	Thick (H)		Thin (L)		Thick (H)		Thin (L)		
Cell number	80	85	83	79	84	77	76	78	
Experiments 1 and 2 CD/GPL	H/L L/H	H/H L/L	L/L H/H	L/H H/L	H/H L/L	L/H H/L	H/H H/L	L/L L/H	Feed pH = Low
Experiments 3 and 4 CD/GPL	L/L H/H	H/L L/H	L/H H/H	L/L H/L	H/L H/H	L/L L/H	L/H H/H	L/L H/L	Feed pH = High
Experiments 5 and 6 CD/GPL	H/H H/L	L/H L/L	H/H H/L	L/L L/H	H/L L/H	L/L H/H	L/H L/L	H/L H/H	Feed pH = High
Experiments 7 and 8 CD/GPL	H/L L/H	L/L H/H	L/H H/H	L/L H/L	L/H H/H	H/L L/L	L/H H/H	H/L L/L	Feed pH = Low

CD = current density, GPL = grams/litre caustic, H = High value, L = Low value, see Table 1.1 for quantitative values of N_m and t.

beginning and the end of the experiment in a manner similar to the procedure described above. The N_m and t values were the same before and after the experiments to within experimental error (±5%). These resistance measurements indicated that N_m did not change because of the experiment. It should be pointed out, however, that no attempt was made to determine N_m while the cells were operating. This would be a useful exercise but was beyond the scope of this work.

Once the diaphragms were mounted, a two-level factorial experiment was performed according to Table 1.2. The independent variables in the experiment were: cell current, HCl feed concentration (NaCl feed concentration was fixed during the experiment), cell differential head, N_m, and t. Also, the temperature of the cells was fixed at 70°C. As shown in Table 1.2 the experiment was designed so that a dependent variable, the effluent caustic concentration (g/l), replaced the cell differential head as an independent variable. (This replacement may have affected the random error in the results.) Many response variables (e.g. diaphragm IR drop, caustic concentration in the effluent, etc.) were recorded for each cell for the randomized combinations of the independent variables.

1.4.1 Results

A total of 121 data sets were obtained and analysed statistically for relationships between, for example, the potential drop across the diaphragm (diaphragm IR drop) and the independent variables. Figure 1.2 presents some diaphragm IR drop data as a function of the product of N_m and t, which is essentially the effective thickness of the diaphragm. It was found that using the product of N_m and t, instead of both separately, improved the empirical fit of the data shown in Fig. 1.2. The line in Fig. 1.2 is the regression line for which $r^2 = 0.9$ and the F ratio is 284. Another correlation is shown in Figs. 1.3 and 1.4 for caustic current efficiency dependence on effluent caustic concentration. The regression coefficient (r^2) for the lines shown in Figs. 1.3 and 1.4 is 0.8 and the F ratio is 94.

1.4.2 Comparison with theoretical model

The data obtained was also used for comparison to a theoretical model of a diaphragm cell which was presented elsewhere [2]. Figure 1.5 shows a comparison between the diaphragm IR drop calculated according to the model and the data. The solid line is the regression equation for this comparison and has a r^2 value of 0.9, an intercept of 6.8 millivolts and a slope of 1.01. The F ratio for this line is 601.

Figure 1.6 shows a comparison of the predicted caustic current efficiency from the model [2] with the experimentally observed efficiencies. Here r^2 is 0.6, the intercept is 7%, the slope is 0.92, and the F ratio is 83.

Finally, Fig. 1.7 shows a comparison between the predicted caustic flow rate and the experimentally observed values. In this figure r^2 is 0.9, the intercept is −0.02, the slope is 0.85 and the F ratio is 605.

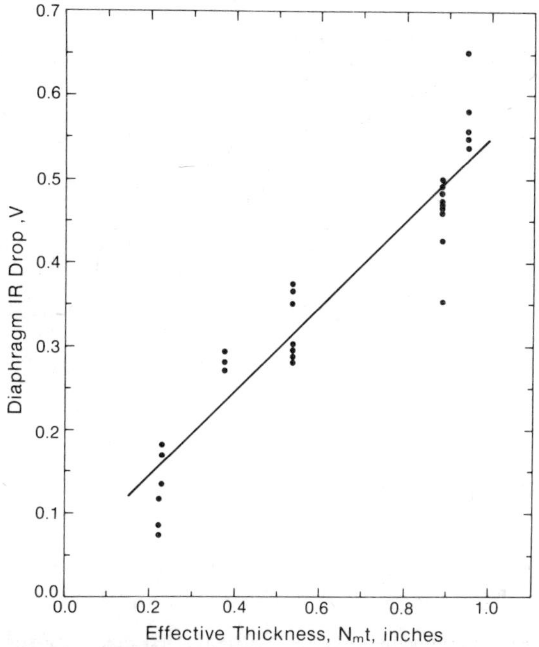

Fig. 1.2 – Correlation of experimental diaphragm IR (voltage) drop with the product of N_m and t at high current densities.

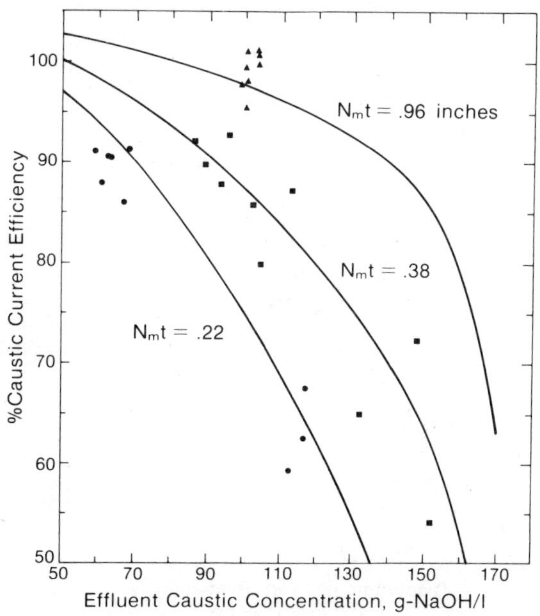

Fig. 1.3 – Experimental effect of caustic effluent concentration and $N_m t$ on caustic current efficiency for low current density and low feed acid concentration.

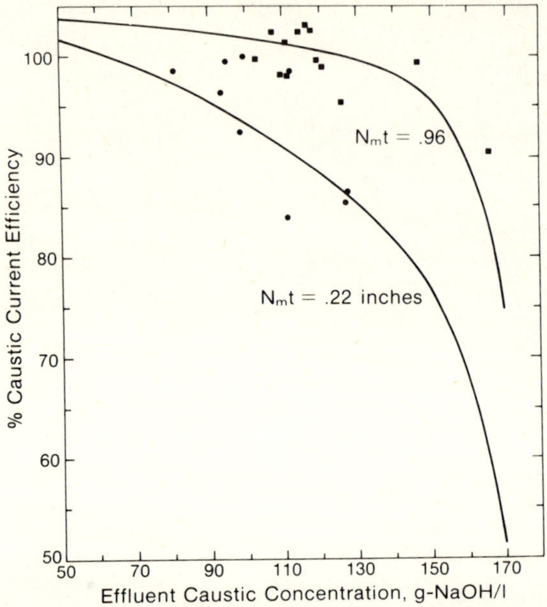

Fig. 1.4 – Experimental dependence of caustic current efficiency on caustic effluent concentration and N_mt at high current density and low feed acid concentration.

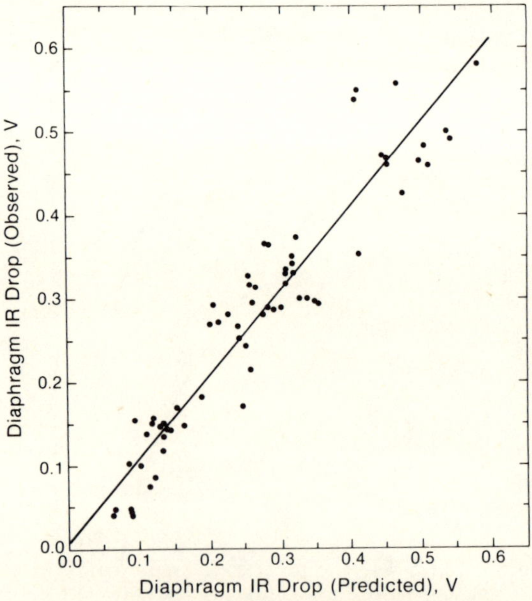

Fig. 1.5 – Comparison of experimentally observed diaphragm IR drops to model predictions.

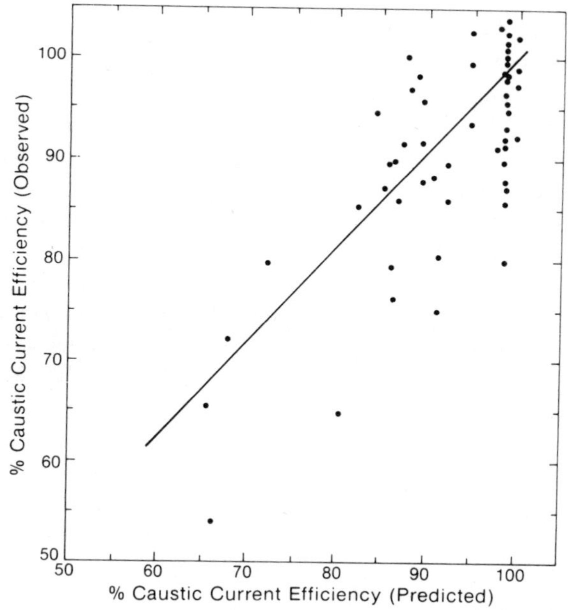

Fig. 1.6 – Comparison of experimentally observed caustic current efficiencies to model predictions.

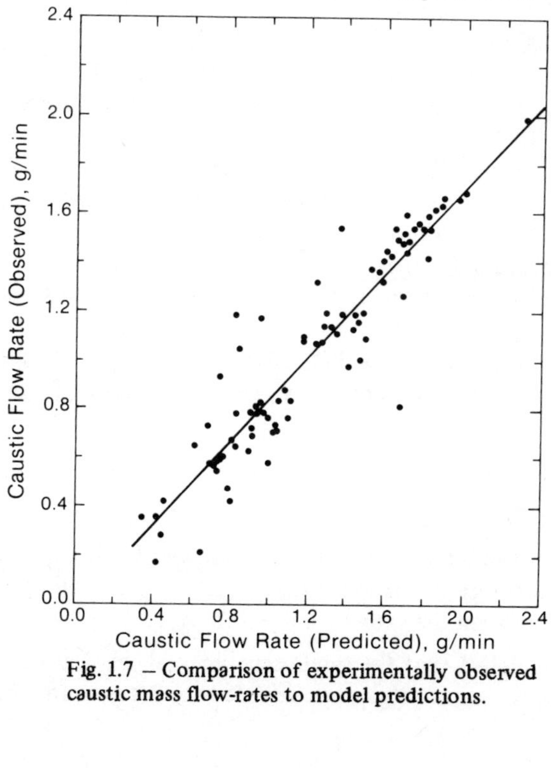

Fig. 1.7 – Comparison of experimentally observed caustic mass flow-rates to model predictions.

1.5 DISCUSSION AND CONCLUSIONS

Measurement errors were present in the independent variables and affected the data analysis. While errors in current density and feed brine acid concentration are systematic and fairly small, the NaOH product concentration analysis is subject to considerable random error, and is felt to be a weak point in the study; that is, the use of head instead of caustic concentration may improve future experiments. The ±5% error in both the thickness and N_m should not affect the results of the data analysis.

The response variable measurement errors and their treatment in the statistical analysis has not been dealt with rigorously in this report. The fact that significantly high correlations between $N_m t$ and the response variables were observed despite fairly high error levels is proof of the importance of N_m in diaphragm characterization. Again, both N_m and t can be measured and their product dictates cell performance.

Considering the large random error that is inherent in a study of this kind, agreement with the predictions of the mechanistic model is good and confirms both the approach taken in developing the model [2] and the significance of the MacMullin number in characterizing the diaphragm.

1.6 REFERENCES

[1] Koh, W-H., *Characterization of Structural Properties of Polymer-Modified Asbestos Chlor-Alkali Separators*, presented at the Electrochemical Society 157th Meeting, St. Louis, MO, May, 1980; *AIChE Symposium Series 204*, 1981, Volume 77, 213.

[2] Caldwell, D. L., Poush, K. A., White, R. E., and Van Zee, J. W., *Mathematical Model of the Chlorine Cell Diaphragm*, presented at the Electrochemical Society 161st Meeting, Montreal, Canada, May, 1982.

[3] Hine, F., Diaphragm Engineering in NaCl Electrolysis, *Soda to enso*, 219, June, 1980; Hine, F., Yasuda, M., and Fujita, K., Effects of the Brine Impurities of the Cell Performance of the Diaphragm-Type Chlor-Alkali Cell, *J. Electrochem. Soc.*, 1981, **128**, 2314.

[4] MacMullin, R. B., and Muccini, G. A., Characteristics of Porous Beds and Structures, *J.A.I.Ch.E.*, 1956, **2**(3), 393.

1.7 OPEN DISCUSSION

1. In opening the discussion from the floor, Dr Strasser, Uhde GmbH, questioned Dr White on the design of his experimental cells – particularly with respect to the distance between the cathode and diaphragm.

   Professor White replied that the small experimental cell that was shown with a finite diaphragm – cathode gap was designed for determining the diaphragm characterization parameter the MacMullin number. The cell was only used for resistance measurements using an a.c. signal of about 1430 c.p.s and pieces of diaphragm about 3 inches square. The overall cell dimension was about 10in × 10in × 10in and no chlorine was ever produced from this cell.

By contrast, the experimental chlorine cells did not have the diaphragm separated from the cathode but the diaphragm was not deposited.

2. A question from Dr D. F. Schlafer, BASF, concerned the reaction between aqueous chlorine and chloride ion to give the Cl_3^- anion. Dr Schlafer asked which analytical technique was used to detect this species and what are its chemical and physical properties.

   In response, Professor White said that the anion itself was not analyzed for in the study.

   Only the total dissolved chlorine was determined analytically. However, there are studies in the literature that have claimed this species, for example, Nagy has presented an equilibrium constant for its formation.

3. M. J. P. Detourney, Solvay, asked whether the diaphragm deposition conditions could be determined to achieve a desired MacMullin number or was the MacMullin number only found after the decomposition.

   One would prefer the former, the latter is obviously possible and studies are progressing to enable the former to be realized, replied Professor White.

4. In reply to a question from Mr S. Herbert, Dead Sea Bromine Co., about the kinetics of side cathodic reactions for the reduction of chlorate to chloride and the ratio of oxidant species in anolyte to catholyte, Professor White regretted that this was proprietary data which he could not release.

5. Mr A. K. Rao, D.C.M. Chemical Works, India, asked how the concentrations of hypochlorite and hypochlorous acid within the diaphragm as shown on one of the slides used in the paper has been determined.

   Professor White answered that these were only predictions and could not be measured directly. The only direct measurement was of effluent liquors from the cell. By varying the diaphragm thickness and MacMullin number a theoretical model was derived which was fully consistent with concentrations actually achieved in the cell liquors.

6. Professor Kahn from India, asked in what concentration chlorate was formed in the cell since the reaction involved seemed trivial.

   Professor White confirmed that the reaction was slight and only became of importance if there was a limit to the quality of chlorate that was allowable in product liquors. Otherwise, it could be ignored.

NOMENCLATURE

A	= Diaphragm active area, cm^2 or in^2
N_m	= MacMullin number, dimensionless
r	= Resistance of an electrolyte saturated porous medium (i.e., resistance of a wet diaphragm), Ω
r_0	= Resistance of a volume of electrolyte which has the same volume as the diaphragm, Ω
R	= Total resistance of electrolyte-filled diaphragm and solution between the electrodes, Ω
$R_{S,\text{after}}$	= Resistance of standard cell after measuring R, Ω
$R_{S,\text{before}}$	= Resistance of standard cell before measuring R, Ω
t	= Diaphragm thickness, cm or in
ρ	= Resistivity of electrolyte, Ω-cm or Ω-in

CHAPTER 2

PPG Bipolar Diaphragm Electrolyzer – New Design for Low Power Consumption

R. J. SCOTT, PPG Industries, Inc., Pittsburgh, Pa 15222, USA

2.1 INTRODUCTION

Saving energy to reduce operating costs is the name of the game in electrochemical technology. To that end, PPG Industries has developed a new bipolar electrolyzer that will reduce by nearly one quarter the amount of energy required to produce 1000 tons per day of chlorine at its Lake Charles, La., chemicals complex, by the diaphragm cell process.

The new electrolyzers, designated as the PPG Bipolar Diaphragm Electrolyzer Model V-1161, are being installed as part of a $100 million-plus modernization at Lake Charles. When completed next year, the new production facilities will conserve the equivalent of 500 000 barrels of fuel oil a year in energy for manufacturing chlorine and caustic soda.

This paper will present the history of the development of the new generation electrolyzer, along with its main design features and operating characteristics. The new technology will be compared to the earlier PPG diaphragm electrolyzer, Model V-1144, with commercial situations presented for each model.

2.2 MODEL V-1144 BIPOLAR DIAPHRAGM ELECTROLYZER

We have presented technical papers on the earlier model V-1144 bipolar electrolyzer at two previous chlorine cell symposiums here. To bring you up to date on PPG diaphragm electrolyzer technology and to allow a comparison with the new model V-1161, I will begin with a brief review of the model V-1144.

The model V-1144, designated the Glanor®† electrolyzer, was developed and commercialized by PPG in the early 1970s, jointly with Oronzio de Nora Impianti Elettrochimici, S.p.A., of Milan, Italy. Previously, PPG had developed and used the bipolar design in sodium chlorate and chlorine cells with graphite anodes, and this experience laid the groundwork for the V-1144 technology.

† Glanor® is a registered trademark of Oronzio de Nora Impianti Elettrochimici, S.p.A.

Development of the Glanor® electrolyzer technology produced the first PPG chlorine cell design that eliminated the use of graphite anodes and rubber linings. Materials used in the construction of the electrolyzer included metal anodes, titanium linings, and fiberglass reinforced plastic (FRP) brine boxes and auxiliaries – all long-life materials when compared to the graphite cells. The diaphragm consisted entirely of slurry-deposited asbestos.

The V-1144 bipolar electrolyzer contains ten bipolar elements clamped together with tie rods between two end electrode elements, resulting in an electrolyzer module of 11 cells in series. Operating in the 70-kiloampere range, each electrolyzer is capable of producing about 26 short tons of chlorine per day. Up to 18 electrolyzers can be installed in series in one electrical circuit, depending on the plant capacity desired.

Development of this electrolyzer provided significant economic savings in the cost of producing chlorine as compared to the graphite cells. Electrolysis power consumption was reduced approximately 14%, from about 2900 to about 2500 d.c.-kWh per short ton of chlorine. Long-life cell materials and long diaphragm life resulted in significant reductions in costs for labor and materials.

The model V-1144 bipolar electrolyzer has achieved significant commercial success, both through licensing and by use within PPG. In the world today, there are seven operating plants using this technology, and four more plants are under construction. These 11 plants have a total rated capacity of over 2 million tons of chlorine per year. This total includes PPG's own installation at Lake Charles, Louisiana, which has a capacity of 1500 tons per day of chlorine. PPG's plant contains four circuits of 16 electrolyzers each. The circuits were placed in operation between 1977 and 1980.

2.3 IMPROVED DESIGN – PROTOTYPE ELECTROLYZER

By the mid-1970s, fabrication and operation experience had been gained on the V-1144 electrolyzers. Also, energy costs were increasing rapidly with no slow-down being anticipated. The need for an improved cell design to provide for PPG's future needs was evident. A program was initiated to improve the bipolar diaphragm cell design, with the following goals in mind:

A. Reduced power consumption.
B. More stable diaphragm.
C. Reduction in the amount of 'critical' titanium and steel welding in the fabrication of the cell elements.

The improved cell design has allowed us to achieve these goals.

Reduced power consumption was achieved in two ways: (a) the metal to metal electrode gap was reduced from 11 mm to 6 mm, and (b) the electrode area was increased by 40% within the same overall dimensions of the cell elements, so that operation at the same current would result in significantly lower current density, and therefore, lower voltage and power consumption. This idea of significantly increasing the electrode area within the same cell dimensions is one of the key features of the new design. Lower current density operation can be achieved in other ways, such as by decreasing amperage and increasing the number of cell units. However, PPG's approach is more economical, since only the electrode area is increased, and not the other parts of the cell, nor the number of cells.

Also, this approach reduces the amount of cell room piping and electrical bus required, since the number of cells is not increased for a given production requirement.

The diaphragm performance was improved by the use of a modified, asbestos diaphragm, which had been under development for some time previously.

The reduction in the amount of 'critical' titanium and steel welding was achieved through use of explosion-bonded titanium steel for the bipolar plate. Previously the bipolar plate was steel only, with a loose titanium liner used on the anolyte side of the cell.

A prototype electrolyzer embodying the new design features was constructed. This electrolyzer consisted of three cells, whose size were about half the size of the projected commercial electrolyzer. It was placed into operation in a commercial graphite cell circuit operating at 30 kA, in June, 1978. The first operation test run lasted 14 months, at which time the electrolyzer was disassembled for inspection. The second test run started in October, 1979, and continues today, after 32 months, with the same diaphragm. It is still operating at a high current efficiency of 95–96%.

2.4 MODEL V-1161 ELECTROLYZER COMMERCIALIZATION

By 1979, PPG perceived the need for an efficient diaphragm cell for use in modernization of its older diaphragm cell area at the Lake Charles complex. This older cell area, containing Hooker S-type cells using graphite anodes, has a production capacity of almost 1000 tons per day of chlorine. It was decided to scale up the prototype electrolyzer to a full size (70–75 kA) electrolyzer design, and to use these bipolar diaphragm electrolyzers in a project to install 1000 tons per day of new capacity, and to scrap the old cell area.

The design of the full size electrolyzer was based upon the features of the prototype as described earlier. The new electrolyzer was designed to the same overall dimensions as the V-1144, with the idea that the electrolyzers would be interchangeable in the circuits, and the V-1161s could replace V-1144s in the future if economics dictated. The new electrolyzer design also contains 11 cells. However, in order to accommodate the additional electrode area, the cell elements contain 61 vertical rows of anodes (and cathodes) as compared to 44 rows in the earlier model. Thus, the new design was designated V (vertical) -1161.

Due to the project schedule, there was insufficient time to build and further to test a full size electrolyzer before making a final decision on the selection of the cell for this project. Engineering of the cell area began in 1980, and in early 1981, an order was placed for the 38 bipolar electrolyzers, model V-1161, required for this project.

The first shipment of electrolyzers arrived at the PPG plant in April, 1982, and two electrolyzers subsequently were assembled and installed in an existing V-1144 operating circuit. This allowed us to check out the mechanical and operational aspects of the units, and to provide training for company personnel. The first full circuit of V-1161 electrolyzers is scheduled for operation in early 1983.

2.5 DESIGN AND OPERATING CHARACTERISTICS – MODEL V-1161 VERSUS MODEL V-1144

Table 2.1 depicts certain design features and operating characteristics of the PPG model V-1161 bipolar diaphragm electrolyzer. A comparison is made with the model V-1144 electrolyzer.

Table 2.1
Design and operating characteristics

Item	Model V-1161	Model V-1144
No. of cells per electrolyzer	11	11
Active anode area per cell (m^2)	49	35
Electrode gap (mm)	6	11
Current load (kA)	72	72
Current density at 72 kA (kA/m^2)	1.47	2.05
Cell voltage (V)	3.08	3.50
Current efficiency (%)	95–96	95–96
Power consumption (d.c.-kWh/STCl_2)	2200	2500
Anode gas		
% Cl_2	97.3–98.0	97.3–98.0
% O_2	1.5–2.2	1.5–2.2
% H_2	<0.10	<0.10
% CO_2	0.40	0.40
Cell liquor		
NaOH (g/l)	135–145	135–145
$NaClO_3$, % A.B.	0.03–0.15	0.03–0.15
Production per electrolyzer		
Chlorine, STPD	26.7	26.7
NaOH, STPD (A.B.)	29.8	29.8

As can be seen from Table 2.1, the goal of a significant reduction in power consumption has been achieved. The power consumption per ton of chlorine for the model V-1161 is about 12% less than that for model V-1144. Compared to the graphite diaphragm cells being replaced at Lake Charles, the model V-1161 electrolyzer effects a reduction in power consumption per ton of chlorine produced of almost 700 kWh, or about 24%.

The technology of the model V-1161 bipolar diaphragm electrolyzer gives an efficient, competitive method of producing chlorine and caustic soda, particularly for relatively large capacity plants.

The diaphragm cell process has a keen competitor in the new membrane cell process, which also is available now. However, the diaphragm cell process remains competitive in situations where some of the following factors exist:

(a) Salt is available as a brine solution.
(b) Power is generated on the premises, with turbine exhaust steam available to feed caustic soda evaporators.
(c) In the replacement of older diaphragm cells, where the caustic soda evaporators are still usable.
(d) Relatively large capacity plants.

Where the diaphragm cell process is competitive, the PPG bipolar diaphragm electrolyzer, model V-1161, is a viable competitor.

2.6 DISCUSSION

1. The first (anonymous) questioner asked about current leakage around individual electrodes in the PPG bipolar electrolyzer.

   Mr Scott replied that because of insulation and attention to detail around the liquor feeds current leakage was negligible.

2. Dr Janssen, Eindhoven University, raised the question of gas release from the anode surface and asked whether gaps or spaces were provided. The PPG electrolyzer is fitted with expanded metal anodes which have been designed for bubble release and also the anode/cathode gap is 6 mm which gave ample space for gas movement.

3. Professor Kahn of India asked what were the requirements for brine purity.

   Mr Scott replied that the brine purity requirements were the same as for any modern diaphragm cell plant, i.e. less than 10 p.p.m. calcium plus magnesium, so conventional brine treatment was all that was required.

4. Mr D. Ravier, Chlorchemie, enquired about the lifetime of the diaphragm in this electrolyzer.

   PPG have operated a prototype test cell which is still operating after 32 months with the same diaphragm, however, they would expect 2 years in a commercial installation.

CHAPTER 3

Synthetic Diaphragms for Chlorine Cells

J. J. BOWSER, W. L. Gore & Associates Inc.

3.1 INTRODUCTION

This chapter discusses the development of a microporous PTFE diaphragm to replace asbestos and modified asbestos in chlor-alkali cells.

In the mid-1970s Gore became interested in electrochemical cell separators as a potential new application area for their Gore-Tex®† expanded polytetrafluoroethylene (PTFE) technology. As a result, a programme of collaboration with ICI began in 1974 to develop a PTFE diaphragm for chlor-alkali cells utilizing Gore's expertise in expanded PTFE systems and ICI's chlor-alkali experience. The collaboration complemented the in-house ICI development of the DMT series of modern diaphragm cells.

3.2 THE ADVANTAGES OF A PTFE DIAPHRAGM

Currently, there are a number of modified asbestos diaphragm techniques which are considered state of the art for diaphragm cell use. Using these as a base case, one can consider what might be gained from the development of a PTFE-based microporous diaphragm. These advantages are listed in Table 3.1.

Table 3.1

Advantages of a new PTFE diaphragm over existing modified asbestos technology

Reduced cell voltage
Good physical uniformity
Improved chemical stability
Extended lifetime
Asbestos free

† Gore-Tex® is a trademark of W. L. Gore & Associates.

3.2.1 Reduced cell voltage

If a synthetic diaphragm is made which is immune to gas bubble erosion, the anode-to-cathode gap can be reduced substantially, resulting in lower IR losses in the brine. Additionally, due to a reduction in the amount of solid material between the electrodes, the voltage drop across the diaphragm will also be less. In these days of high energy costs, the financial savings can be significant.

3.2.2 Good physical uniformity

The technique of slurry deposition of asbestos fibre onto cathode boxes can lead to non-uniform laydown of the fibre. The synthetic diaphragm gives the opportunity for selection and quality control off the box and prior to fitting, particularly if the diaphragm is in the form of a sheet.

3.2.3 Improved chemical stability

The performance of the present modified asbestos diaphragms depends to a large degree on the unique and sensitive chemistry of silica and magnesium hydroxide gels. Sustained intrusion of acid brine or shifts in pH within the diaphragm due to changes in cell running conditions cause erosion of the diaphragm and subsequent loss in current efficiency. A synthetic, such as a PTFE, diaphragm whose properties are not dependent upon active chemistries such as gels, can withstand large fluctuations in running conditions, including complete power loss and subsequent restoration without deterioration in diaphragm performance.

3.2.4 Extended lifetime

The PTFE diaphragm can have an extended lifetime over modified asbestos as a result of the improved chemical stability and also better physical durability to gas bubble erosion.

3.2.5 Asbestos free

The handling of asbestos fibre for diaphragm deposition is carried out using safe working methods. However, general environmental concerns about the use of asbestos and the longer term availability of suitable quality asbestos fibre for diaphragm use will bring pressure to bear on its continuing use in chor-alkali cells.

3.3 ALTERNATIVE PTFE SYNTHETIC DIAPHRAGM

There is a signficant economic incentive embodied in the above advantages and therefore there have been many attempts to produce such a separator in the last decade. The patent literature is a good indicator of the extensive activity in this area. The main alternative types of separator which have been patented are listed in Table 3.2 and include the following examples.

Table 3.2
Patented synthetic diaphragm development

Deposited fibrous fluoropolymers
Calendered PTFE
PTFE fabrics and felts
Expanded PTFE

3.3.1 Deposited fibrous fluoropolymers

Under this first heading, fluoropolymer fibres are vacuum deposited from a slurry onto the cathode box in the conventional asbestos diaphragm manner. A wetting agent is either directly incorporated into the fibre during its manufacture or added into the fibre slurry. ICI, Solvay and BASF Wyandotte have all exhibited activity in this area.

3.3.2 Calendered PTFE

DuPont seems to have been the first to publicize its work with PTFE which is formed into a sheet by calendering. An inorganic filler provides the necessary porosity. The sheet is formed in the presence of a lubricant by rolling, followed by repeated folding and rolling at 90° increments. ICI and Hooker improved on this technique by adding removable fillers such as calcium carbonate or starch to increase the porosity of the diaphragm.

3.3.3 PTFE fabrics and felts

Fabrics and felts would typically be used as support media for materials such as silica or zirconia gels. The gel would be the active component in determining permeability and current efficiency rather than the PTFE structure. Olin, and to a limited extent, PPG, have been active in this area.

3.3.4 Expanded PTFE

This is the type of diaphragm which Gore and ICI have jointly concentrated on because it is considered that expanded PTFE has certain advantages over these other types.

3.4 THE GORE-TEX® DIAPHRAGM MANUFACTURING PROCESS

The process for manufacturing Gore-Tex® expanded PTFE products and the products themselves are covered in a number of Gore patents. The technology has been used by Gore since 1970 and is the basis for many well established product lines. For the electrochemical diaphragm, a proprietary chemically stable wetting agent is used which is blended with PTFE at a very early stage in processing. This allows the agent to be incorporated into the PTFE structure and avoids the necessity for what may be less permanent treatments at a later time. Since PTFE is not normally melt processible, the PTFE compound is mixed with a lubricant to allow it to be paste-extruded. This mixture is preformed in a pellet, then extruded through a die to give a flat sheet of material. Lubricant is removed by solvents or drying.

Porosity is developed through expansion of the sheet. The sheet is heated to a temperature below 327°C, the commonly accepted melt point of PTFE, and stretched at a high rate. Because this is also an orienting process, the matrix tensile strength of the polymer is increased through expansion. By adjusting certain process parameters, a wide range of porosities, tortuosities, and strengths can be obtained.

To add further strength and physical stability, the diaphragm structure is locked by heating above the melt point. This increases the amorphous content of the polymer, causing molecular entanglements and discouraging any further dimensional changes in the diaphragm.

These basic process details are shown schematically in Fig. 3.1. Diaphragms are produced in sizes up to 1.5 × 1.0 m, in sheet form. This is sufficiently large to clad all the alternative models of production diaphragm cells in operation today. The reproducibility of the process is good and the uniformity in properties is superior to a calendered sheet or a fibrous mat formed on a cathode box by vacuum deposition. Because the porosity is formed by expansion, there are no pore-forming agents to be leached out.

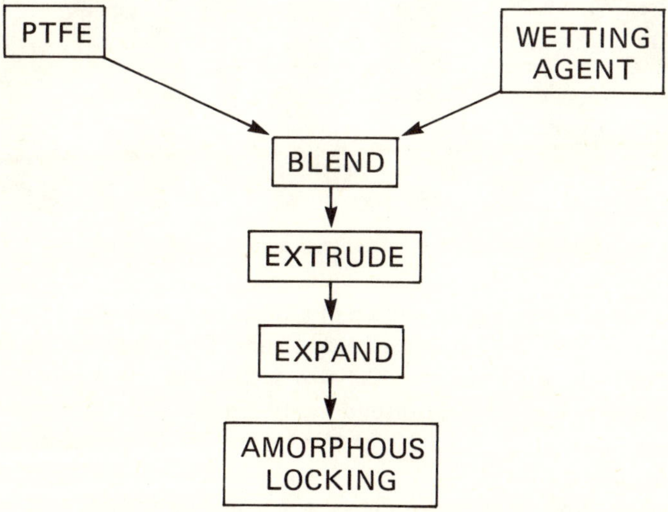

Fig. 3.1 – Gore-Tex® diaphragm manufacturing process.

3.5 PROPERTIES OF THE GORE-TEX® DIAPHRAGM

An electron micrograph of the surface of one of the diaphragms is shown in Fig. 3.2. The original magnification is ×1500, and the small white bar represents 5 μm. During expansion, fibrils of PTFE are formed, which are anchored by nodes. The strength of the polymer increases significantly during this process. In fact, matrix tensile strengths in the range of 10 000–20 000 p.s.i. are common. Despite this physical robustness, Gore-Tex® diaphragms are flexible, which aids in fitting them to existing cathode boxes. The diaphragms do not have any exotic storage requirements and shelf-life will not be a concern.

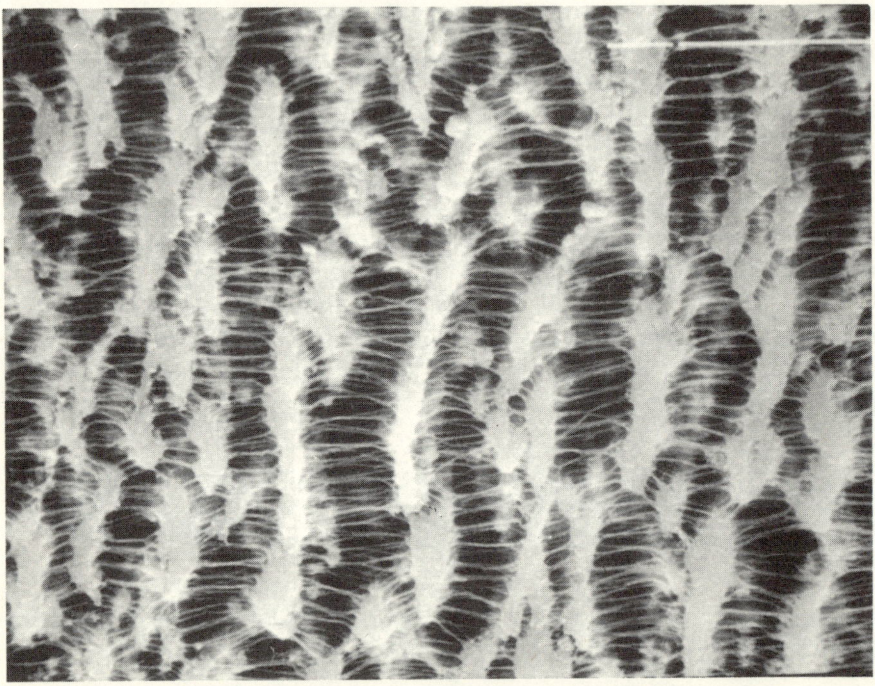

Fig. 3.2 – Gore-Tex® diaphragm surface (mag. ×1500).

In Fig. 3.3 is shown a cross-section of a Gore-Tex® diaphragm at an original magnification of ×50, the white bar being 50 μm. The magnification is too low to discern the microstructure, but the two surfaces of the diaphragm can be seen, and there is a good sense of uniformity, on this scale. A cross-section at a higher magnification of ×2000 is shown in Fig. 3.4. The microstructure can now be seen quite distinctly.

Electrolyte flow through the diaphragm would be vertical with respect to Figs. 3.3 and 3.4. The diaphragm is highly porous, about 75%, in fact, which allows for a low electrical resistance.

Fig. 3.3 – Gore-Tex® diaphragm cross-section (mag. ×50).

Fig. 3.4 – Gore-Tex® diaphragm cross-section (mag. ×2000).

Since the diaphragms are intended largely for retrofitting to existing cells, they must be designed within certain permeability limits. This is accomplished by varying thickness, tortuosity and porosity, each of which can be controlled in the Gore process.

Choosing the thickness of the diaphragm is governed by the trade-off between electrical resistance and current efficiency.

Tortuosity is sometimes thought of as a measure of effective path length through a porous medium. The nodes and fibrils act to divert hydraulic flow through a series of orifices and increase path length and resistance to flow. As a result the back migration of hydroxyl ions can be effectively controlled.

The effective size of pores such as these depends on the method by which they are measured. If one accepts that seeing is believing, then one might physically measure pore size from an electron micrograph. From the micrographs it is seen that the largest pore, or space between fibrils, is roughly $5 \times 1\,\mu m$. Most are smaller. Mercury porosimetry, while a well-established technique, is unsatisfactory with Gore-Tex® diaphragms. The porosity is so high that the hundreds of p.s.i. pressure necessary to force mercury into the pores crushes the structure. Methanol bubble point porosimetry, on the other hand, seems to give more useful numbers, of the order of $0.8\,\mu m$ maximum continuous pore size and a $0.4\,\mu m$ flow-weighted average. These results indicate a small effective pore size and also a narrow pore size distribution, both of which contribute to high current efficiencies.

3.6 GORE-TEX® DIAPHRAGM CELL PERFORMANCE

The initial stages of the diaphragm testing programme with ICI Mond Division were undertaken at the laboratory scale with micropilot cells of a 25 A capacity as shown in Fig. 3.5. From this scale the evaluation has proceeded through a series of pilot cells to full-scale commercial DMT43 units of 120 kA capacity.

During the course of the collaboration several alternative Gore-Tex® diaphragm systems have been developed and evaluated.

The laboratory cell experience based on selected data for the preferred system is summarized in Fig. 3.6.

The data shows that the diaphragm goes through a settling out period during the first 2 months of operation, after which it typically achieves a relatively stable condition.

The midline of the voltage plot, at 3.2V after it has stabilized, represents an energy saving over a modified asbestos cell of about 8%. This will be transferable to production cells if expandable anodes are used. For instance, an anode to cathode gap reduction from 6 to 3 mm gives a 3.5% voltage saving. This is augmented by a 4.5% reduction in IR drop across the diaphragm itself, for a total realized saving of 8%. This can be equivalent to about 250 kWh per t Cl_2.

Fig. 3.5 – 25 A micropilot cell.

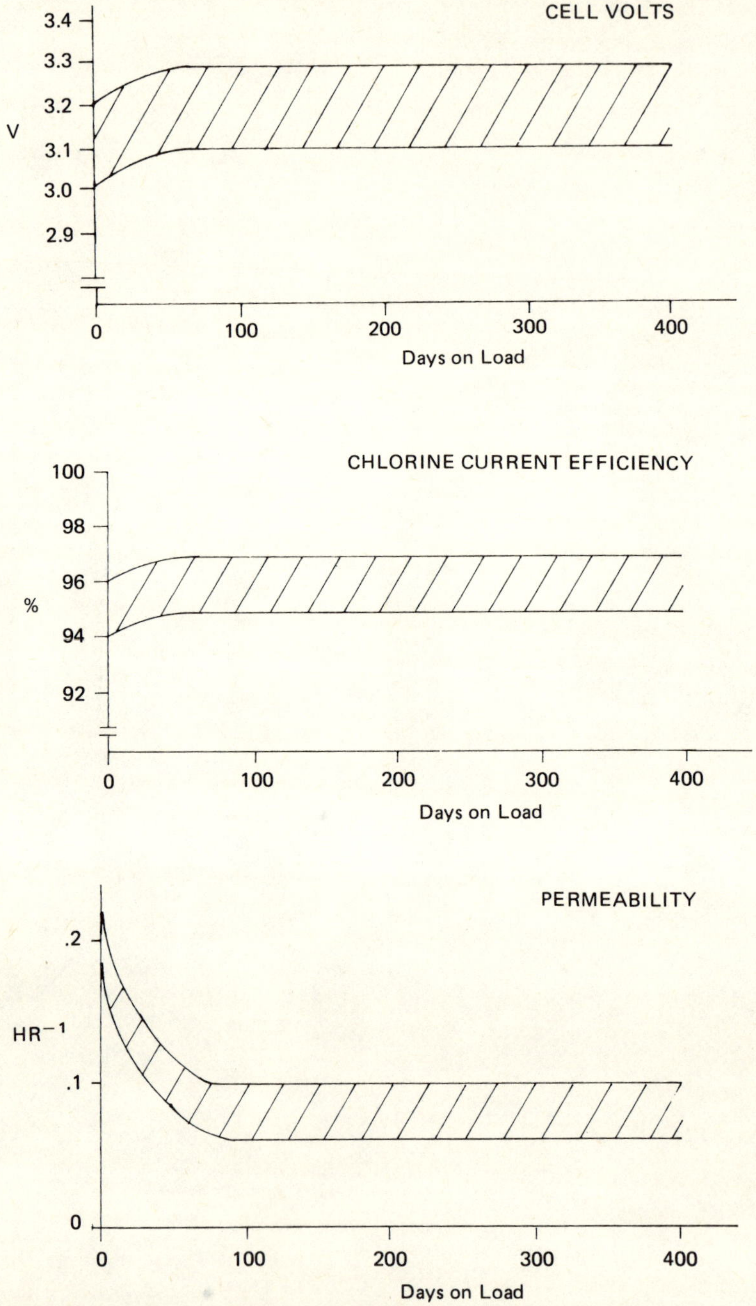

Fig. 3.6 – Micropilot cell data for Gore-Tex® diaphragm.

Current efficiency at 10–11% w/w NaOH production is found to be as good as with modified asbestos, with the added feature that it does not drop off with running time of the cell.

Permeability is stable after the initial 2 months and remains within the required limits for ecomonic operation of a production unit.

A working lifetime of 2 years is projected as a result of the good chemical and physical stability of the diaphragm. This may make lifetimes of other cell components, such as gaskets, more important than they are presently. The diaphragm is operated on prevailing modified asbestos quality brine. In the event of temporary extreme upsets in brine quality, the Gore-Tex® diaphragm can be successfully brine washed to remove impurities. The procedure is to remove power for a short time to allow the brine to lower the pH in the separator and flush deposits from the diaphragm.

An additional feature of this synthetic diaphragm arises from its insensitivity to changing cell chemistry and current loads. In certain situations, it may be economical to reduce load to the cell room during hours of high marginal electrical power cost and increase load during the corresponding inexpensive period, the magnitude of these load changes being limited by the local rectification capabilities. Although modified asbestos separators would suffer from this practice, Gore-Tex® diaphragms would not.

3.7 CLADDING OF CATHODE BOXES WITH GORE-TEX® DIAPHRAGM

To fit the Gore-Tex® diaphragm to existing cathode structures, a number of techniques have been devised. One of the preferred methods for a DMT43 cathode box is shown in Fig. 3.7.

Brine flow through the inactive top and bottom of the box is prevented by a melt processible fluoropolymer film. This has been moulded as shown to form what might be called a top hat. The diaphragm itself is formed into a continuous loop and lowered into the anode slot after supporting bridges in the top hat have been cut away. A liquor-tight seal is made between the top hat and the diaphragm with two titanium loops and a series of clips. Performance of this cladding system has been good, indicated by extremely low levels of H_2 in the chlorine cell gas.

Similar cladding methods have been developed for three major types of cells to date. These are the ICI DMT, Hooker H series and the Diamond-Shamrock MDC series.

During the course of the collaboration ICI has developed a number of cladding methods.

The patents which pertain to these techniques are given in Table 3.3. The first two refer to the fluoropolymer top hat. The third claims a procedure in which the diaphragms are stitched to join them together. The fourth describes

Fig. 3.7 – Cladding of DMT43 cathode box with Gore-Tex® diaphragm.

the clips as shown in Fig. 3.7. The last, a South African patent, claims a method which eliminates the clips in favour of a fluoropolymer weld.

Table 3.3
Relevant issued cell cladding patents

US4153530	'Top Hat'
US4156639	'Top Hat'
US4236952	Stitching
US4278523	Clipping
SA80-4021	Welding

3.8 THE PRESENT DEVELOPMENT POSITION

ICI is proving the Gore-Tex® diaphragm both in its cell development facility and at its 180 000 t.p.a. diaphragm cell plant at Wilton, England. This plant contains state of the art DMT43 cells. The Wilton evaluation is being conducted on a cell-by-cell replacement basis.

Gore and ICI plan for a joint commercialization with this product in the first quarter of 1983.

Although new technologies, such as ion exchange membranes are beginning to find their way into the chlor-alkali industry, many millions of tons of chlorine are, and will continue to be, produced in diaphragm cell installations. Gore-Tex® microporous diaphragms as a replacement for asbestos and modified asbestos can offer those locations the benefits of lower energy costs, improved diaphragm lifetime and operational stability, and avoidance of asbestos.

3.9 ACKNOWLEDGEMENT

The Gore-Tex® diaphram evaluation programme at ICI Mond Division has been co-ordinated throughout the development by Dr G.W. Cowell and his assistance in the provision of data and the preparation of this chapter is gratefully acknowledged.

SECTION B

The Development of Membranes and Membrane Electrolyzers

CHAPTER 4

High Performance Membranes for the Chlor-alkali Industry

S. M. IBRAHIM, E. H. PRICE, R. A. SMITH, E. I. Du Pont de Nemours & Co., Inc., Wilmington, Delaware 19898

4.1 INTRODUCTION

This paper will present some of the performance experience which has been gained on Du Pont's new Nafion®† perfluorinated membrane, designated Nafion® 901. This membrane has been extensively tested in laboratory and full-sized chlor-alkali cells around the world and recently has been chosen for use in new plant installations. Nafion® 901 is the first member of a new family of membranes which we are confident will be recognised as a major contribution to chlor-alkali technology. Data will be presented on efficiency, power consumption, durability, life, product quality and practical aspects of use.

4.2 NAFION® MEMBRANE PRODUCTS

High performance membranes for the chlor-alkali industry are the result of more than a decade of product development by Du Pont. The Du Pont Nafion® product line (Table 4.1) reflects the history of the development of membrane technology, and Du Pont's contributions to it.

Table 4.1
Nafion® products

100 series	Film – chemical stability
400 series	Reinforced – durability
300 series	Dilute caustic with high current efficiency
900 series	Combines high current efficiency, low voltage, high caustic strength and durability

† Registered trademark of E. I. Du Pont de Nemours & Co. Inc. for its perfluorinated membranes.

The first breakthrough was the availability of chemically stable membrane, introduced commercially by Du Pont in 1969.

Next came reinforced membrane for greater physical durability – our Nafion® 400 series.

The 300 series introduced the first reinforced *composite* membranes – a 'tight', low water absorption cathode-side layer for high current efficiency backed by a high water absorption anode-side layer for low voltage. This gave low power consumption for dilute caustic.

Low power consumption at high caustic strength was subsequently achieved by using a weak acid on the cathode side.

The 900 series, the first member of which is now fully commercial, combines truly high current efficiency with low voltage at high caustic strength in a durable product capable of sustained stable performance in commercially available membrane electrolyzers.

4.3 NAFION® 901

Nafion® 901, the first member of this new family of membranes, is a carboxyl-sulfonate bimembrane with a unique reinforcement.

Its high performance has been proven in several types of commercial size electrolyzers under a variety of operating plant environments around the world.

Importantly, it is now available in commercial quantities.

4.3.1 Current efficiency

Nafion® 901 gives high current efficiency 94.5–96.5% at high caustic strength, with a broad maximum at about 33% caustic (Fig. 4.1).

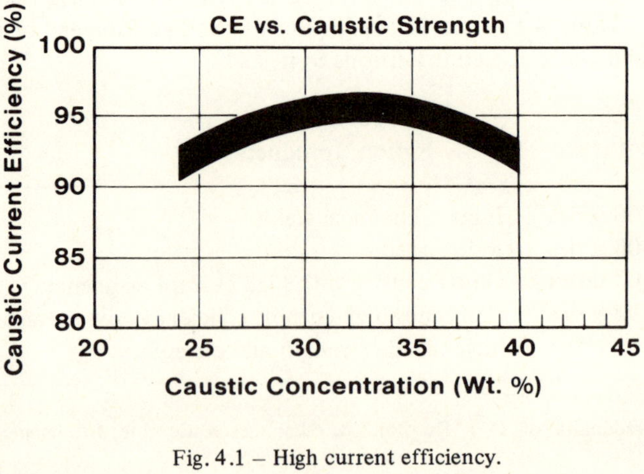

Fig. 4.1 – High current efficiency.

4.3.2 Voltage

Nafion® 901 cell voltage ranges from 3.3 to 3.9V at 3 kA/m^2.

Cell voltage is highly dependent on operating conditions: temperature, pressure, electrolyte concentrations, and current density. Cell voltage is also strongly influenced by cell design: structural IR drop, gap, membrane position, electrode overvoltage (especially the cathode, since the Beer's type anode has low overvoltage and is used widely), internal circulation of electrolyte, and electrode design.

4.3.3 Power consumption

Power consumption achievable with Nafion® 901 has ranged from 2200 kWh/MT at 2 kA/m^2 to 2900 kWh/MT at 5 kA/m^2 (Fig. 4.2). It should be pointed out that the voltages used in computing these power consumptions were electrode-to-electrode.

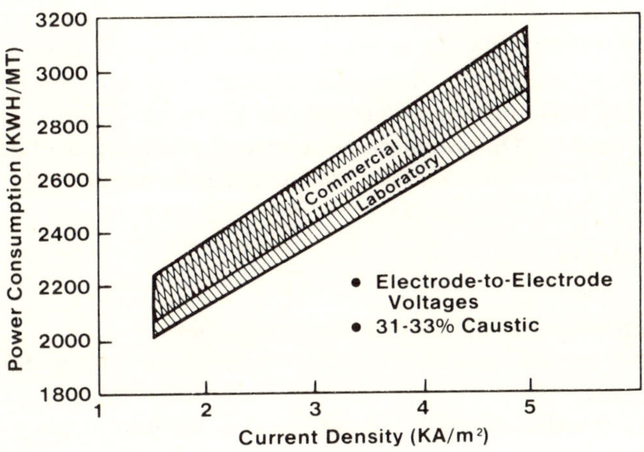

Fig. 4.2 – Power consumption.

Most experience has been gained at 3 kA/m^2 where power consumptions in the range of 2400–2600 kWh/MT have been realized in commercial sized electrolyzers. Laboratory cell experience has given results down to 2300 kWh/MT at 3 kA/m^2.

It is clear that Nafion® 901 is capable of high current efficiency, low voltage and consequently low power consumption.

What about stability of this high performance?

4.3.4 Performance stability

The stability of performance is excellent.

Figure 4.3 is an example. After 787 days, current efficiency remains at 95%,

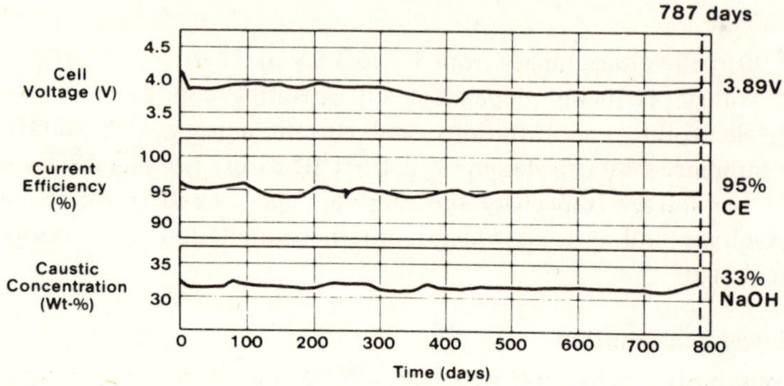

Fig. 4.3 – Stability of performance.

cell voltage is stable at 3.89 V (3.74 V electrode-to-electrode) and power consumption has been essentially flat for the entire run. This run in a laboratory cell was interrupted for re-gasketing. It was still running as this paper was printed.

This kind of performance, seen repeatedly in our hands and, more importantly, in the hands of our customers, has provided a high degree of confidence in the ability of Nafion® 901 to give *sustainted* high peformance.

4.3.5 Membrane life

How long will it last?

There is no numerical, universal answer to this question.

From what we have seen we think the useful life of Nafion® 901 will be measured in years.

It has the chemical stability and the physical stability to last for several years.

Actual life will depend strongly on conditions of use. The main factors will determine how long high performance can be realized are membrane handling and contact with metal, cell design, brine hardness, brine sulfate, corrosion, electrolyte concentrations, and temperature.

Pains to avoid mechanically damaging the membrane and to maintain careful process control are the most critical factors in realizing long membrane life.

4.3.6 Durability

The excellent durability of Nafion® 901 has been demonstrated.

In a $4 m^2$ cell which operated as an adjunct to a commercial chlor-alkali plant, 901 ran for a year with an *average* caustic current efficiency of 94.1% and stable voltage. This despite at least 28 shutdowns – scheduled and unscheduled, for the whole variety of reasons which chlor-alkali plant operators know all too well. Recovery from each of these upsets was rapid, and the membranes returned in each case to the level of performance seen prior to the upset.

In a 70m^2 electrolyzer, to cite a second example current efficiencies of about 96% and voltages in the 3.6 range were seen after 365 days of operation.

Nafion® 901 has been successfully removed for re-gasketing and restarted in commercial sized electrolyzers.

Perhaps the best way of describing the durability of Nafion® 901 is to use the words of one of our customers who is most familiar with this membrane. He calls Nafion® 901 an 'operator's membrane'.

An example of 901's ability to recover from an upset is seen in the two runs shown in Fig. 4.4. Caustic concentration was deliberately allowed to go to 40% in one case, and to 15% in another case. After the 10-day excursion, caustic strength was restored to 32%.

Recovery from each excursion was complete. It took less than 2 days for current efficiency to recover from the 40% excursion. It took about 4 days to recover from the dilute caustic excursion, a fact we attribute to the slower process of membrane dehydration. Recovery of voltage followed the same pattern.

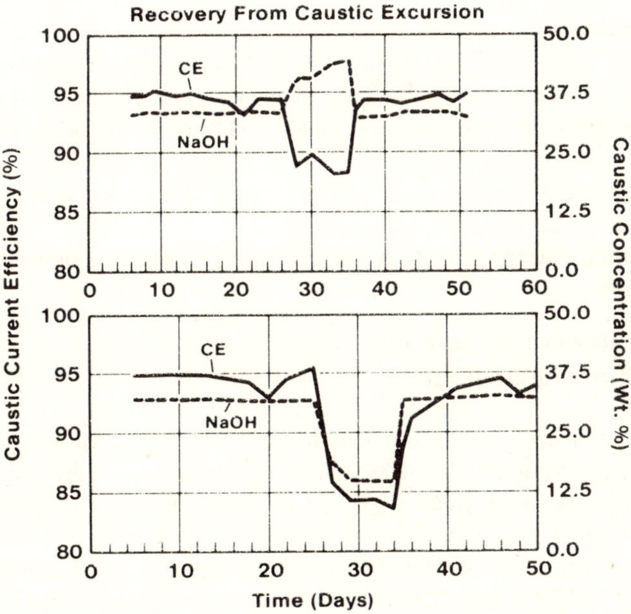

Fig. 4.4 – Durability.

4.3.7 Current density

Most of our experience with 901 has been at 3 kA/m^2.

We began testing at 5 kA/m^2 beacuse we thought it would give us an accelerated aging picture. As it turned out, however, Nafion® 901 runs fine at 5 kA/m^2. After 660 days it was still running at 94.8% current efficiency and the same voltage, $V_{ee}{}^{\dagger} = 4.08$, with which it started out.

Because water transport is higher at 5 kA/m^2, this 660 days is the approximate equivalent of 3.5 years of 'membrane work' at 3 kA/m^2.

A test at 10 kA/m^2 was run for a month to determine what physical damage might occur. Current efficiency of 93.0 and $V_{ee} = 4.30$ were observed. It was shut down for membrane inspection. There was no evidence of physical damage to the membrane from this brutal exposure.

To assess the membrane's ability to perform under daily current density change, Nafion® 901 was run for 490 days with daily cycling of current density between 2 kA/m^2 and 3 kA/m^2. Current efficiency remained high throughout; 96.6% was measured on day 490. Voltage dropped over the first few months to a level on the low side of our over-all experience with 901 in this type of cell, $V_{ee} = 3.38$ at 2 kA/m^2 and 3.56 at 3 kA/in^2. Also, the voltage coefficient was lower than is seen in normal operation.

The main point is that the membrane will tolerate load shedding and remain stable in performance.

4.3.8 Temperature

Most of the early work was done at 80°C. Testing at 90°C showed essentially no current efficiency penalty and no evidence of membrane instability. After 602 days at 90°C, performance was stable and current efficiency remained high at 95.7%.

A run at 95°C has gone for 517 days, again without any evidence of fall-off in performance. Current efficiency was 94.7%.

The voltage drops linearly in the 70–110°C range at a rate of 100 mV per 10°C increase.

4.3.9 Caustic and chlorine quality

There is a new term creeping into the chlor-alkali industry lexicon. It is 'membrane grade caustic'.

On a 50% caustic basis, sodium chloride is 100 p.p.m. or less. This has been determined for Nafion® 901 after 2 years of operation in laboratory cells; and after 1 year in a commercial sized cell, sodium chloride was measured at less than 50 p.p.m.

Chlorine typically contains less that 1.5% oxygen if neutral brine feed is used. If hydrochloric acid is added to the brine feed, the oxygen can be cut to less than 0.5%.

† V_{ee} = electrode-to-electrode voltage.

4.3.10 Convenience

To be practical in a chor-alkali plant, a membrane must be convenient to use.

Nafion® 901 *is* convenient.

It is fully conductive as received in the K^+ ionic form.

Pre-treatment is simple. Soak in 1–2% caustic for a few hours or overnight. Rinse, or mount directly.

Mounting the membranes wet, in their expanded state, insures that they will be flat under operating conditions.

If the electrolyzer is to be stored for more than a day or so before it is filled, simply add a little water to each cell as you build it, and stopper up the ports. The membranes will remain moist from the humidity in the electrolyzer until startup.

Membranes removed for subsequent re-use may be rinsed off, air dried, an stored until needed.

4.3.11 Availability

Nafion® 901 is available in standard widths of 0.61, 0.75, 1.22 and 1.50 m.

Standard length is 0.61 to 2.5 m.

Non-standard widths and lengths are available and will be quoted upon request.

4.3.12 Supply dependability

Nafion® 901 is manufactured at Du Pont's Fayetteville, North Carolina, plant. The intermediates for Nafion® are also manufactured there from basic fluorochemicals made at other Du Pont plant sites going all the way back to Du Pont-owned fluorspar in the ground.

The Nafion® plant incorporates the latest Du Pont fluorocarbon process technology, and it represents over a decade of perfluorinated membrane manufacturing experience.

This plant has the capacity to meet the total world demand for membrane for at least the next 15 years.

4.4 SUMMARY

In summary, Nafion® 901 is now fully commercial. Its high performance has been proven widely. It is a practical, durable membrane – an 'operator's membrane'. And supply of commercial quantities is assured from our plant at Fayetteville.

Nafion® 901 is the first membrane of a new family of high performance membranes which you can count on from Du Pont as you make your plans to adopt the membrane cell process. Nafion® 901 has been developed to complement electrolyzers designed by captive users and electrolyzers offered by the leading technology sellers worldwide.

Du Pont intends to play a leading role as membrane cell technology matures in the years ahead.

To this end we have an ongoing aggressive membrane product development program to provide optimum membranes for present and future electrolyzer designs.

Du Pont's developments aim at high performance which can be realized in available electrolyzers operating in the real world environment of a chlor-alkali plant. And we are committed to continue to supply membrane reliability at the lowest possible cost. This combination of high performance, product convenience and durability, and low cost will assure users of the highest possible value-in-use.

4.5 DISCUSSION

1. Mr V. Ramadurai from Standard Mills Co. Ltd., India, questioned the power differences quoted in Fig. 4.2 and also why current efficiency decreased in both cases shown in Fig. 4.4.

   Mr Smith replied that the power differences resulted from differences in cell design, both laboratory and commercial scale results were quoted. The reason for the decrease in current efficiency shown in both cases in Fig. 4.4 was illustrated in Fig. 4.1. Since current efficiency is a maximum at between 33 and 35% caustic strength, then an excursion into either a region of lower or of higher strength caustic would produce a reduction in current efficiency.

CHAPTER 5

AZEC System – New Ion-exchange Membrane Chlor-alkali Process

MASAO NAGAMURA, HIROSHI UKIHASHI, OSAMU SHIRAGAMI, Asahi Glass Co., Ltd.

5.1 INTRODUCTION

A membrane chlor-alkali process is now recognized worldwide as the most energy saving process for the production of chlorine and caustic alkali.

Asahi Glass, one of the leading companies in the Japanese chlor-alkali industry, has long experience in the development and manufacture of ion exchange membranes for desalination and concentration of sea water and other separation processes, along with electrodialysis cell technology. Furthermore, as one of the two major companies producing organic fluorine compounds in Japan, we have successfully commercialized various fluorochemical products which were the direct outcome of our research and development.

Drawing on our experience in these fields, we have been deeply involved in developing a new ion-exchange membrane for chlor-alkali production since 1974.

In the late spring of 1975, we announced our success in the development of high performance membrane made of an entirely new carboxylated perfluoropolymer. A pilot plant for manufacturing the membrane was brought into operation in the same year.

A chronology of our membrane chlor-alkali process development is summarized in Table 5.1. In May 1978, we started the commercial production of out membrane named Flemion. Along with the improvement of membrane itself, the development of an electrolyzer was forwarded which was to enable Flemion to function most effectively. Commercial Flemion chor-alkali plants have been operating successfully in our Osaka factory since September 1978, Nippon Carbide Co. since January 1981, and Thai-Asahi Caustic Soda Co. since July 1981.

Meanwhile, our membrane technology has been transferred on a contract basis successively to the competent world chlor-alkali producers. Last year, a contract with Du Pont Co. concerning the cross-licensing of the patents on ion-exchange membrane was announced, which paved the way for our mutual cooperation in this important technological field.

Table 5.1

A chronology of Flemion chlor-alkali process.

Year		Progress of technology		Technical disclosure
1975	June	High performance perfluorocarboxylate membrane (pat. appl.)		
	June	Monopolar cell		
1976	Sept.	Demonstration cell		
1977	Apr.	Uniquely reinforced membrane	Nov.	PPG (joint develop. bipolar cell)
1978	May	Commercial production of Flemion	Mar.	Olin
	Sept.	Start up of Osaka plant		
1979	Jan.	Flemion 330 for KOH		
	Feb.	Flemion 430 for less conc. NaOH	June	ICI
	Nov.	Pat. appl. for Flemion DX and AZEC		
1980	Nov.	AZEC system established	Jan.	Uhde, Hoechst, Bayer (membrane and cell)
1981	June	AZEC demonstration cell	Jan.	Nihon Carbide plant
	June	Expansion of membrane plant	July	THASCO plant
	Nov.	Commercial production of Flemion DX	Sept.	Du Pont (pat. cross licencing)
			Sept.	Dow Chemical
1982	Mar.	Advanced Flemion DX	May	Solvay

5.2 THE GENERAL ASPECT OF FLEMION

Flemion is made of an unique perfluoropolymer. The essential features of the polymer are summarized as follows:

(1) A carboxylated type ion-exchange group.
(2) A high ion-exchange capacity.
(3) A high molecular weight.

Such a structure leads to a low water content despite a high ion-exchange capacity, as shown in comparison with a sulfonated type in Fig. 5.1, which endows a high fixed ion concentration within the membrane resulting in a high current efficiency during electrolysis producing a strong concentrated caustic solution such as 35 to 40%. The electric conductivity of the membrane is as high as that of a sulfonated type as shown in Fig. 5.2.

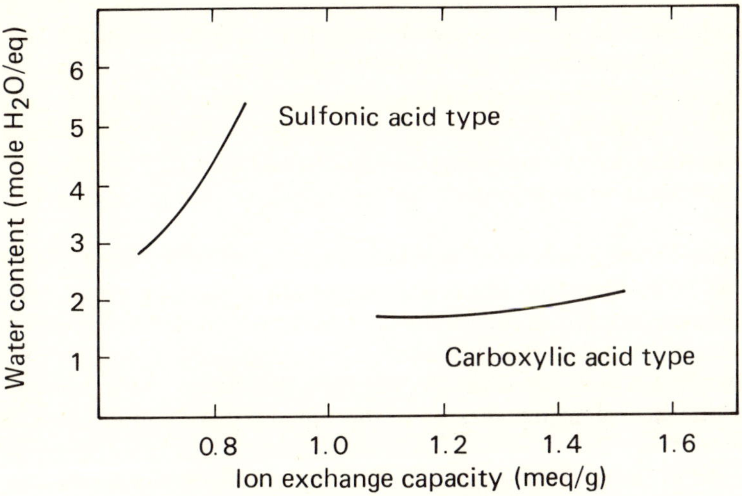

Fig. 5.1 – Water content of membranes.

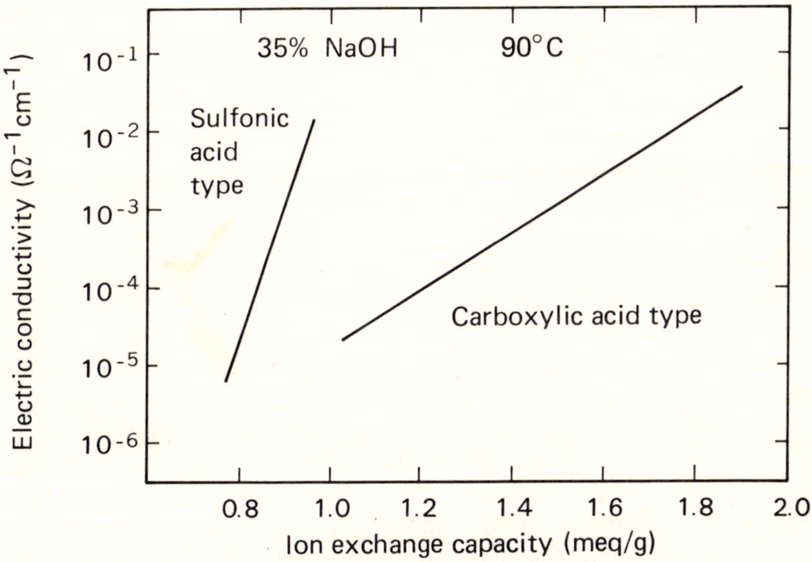

Fig. 5.2 – Electric conductivity of membranes.

The whole family of Flemion membranes produced from basic fluorochemicals, which are also our products, through a process comprising three main steps: (1) synthesis of functional monomers; (2) copolymerization of functional monomers with tetrafluoroethylene, and (3) membrane fabrication.

Owing to our extensive experience and the technical background in the synthesis of fluoro-organic compounds, we have developed various routes for synthesis of carboxylated perfluorovinylethers.

By a proper selection of the copolymerization process such as bulk, solution, or emulsion polymerisation, any copolymer of required ion-exchange capacity and adequate molecular weight can be produced without difficulty. The processability of the polymer is modulated not only by the adjustment of molecular weight, but also by the incorporation of non-functional perfluorovinylether as a third monomer so as to reduce the crystallinity of the resulting polymer, if necessary.

By use of these polymers, membranes reinforced by a unique method are fabricated by an advanced extrusion technique with even and accurate thickness. The membranes reinforced conventionally with woven fabric are also available.

Moreover, the lamination technique of two membranes of different thickness and different ion exchange capacities has been established. Through this means, the performance of membranes could be optimized in accordance with the conditions of use.

Recently, our further effort resulted in the accomplishement of a new high performance membrane, Flemion DX, and an innovative cell system using Flemion DX, named AZEC, which realized a cell voltage close to ideal.

5.3 AZEC SYSTEM

AZEC is Asahi Glass' proprietary membrane cell system where a new high performance membrane, Flemion DX, is applied in conjunction with a new electrode system. In this system, the distance between the membrane and electrode can be maintained at nearly zero without any inconvenience.

5.3.1 Flemion DX

Flemion DX has organic and inorganic composite texture. With this membrane, gas bubbles can be removed quite easily from the membrane surfaces. The goal has been attained by increasing the hydrophilicity of surface with a porous, non-electrode inorganic layer. The characteristics of this membrane are shown in Fig. 5.3, where the dependence of cell voltage upon electrode gap is shown in comparison with that for standard Flemion 230. A lower cell voltage than with Flemion 230 and its regular decrease with decrease of the gap results in little stagnation of gas bubbles on the surface of Flemion DX. This indicates the feasibility of the use of Flemion DX with zero gap between membrane and electrode.

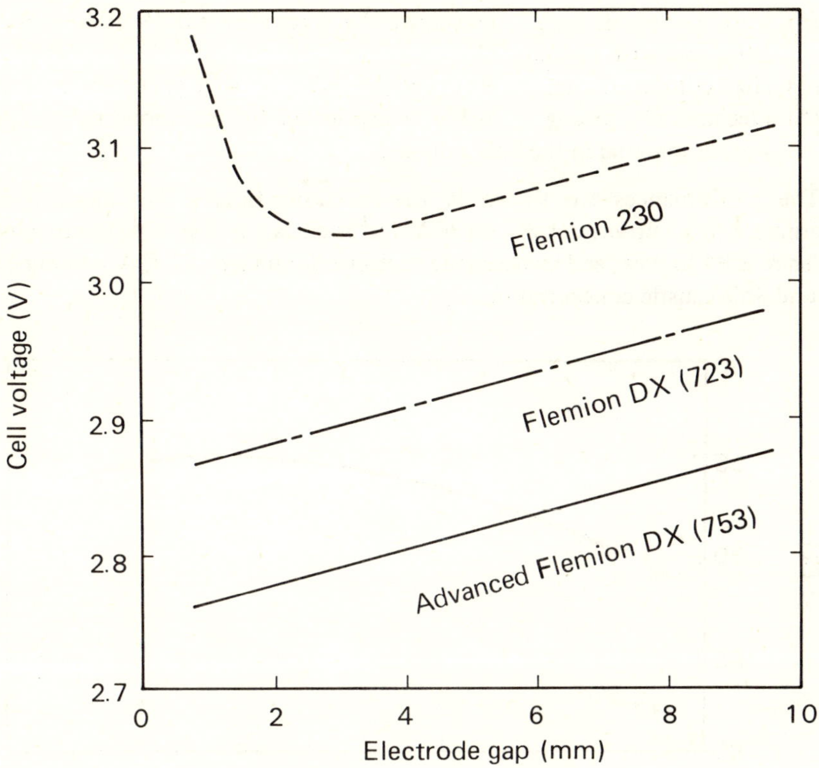

Anolyte	:	200 g/l NaCl
Catholyte	:	35 wt% NaOH
Current density	:	20 A/dm^2
Temperature	:	90°C

Fig. 5.3 – Comparison of dependence of cell voltage upon electrodes gap.

Flemion 723, which is shown in Fig. 5.3, has a similar chemical composition to Flemion 230 and can be used to replace the latter in the ordinary membrane cell to save the voltage drop.

Recently we have developed a new DX type of membrane, Flemion 753, which is specially designed for the use in AZEC system. The features of this membrane as as follows:

(1) Chemically homogeneous laminated membrane – functional group is consistently carboxylic type.

(2) Hydrophilic surface – easy bubble release – coated with non-corrosive and non-electroconductive inorganic materials.

(3) Very low ohmic drop – reducing cell voltage almost 0.2 V at 40 A/dm^2 compared with Flemion 723.
(4) High current efficiency – 94 to 95% over a broad range of current density.
(5) Mechanically strong – highly strengthened by incorporating special fibrous material in the bulk polymer.

The performances are shown in the following figures. The electrolytic performance at a current density of 30 A/dm^2 is shown in Fig. 5.4. The current efficiency is 94 to 95%, and an electrode to electrode voltage of 2.89 V is obtained at round 35% caustic concentration.

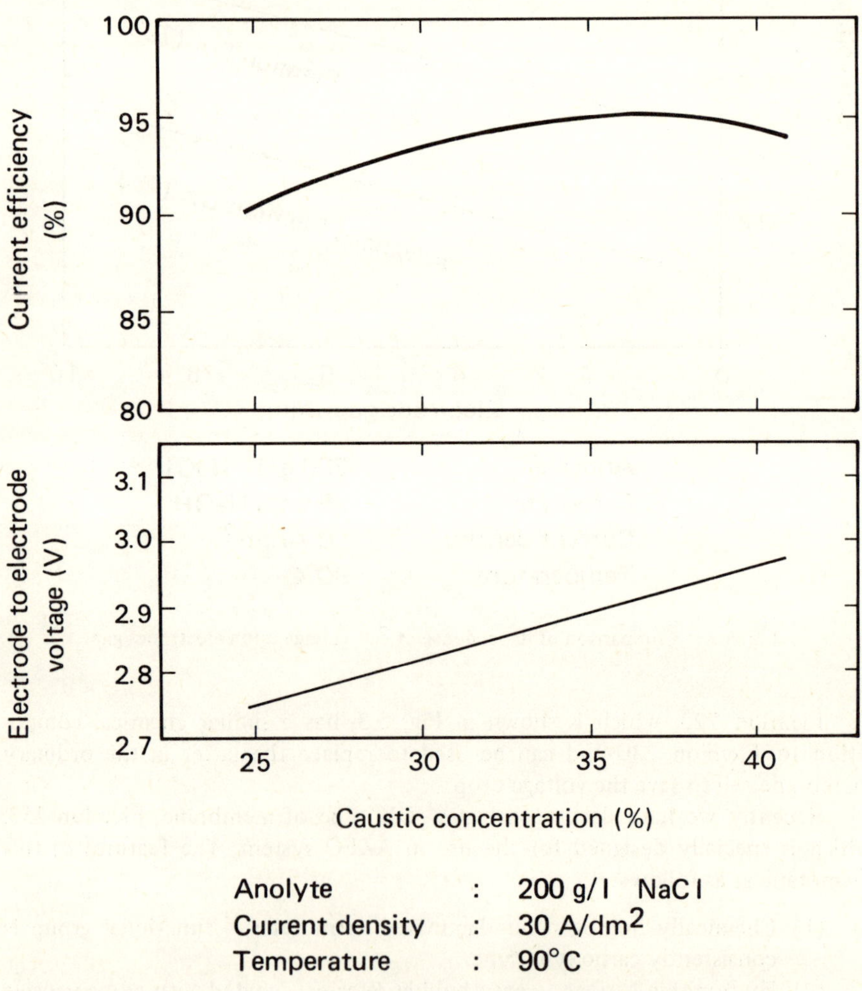

Fig. 5.4 – Electrolytic performance of Flemion 753.

The influence of current density on the performance of 753 is shown in Fig. 5.5, with an optimum caustic concentration of 35%. The high level of current efficiency is maintained over a broad range of current density. The increase of voltage with current density for 753 is much less than that for 723. At 40 A/dm^2, the voltage for 753 is as low as 3.02V, which is about 0.2V lower than 723. As shown in Fig. 5.6, the content of sodium chloride in caustic is less than 50 p.p.m. at 20 A/dm^2 and decreases with the increase of current density. All these facts indicate the advantage of using Flemion 753 at a high current density.

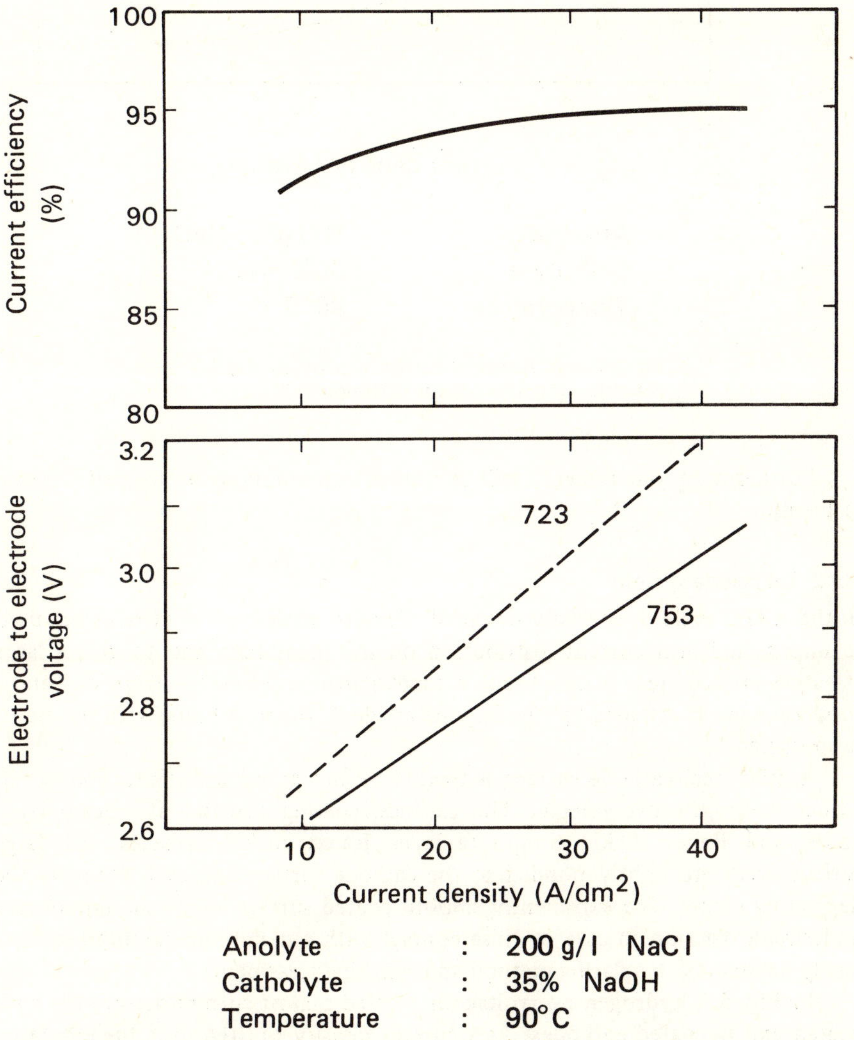

Fig. 5.5 – Influence of current density on performance of Flemion 753.

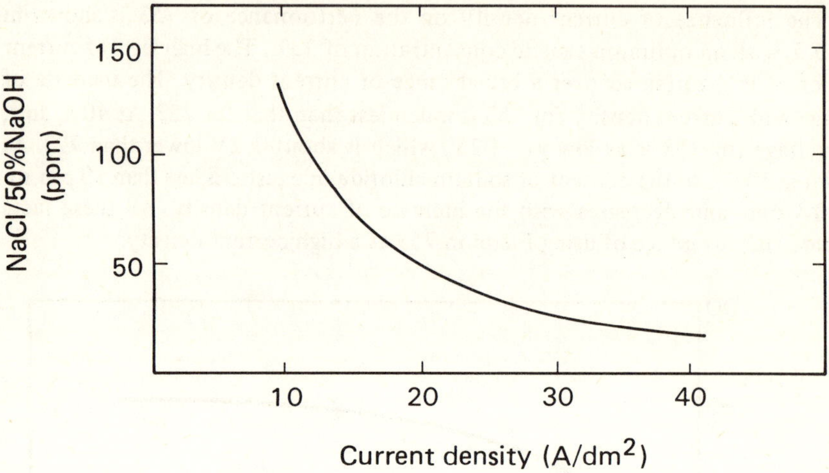

Anolyte	:	200 g/l NaCl
Catholyte	:	35% NaOH
Temperature	:	90°C

Fig. 5.6 – Influence of current density on sodium chloride content in caustic (Flemion 753).

Furthermore, our research still continues into obtaining better performance of Flemion.

5.3.2 Electrode system

In the AZEC system, specially designed, flexible mesh type electrodes are used to impose uniform current distribution on the membrane and to give a large effective surface area. A membrane is sandwiched between the electrodes, and a good zero-gap is attained by pressing the anode to the membrane with the aid of support ribs.

A mild steel cathode surface is treated by our active coating technology to reduce hydrogen overvoltage. The cathode coating consists of electroplated nickel and Raney nickel composite layer. Raney nickel particles with large surface areas are tightly bonded to the cathode surface and each other by the depositing nickel. The sufficiently smooth coated surface which should not give undesirable friction in case of close contact with a membrane is obtained by a newly devised electroplating method and finishing treatment.

In Fig. 5.7, hydrogen overvoltage is plotted against current density for both coated and uncoated cathodes. At a current density of 30 A/dm^2, the reduction of overvoltage is as much as 0.2 V, and is kept constant as shown in Fig. 5.8.

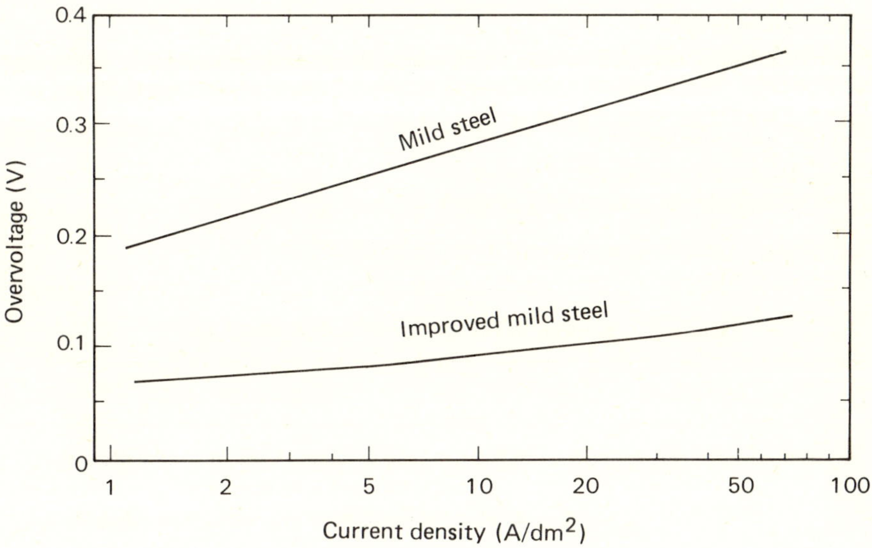

Electrolyte : 35% NaOH
Temperature : 90°C

Fig. 5.7 – Hydrogen overvoltage versus current density.

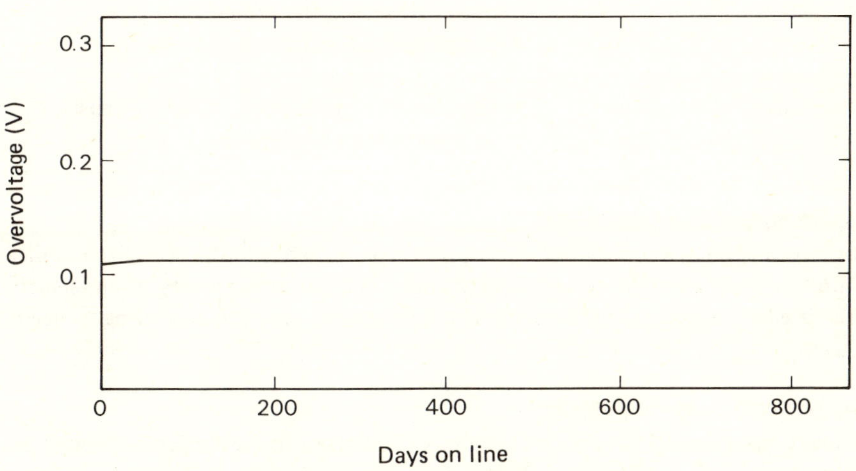

Electrolyte : 35% NaOH
Temperature : 90°C
Current density : 30 A/dm^2

Fig. 5.8 – Durability test of the improved mild steel cathode.

5.3.3 Electrolyzer

Our fundamental concept on the AZEC electrolyzer design lies on the following:

- Simple and compact structure
- Operational reliability
- High electrolytic performance
- Flexibility in production capacity

The factors examined are as summarized in Table 5.2.

Table 5.2

Factors examined in the design of AZEC electrolyzer

Factors	Remarks
Monopolar filter-press type	• Composed of assemblies of unit cells of 26 cm wide and 1.45 m high • Having flexibility in production capacity
Vertically long electrolyzer	• Gives rise to easy gas bubble release from electrodes and complete mixing of electrolytes due to a sufficient natural circulation
Short current path	• Electrode structure bringing on minimum ohmic drop and the skilled zero-gap • Simplified cell structure
Materials of cell structures	• Capable of a production of strong caustic solution with a high purity

(1) ***Monopolar filter-press type***

Our electrolyzer is a monopolar filter-press type, which is composed of assemblies of unit cells of 26 cm wide and 1.45 m high. The current capacity of an electrolyzer is easily adjusted by changing the numbers of unit cell according to needs of users.

(2) ***Vertically long electrolyzer***

Evolving gases induce a sufficient natural circulation of electrolytes through an electrolyzer. Gases are easily released from the electrodes, and the electrolytes are completely mixed.

(3) ***Short current path***

The electrode structure resulting in minimum ohmic drop and the skilled zero-gap has been devised, and the structure of an electrolyzer is simplified.

(4) ***Materials of cell structures***

The materials in contact with the electrolytes are carefully selected so as to produce a strong caustic solution with a high purity.

Figure 5.9 shows a schematic configuration of an AZEC electrolyzer. Electrolytes and gases pass through inner ducts of an electrolyzer to gas–liquid separators. Electrolytes, after being separated from gases are partially recirculated to the electrolyzer. Fresh feed of brine or water is added into the recirculated anolyte or catholyte respectively, and is mixed completely. The electrolyzer is tall and an electric connection between adjacent electrolyzers is easily conducted from side to side, which makes the length of bus bar short. Thus an ohmic drop becomes small and a floor space is utilized efficiently for installing electrolyzers.

A first monopolar AZEC prototype electrolyzer having a current capacity of 64kA was constructed in our Chiba factory and put into operation in June 1981. Since then, several modified electrolyzers have been operated in parallel as shown in Fig. 5.10 with expected performances covering a wide range of operational conditions. Figure 5.11 shows the dependence of the electric power consumption on current density. A high purity caustic soda is produced as shown in Table 5.3.

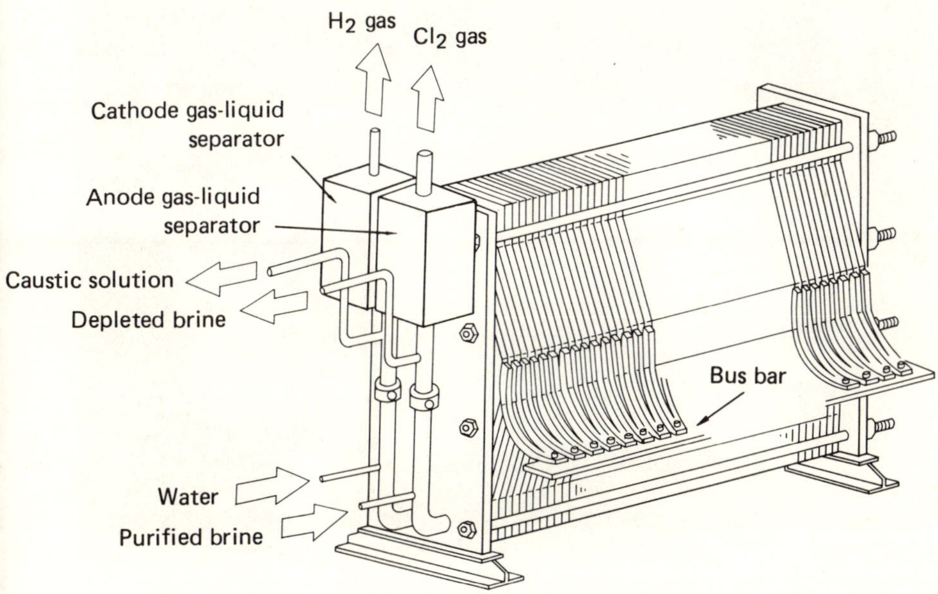

Fig. 5.9 – Schematic configuration of an AZEC electrolyzer.

Fig. 5.10 – Monopolar AZEC prototype electrolyzers.

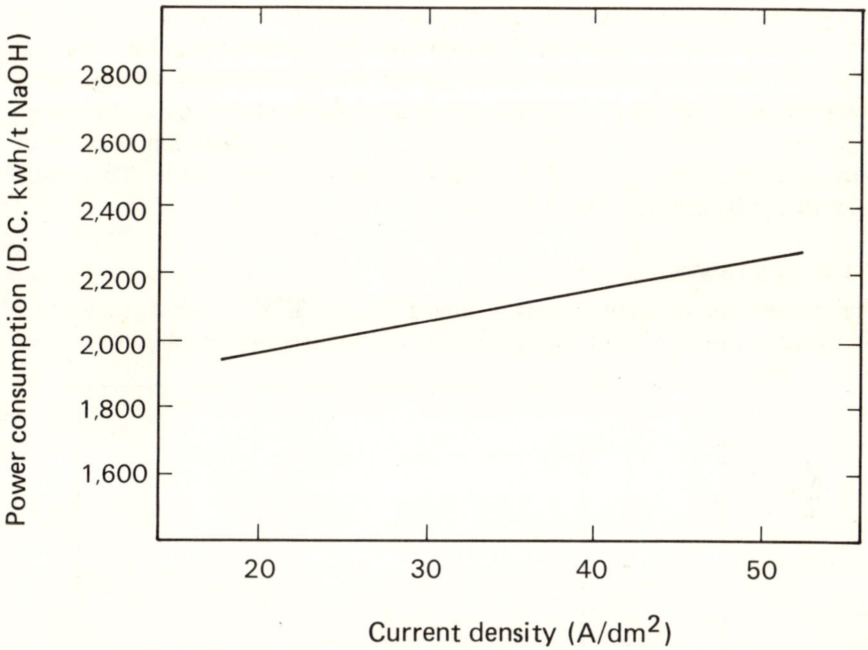

Fig. 5.11 – Consumption of electrolytic power of AZEC.

Table 5.3
An example of analysis of caustic soda

Brine:	Solar salt brine
Membrane:	Flemion 753
Current density:	30 A/dm^2

Component	Content
NaOH	35.1 %
Na_2CO_3	0.06%
NaCl	19 p.p.m.
$NaClO_3$	<10 p.p.m.
Na_2SO_4	<1 p.p.m.
SiO_2	5 p.p.m.
Al_2O_3	0.3 p.p.m.
CaO	<0.1 p.p.m.
Fe_2O_3	<0.1 p.p.m.

We are also applying AZEC to the electrolysis of potassium chloride. For this purpose, we have developed another type of DX membrane, Flemion 733. With this membrane in AZEC, 35% caustic potash solution is obtained at a current density of 30 A/dm^2 with an electrode to electrode voltage of 3.13 V and a current efficiency of 96 to 97%. It corresponds to a power consumption of only 1560 d.c.-kWh/t KOH. The content of potassium chloride in 50% caustic solution can be kept less than 50 p.p.m.

5.4 CONCLUSION

The progress of Flemion process is summarized in Fig. 5.12, which shows that the power consumption of the electrolysis was reduced by one-third over these

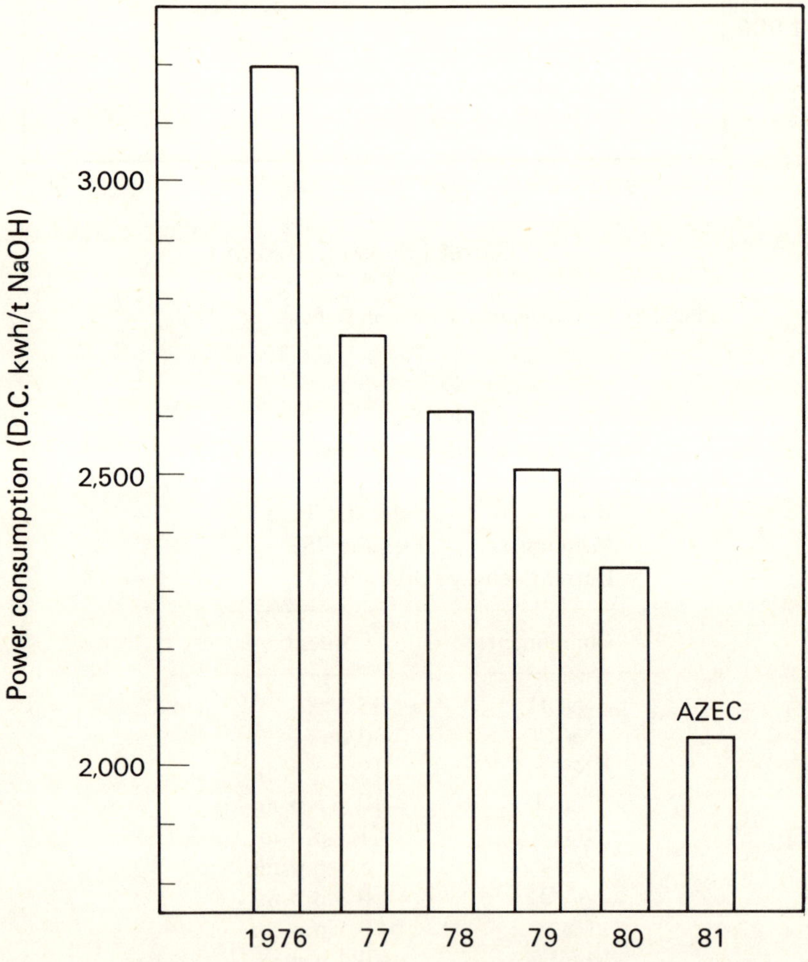

Fig. 5.12 – Progress of Flemion process.

5 years. We believe that the AZEC system, which is the collective outcome of our membrane, electrode and cell technology, may serve as a most energy saving and economic method of chlor-alkali production.

5.5 REFERENCES

[1] Ukihashi, H., and Asawa, T., *151st Meeting, Electrochem. Soc.*, 1977, **1**, 634.
[2] Ukihashi, H., *CHEMTECH*, 1980 (February), 118.
[3] Nagamura, M., Ukihashi, H., and Shiragami, O., *Modern Chlor-alkali Technology*, pp. 211–221, Ellis Horwood, Chichester, 1980.
[4] Nagamura, M., Ukihashi, H., and Shiragami, O., *Chem. Age of India*, 1980, **31**, 442.
[5] Ukihashi, H., *Chem. Economy & Eng. Review (Japan)*, 1981, **13**, No. 5, 29.
[6] Ukihashi, H., and Yamabe, M., *Perfluorinated Ionomer Membranes*, ACS Symposium Series No. 180, Edited by Eisenberg, A., and Yeager, H. L., 1982, pp. 427–451.

5.6 DISCUSSION

1. Dr G. J. Moody from Cardiff asked what plant modifications were needed in the case of potassium hydroxide production and also whether lithium hydroxide could be similarly produced.

   Dr Ukihashi explained that the same equipment could be used for potassium hydroxide as for sodium hydroxide, only the membrane was different. Asahi Glass have no experience of lithium hydroxide and do not see this as an area of future development.

2. Mr R. Williams from Canadian Occidental asked for a capital cost comparison between the AZEC cell and the older commercial MH4 cell.

   The AZEC system is a combination of new electrodes and electrolyzer with zero gap using Flemion DX membrane. The power consumption is lower but the capital costs of the two cells are more or less the same.

CHAPTER 6

Asahi Chemical's Perfluorocarboxylic Acid Membrane

MAOMI SEKO, HIROTSUGU MIYAUCHI, JUKICHI OHMURA, KYOJI KIMOTO, Asahi Chemical Industry Co., Ltd., Tokyo, Japan

Monolayer perfluorocarboxylic acid membrane has a high current efficiency in chlor-alkali electrolysis, but rather a high electric resistance. Asahi Chemical Industry Co., Ltd. has developed a unique multilayer membrane of perfluorocarboxylic and perfluorosulfonic acid which has both a current efficiency of about 95% and a low electric resistance by forming a carboxylic acid layer on one surface of the perfluorosulfonic acid membrane through chemical reaction.

The performance of this type of membrane is influenced by the chemical structure of the monomer from which the starting material is prepared. New perfluoro monomers containing sulfonic acid or its precursors were synthesized and utilized for the preparation of the multilayer membrane. The resulting membrane has high current efficiency in a wide range of caustic soda concentrations and excellent service life.

6.1 INTRODUCTION

Asahi Chemical has engaged in research on chlor-alkali electrolysis utilizing ion-exchange membranes of fluorocarbon polymer since the early 1970s, and discovered that the perfluorocarboxylic acid membrane provides the only possible means of obtaining a high current efficiency in the electrolysis of concentrated caustic soda solution.

In accordance with the results of this research, Asahi Chemical filed in 1974 the basic patent applications covering the material composition of membranes containing carboxylic acid and the use of such membranes in electrolysis throughout the world. These patents have been granted in various countries, and cover fluorocarbon carboxylic acid membranes and fluorocarbon membranes containing both carboxylic acid and sulfonic acid [1–9].

The world's first membrane chlor-alkali plant was started by Asahi Chemical in April 1975 at a capacity of 40 000 metric tons per year of caustic soda. This plant has since been expanded to a capacity of 120 000 metric tons per year of caustic soda.

Since the announcement of this plant operation [9] it has become the consensus of expert opinion that membrane electrolysis will be the predominant process for chlor-alkali production in the future. This is based on the following advantages:

1. Reduced energy consumption in the membrane chlor-alkali process through the utilization of perfluorocarboxylic acid membrane.
2. Lower investment cost due to high current density.
3. Ease of operation.
4. Lower operating cost.
5. High product purity.

Asahi Chemical has licensed around the world the entire package of technology including membrane, electrolysis system, electrolyzer, and heat recovery evaporator system [10–13]. In 1982, the total worldwide capacity of caustic soda production utilizing Asahi Chemical technology will be 520 000 metric tons per year.

This report describes some of the features of Asahi Chemical's perfluorocarboxylic acid membrane, including the basic design concept and the synthesis, structure, and performance of the membrane.

6.2 MEMBRANE CONTAINING CARBOXYLIC ACID

Perfluorocarboxylic acid membrane is the most desirable for chlor-alkali production, as it is the only fluorocarbon membrane to exhibit both a high chlorine resistance and a high current efficiency concentrated caustic soda. The high current efficiency is possible because perfluorocarboxylic acid membrane has a low water content and thus effectively prevents back migration of hydroxide ion. Perfluorosulfonic acid membrane, by contrast, is unable to prevent such back migration due to its low pK_a value and its high water content [14]. Figure 6.1 shows the relation between water content of membrane and caustic soda concentration. Figure 6.2 shows the relation between electrical resistance and caustic soda concentration.

The general methods for preparation of perfluorocarboxylic acid membranes are indicated in Table 6.1. Membrane may be prepared by copolymerization of tetrafluoroethylene with perfluorovinyl ether monomer containing carboxylic acid, its ester or precursor convertible to carboxylic acid [1, 5, 7], or by the introduction of a carboxylic acid group through chemical reaction causing the scission of C–S bond contained in the pendant of the perfluorosulphonic acid polymer [15–20]. Another method involves the coating of perfluorocarboxylic acid polymer of perfluorosulfonic acid membrane or the lamination of two different films of perfluoro polymer, one containing carboxylic acid group and the other containing sulfonic acid groups [2–4, 6, 8, 21, 22].

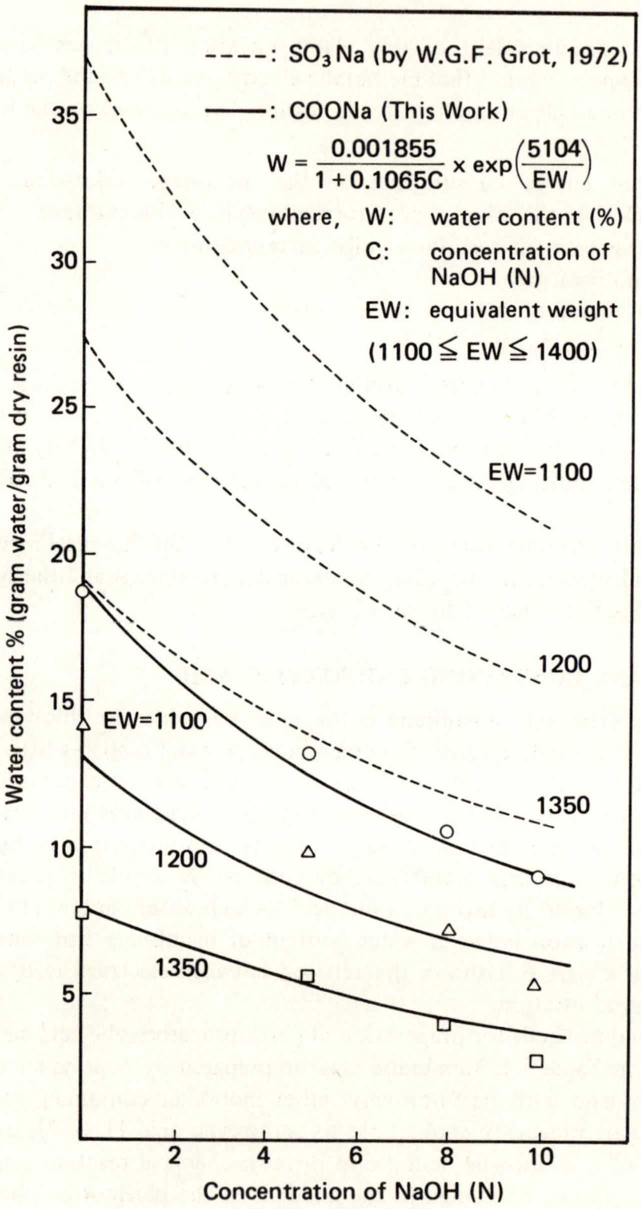

Fig. 6.1 – Water content, equivalent weight and concentration of NaOH, for perfluorocarboxylic and perfluorosulfonic acid membranes.

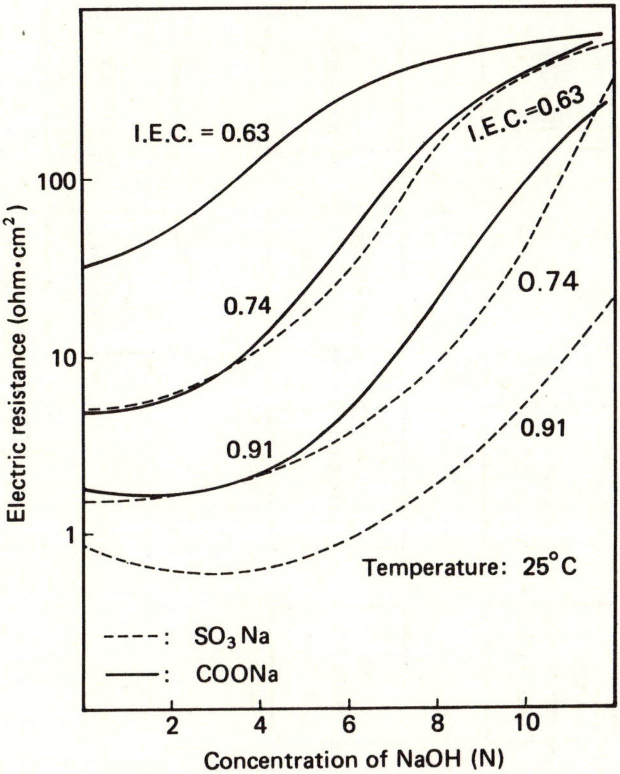

Fig. 6.2 – Electric resistance, ion exchange capacity (meq/gram dry resin) and concentration of NaOH, for perfluorocarboxylic and perfluorosulfonic acid membranes.

Table 6.1
Process for the preparation of perfluorocarboxylic acid membrane

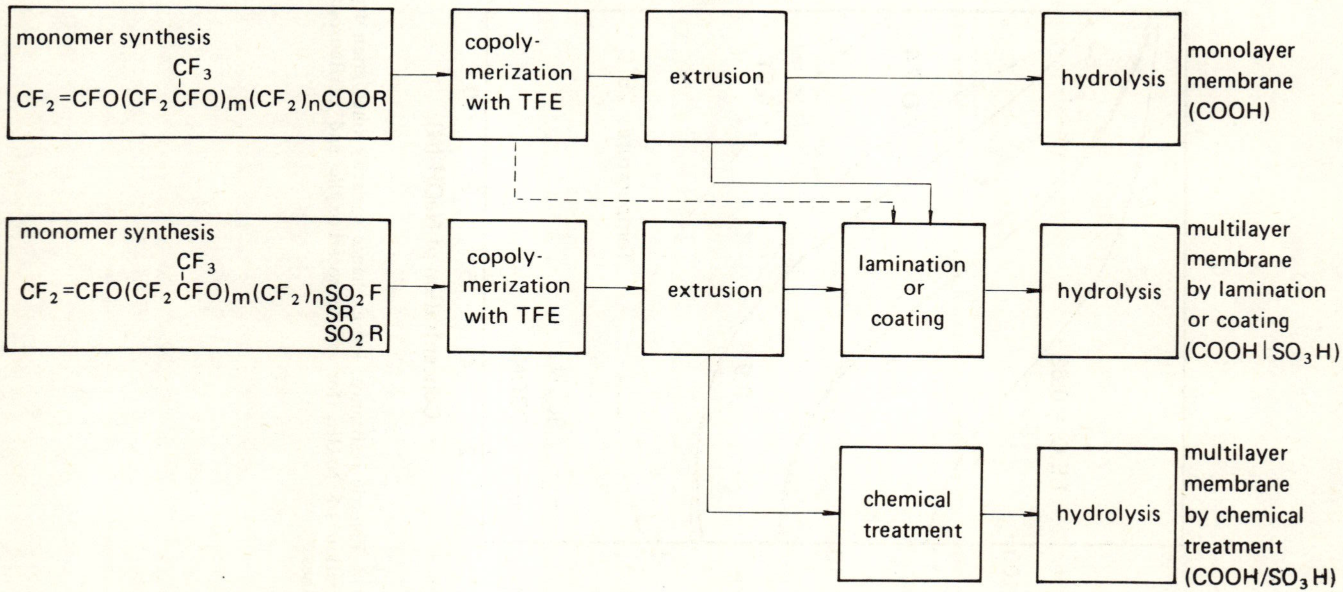

6.3 ION-EXCHANGE GROUPS AND MEMBRANE STRUCTURE

Perfluorocarboxylic acid membrane is prepared in monolayer or multilayer form. Table 6.2 shows a comparison of the properties of perfluorosulfonic acid membrane, perfluorocarboxylic acid membrane and multilayer perfluorocarboxylic and perfluorosulfonic acid membrane in which the carboxylic acid layer faces the catholyte.

As shown in Table 6.2, the multilayer membrane of perfluorocarboxylic acid and perfluorosulfonic acid has a current efficiency of about 95% or more and the lowest electric resistance, and is therefore the most desirable for use in electrolysis. With this membrane, hydrochloric acid can be injected into the brine stream in order to neutralize hydroxide ion migrating from catholyte, thus reducing the generation of oxygen and the formation of chlorate to very low levels. Consequently, chlorine of high purity containing less than 0.5% oxygen can be readily obtained with this membrane, and the anode life can be greatly extended over that obtained with other membranes [23]. The low electric resistance and high current density of this membrane allow significant reductions in the number of cells required and in electrolyzer investment cost.

Table 6.2
Membrane properties classified by ion-exchange group and membrane structure

	Ion-exchange group		
	R_f-SO_3H	R_f-COOH	R_f-COOH/ R_f-SO_3H
pK_a	<1	2–3	2–3/<1
Hydrophilicity	high	low	low/high
Water content	high	low	low/high
Current efficiency % (8N NaOH)	75	96	96
Electric resistance	low	high	low
Chemical stability	very high	high	high
Handling condition (pH)	>1	>3	>3
pH of anolyte	>1	>3	>1
Neutralization of OH^- by HCl	applicable	impossible	applicable
O_2 in product Cl_2	<0.5%	>2%	<0.5%
Life of anode	long	short	long
Current density	high	low	high

Membrane of 100% perfluorocarboxylic acid has high electric resistance and does not dissociate at lower pH, thus making it difficult to neutralize hydroxide ion migrating from catholyte by injecting hydrochloric acid. Consequently, the oxygen content of the chlorine is high and the anode life is reduced. Perfluorosulfonic acid membrane has relatively low current efficiencies, and thus cannot be utilized economically.

6.4 MULTILAYER MEMBRANE PREPARED BY CHEMICAL TREATMENT

Multilayer membrane of perfluorocarboxylic and perfluorosulfonic acid can be prepared either by coating perfluorocarboxylic acid polymer on perfluorosulfonic acid membrane or lamination of two different polymer films containing carboxylic and sulfonic acid [2–4, 6, 8, 21, 22] or by chemical conversion of perfluorosulfonic acid to perfluorocarboxylic acid [15–20]. A comparison of these membranes is shown in Table 6.3.

In the coating of perfluorocarboxylic acid polymer and the lamination of two polymer films, the minimum thickness of the carboxylic acid layer which can be obtained is about 25–50μm. This limits the degree to which the electric resistance can be lowered. A further disadvantage of this process is the necessity of separately preparing both sulfonic acid and carboxylic acid polymers.

Table 6.3
Comparison of monolayer and multilayer perfluoro ionomer membranes

	Type of membrane		
	Monolayer perfluoro-carboxylic acid membrane (COOH)	Multilayer membrane of perfluoro-carboxylic and sulfonic acid by lamination or coating (COOH SO_3H)	Multilayer membrane of perfluoro-carboxylic and sulfonic acid by chemical treatment (COOH/SO_3H)
Thickness of COOH layer (μm)	>125	25–50	⩽10
Current efficiency % (8N NaOH)	96	96	96
Electric resistance	high	medium	low
Necessary number of functional monomers	1	2	1
Membrane fabrication	extrusion	extrusion and lamination or coating	extrusion and chemical treatment

Chemical conversion is considered to be the best method for the preparation of membrane for the chlor-alkali process. With this method, the thickness of the carboxylic acid layer which forms on the surface of perfluorosulfonic acid membrane can be easily reduced to less than 10 μm, resulting in a corresponding reduction in the electric resistance of the membrane. This process also effectively prevents peeling of the carboxylic acid layer, as it is formed from the original polymer matrix with a gradual gradient in composition between the two layers. In this process, moreover, only perfluorosulfonic acid polymer need be prepared beforehand to obtain the multilayer membrane of sulfonic acid and carboxylic acid.

6.5 THICKNESS OF CARBOXYLIC ACID LAYER, CURRENT EFFICIENCY AND MEMBRANE RESISTANCE

The current efficiency and the electric resistance are shown in Fig. 6.3 for the multilayer membrane prepared by chemical treatment. As shown in this figure, the electric resistance increases linearly with the thickness of carboxylic acid layer. The current efficiency increases greatly until the thickness of the carboxylic acid layer reaches 7 μm where saturation tends to occur. From the point of view of power consumption, the optimum thickness of the carboxylic acid layer is in the range of 2–10 μm. Thus chemical treatment is the most appropriate method for preparation of carboxylic acid membrane, because the highest current efficiency can be obtained with a minimum increase in electric resistance.

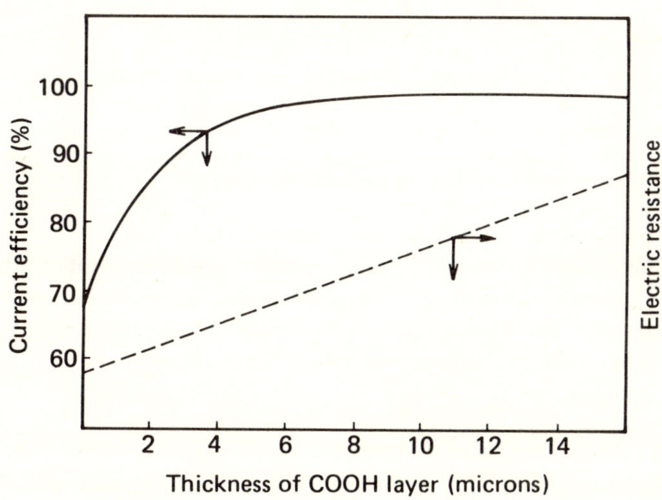

Fig. 6.3 – The thickness of COOH layer, current efficiency, and electric resistance of a multilayer perfluorocarboxylic and sulfonic acid membrane prepared by chemical treatment ($COOH/SO_3H$).

Figure 6.4 shows the relation between the current efficiency, caustic soda concentration and ion exchange capacity of multilayer membrane prepared by chemical treatment of perfluorosulfonic acid membrane. Each membrane shows maximum current efficiency at a certain caustic soda concentration. This is because with increasing concentration of the external caustic solution up to a certain value, the membrane tends to shrink and the COO^- fixed ion concentration in the membrane tends to increase. However, if the concentration of the caustic soda increases beyond this value, hydroxide ion diffusion tends to increase due to increased electrolyte sorption, thus tending to decrease current efficiency.

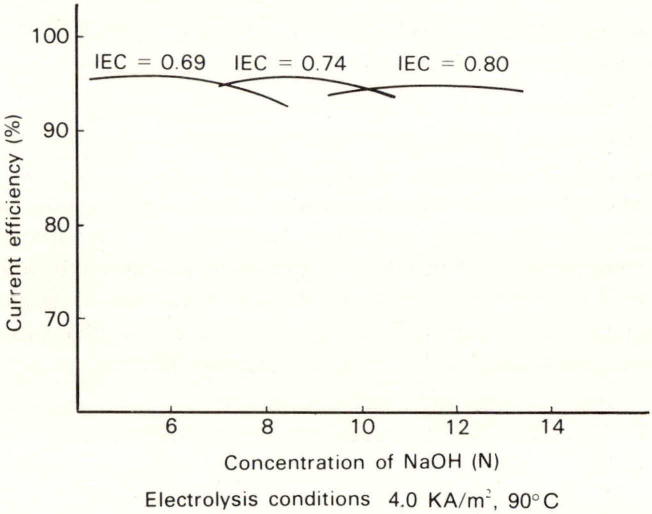

Fig. 6.4 – Current efficiency, ion exchange capacity (meq/gram dry resin) and concentration of NaOH, for multilayer perfluorocarboxylic and sulfonic acid membrane prepared by chemical treatment ($COOH/SO_3H$).

Increasing the ion-exchange capacity of the membrane results in the occurrence of the maximum current efficiency at a higher caustic soda concentration. Therefore in order to obtain high concentration of catholyte, the membrane should have carboxylic acid groups of high exchange capacity.

6.6 CARBOXYLIC ACID FORMATION

Table 6.4 shows the most appropriate method of chemical treatment to convert perfluorosulfonic acid membrane or its precursor to perfluorocarboxylic acid membrane [15, 16]. In this process the sulfonyl halide group in the membrane is converted to sulfonic acid which is then converted to the perfluorocarboxylic acid group through desulfonylation reaction. The perfluorocarboxylic acid pendant has one CF_2 less than the original chain of perfluorosulfonic acid. The extremely high selectivity of this reaction effectively prevents the formation of impure products, which would be impossible to remove from the membrane.

Table 6.4

Chemical treatment to prepare perfluorocarboxylic acid membrane from pefluorosulfonic acid membrane or its precursor

$$-O(CF_2)_nSO_2F \xrightarrow{\text{reduction}} -O(CF_2)_nSO_2H$$

$$-O(CF_2)_nSO_3H \longrightarrow -O(CF_2)_nSO_2Cl \xrightarrow{\text{reduction}} -O(CF_2)_nSO_2H \xrightarrow{-SO_2} -O(CF_2)_{n-1}COOH$$

The following alternative reactions for conversion of perfluorosulfonic acid to perfluorocarboxylic acid have been reported [17–20, 24], but have not been found practical or appropriate:

1. Conversion through CF_2I.
2. Direct preparation of carboxylic acid group from SO_2Cl or SO_3H group.
3. Addition reaction of the pendant having CF_2I or SO_2Cl group.

6.7 MONOMER STRUCTURE

Perfluorosulfonic acid membrane is utilised as a raw material for preparing multilayer membrane by chemical treatment. Table 6.5 shows the conventional method of Du Pont de Nemours & Co. for preparing perfluorosulfonic acid membrane which utilizes perfluorosulfonyl fluoride monomer copolymerized with tetrafluoroethylene [25, 26]. However, short chain sulfonyl fluoride monomer of $m = 0$ cannot be synthesized with this method, since its production is made impossible by ring formation [27]. It is therefore necessary to utilize long chain sulfonyl fluoride monomer of $m = 1$. With this monomer, however, it is impossible to obtain the ion exchange capacity of more than 0.9 meq per gram dry resin. Moreover, the current efficiency of multilayer membrane prepared from this perfluorosulfonic acid membrane tends to decline during long term electrolytic operation.

Table 6.5

Synthesis of $CF_2{=}CFO(CF_2CF(CF_3)O)_mCF_2CF_2SO_2F$

(where $m = 0$ or 1)

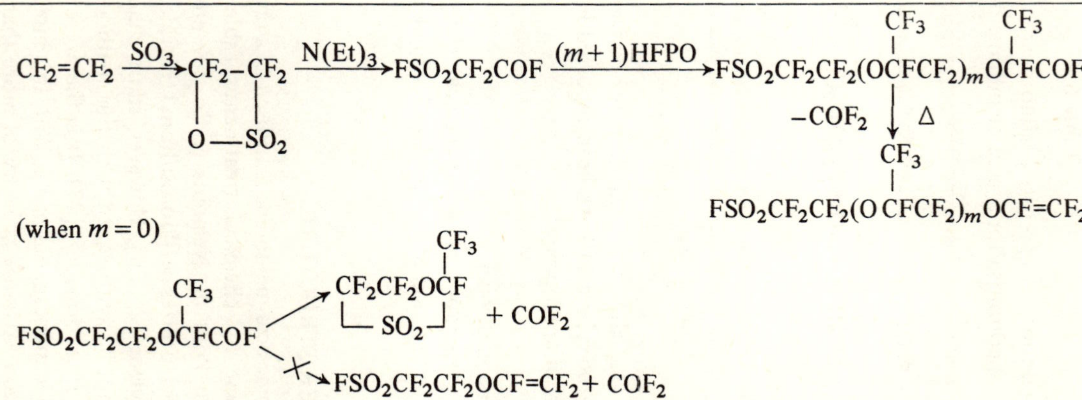

Table 6.6

Synthesis of $CF_2{=}CFO(CF_2CF(CF_3)O)_m(CF_2)_n{-}SO_2F$

(where $m = 0$ or 1, $n = 3$–5)

$n = 3$

$n = 3$–5

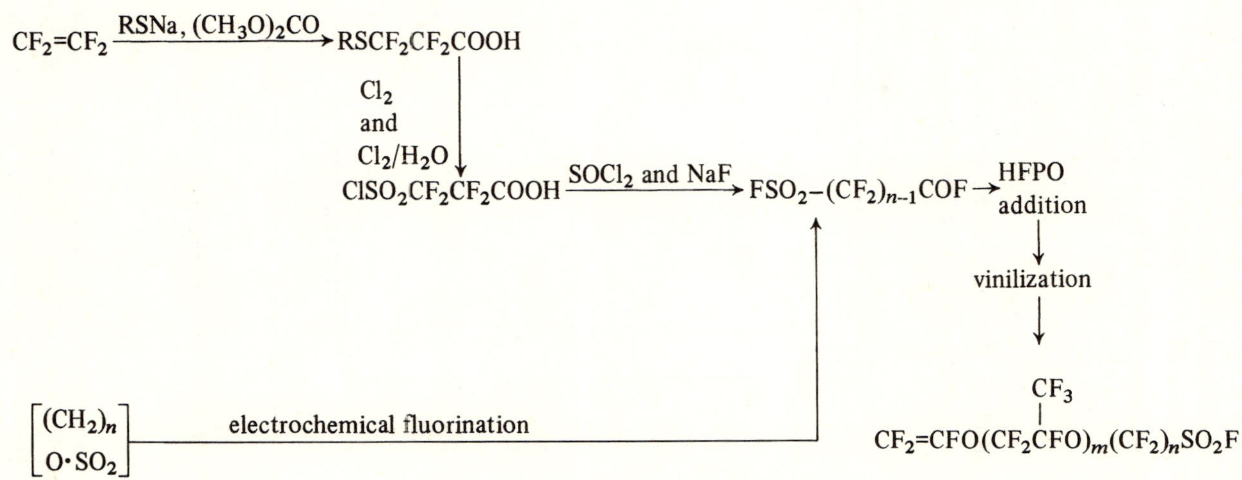

In order to avoid these disadvantages of the conventional perfluorosulfonic acid membrane, Asahi Chemical developed new perfluoroether monomers, which are synthesized as indicated in Table 6.6 [16, 28–35]. These monomers are unique in the number of CF_2 groups bonded to the functional group. This effectively prevents cyclization and thus allows the synthesis of monomer with $m = 0$.

Figure 6.5 illustrates copolymer composition curves for the perfluorovinyl ether monomer with $n = 3$ and $m = 0$ or 1 when it is copolymerized with tetrafluoroethylene. Polymer can be obtained in a wide range of ion-exchange capacities by adjusting the initial composition.

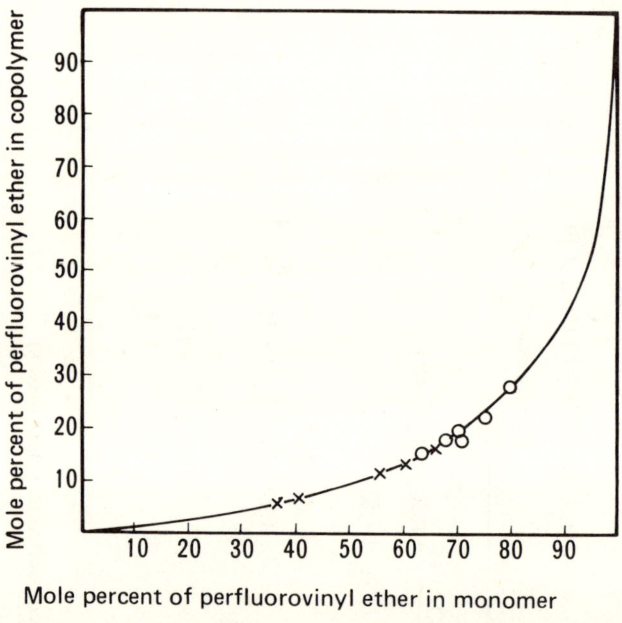

Fig. 6.5 – Copolymer composition curves. $n = 3$, $m = 0$ (○) or 1 (×).

Figure 6.6 shows the relation between the electric conductivity and the ion-exchange capacity of perfluorosulfonic acid membrane prepared from the monomer. The membrane of monomer with $m = 0$ has an ion-exchange capacity of more then 1 meq per gram dry resin, and high physical strength. The membrane of monomer with $m = 1$ has a lower ion-exchange capacity but is still highly flexible. These properties are different from those of conventional perfluorosulfonic acid membranes.

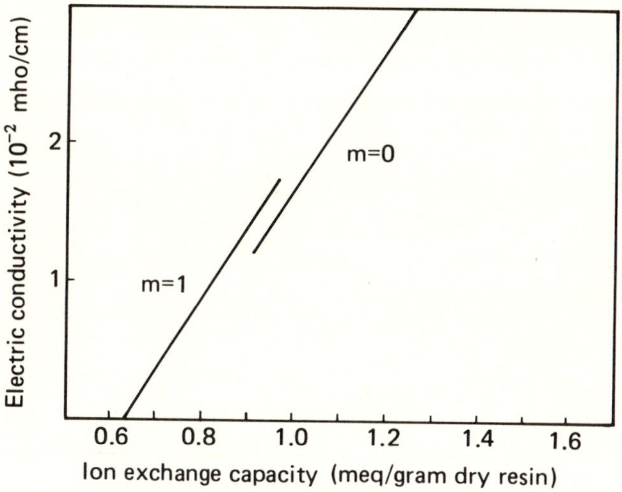

Fig. 6.6 – Electric conductivity in 0.1 N NaOH at 25°C and ion exchange capacity for perfluorosulfonic acid

$$-O(CF_2\overset{\overset{\displaystyle CF_3}{|}}{C}FO)_m(CF_2)_3-SO_3H \text{ membrane.}$$

In the Asahi Chemical process, membrane can be prepared by utilizing combinations of monomers with $m = 1$ and $m = 0$ in any given composition. Therefore membrane can be tailored for any specific concentration of caustic soda from 5% to 40% in electrolysis. Because of the structure of the perfluorosulfonic acid pendant, which differs from that in conventional membranes, Asahi Chemical's multilayer membrane retains a stable, high current efficiency regardless the length of electrolytic operation.

6.8 HYDROGEN GAS ADHESION ON THE SURFACE OF THE MEMBRANE

In the electrolysis conditions, a certain amount of hydrogen gas bubbles tends to adhere to the surface of the membrane. The bubbles can not be removed easily when the distance between the electrodes is below 1mm, resulting in an increase of cell voltage in this region with descreasing electrode distance, as the dotted line in Fig. 6.7 shows. Asahi Chemical developed a unique surface treatment which gives a roughened surface to the membrane during its manufacture to prevent the adhesion of hydrogen gas bubbles to the membrane [36]. With this rough surface membrane, cell voltage decreases with the electrode distance even when it is below 1mm as shown in Fig. 6.7, which allows a further reduction in electrolytic consumption.

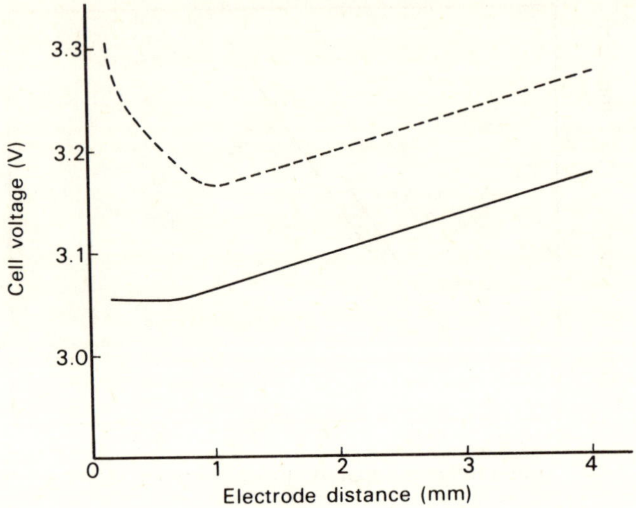

Fig. 6.7 – Electrode distance versus cell voltage.

------- Conventional membrane
——— Rough surface membrane
1.2m × 2.4m cell

Current density: 4.0kA/m²
Temperature: 90°C.

6.9 TYPICAL ASAHI CHEMICAL MEMBRANE

Table 6.7 shows a typical example of the structure of an Asahi Chemical multilayer membrane of perfluorocarboxylic acid and perfluorosulfonic acid made by chemical treatment which has a carboxylic acid layer of about 2 to 10μm thickness. This membrane exhibits high current efficiency and low electrical resistance and can be tailored to any concentration of caustic soda. It yields high performance and especially long service life.

Table 6.7

Structure of Asahi Chemical's membrane

(1) Multilayer membrane prepared by chemical treatment
(2) Thickness of COOH layer: 2–10μm
(3) Chemical structure

$$-(CF_2-\underset{|}{C}F)- \quad \underset{|}{CF_3}$$
$$O-(CF_2CFO)_m-(CF_2)_{n-1}-COOH$$

$$-(CF_2-\underset{|}{C}F)- \quad \underset{|}{CF_3}$$
$$O-(CF_2CFO)_m-(CF_2)_n-SO_3H$$

$(m = 0 \text{ or } 1,\ n = 3\text{–}5)$

Table 6.8 shows typical examples of electrolysis performance in small laboratory cells. These membranes are designed for use with caustic soda in concentrations of 21%, 25%, 30% and 35%.

Tables 6.9 and 6.10 show typical mechanical properties of Asahi Chemical's membrane.

Table 6.8
Performance of Asahi Chemical's membrane

Membrane	Current efficiency (%)	Cell voltage† (V)
Membrane for 21% NaOH	96	2.98
Membrane for 25% NaOH	96	3.03
Membrane for 30% NaOH	96	3.08
Membrane for 35% NaOH	95	3.14

† In laboratory all at 4 kA/m² and 90°C.

Table 6.9
Typical mechanical properties of unreinforced membrane†

Tensile strength (kg/cm²)	390
Elongation (%)	140
Tensile modulus (kg/cm²)	3620

Sample was measured in Na type and in 50% relative humidity at 25°C

† Equivalent weight of sample is 1250.

Table 6.10
Typical mechanical properties of reinforced membrane

Tensile strength (kg/cm²)	780
Elongation (%)	50
Tensile modulus (kg/cm²)	4500

Sample was measured in Na type and in 50% relative humidity at 25°C

6.10 POWER CONSUMPTION OF THE MEMBRANE CHLOR-ALKALI PROCESS

Figure 6.8 shows the power consumption of recently installed commercial cells with steel and activated cathodes. Asahi Chemical has conducted extensive work in improving the characteristics and performance of anode and cathode.

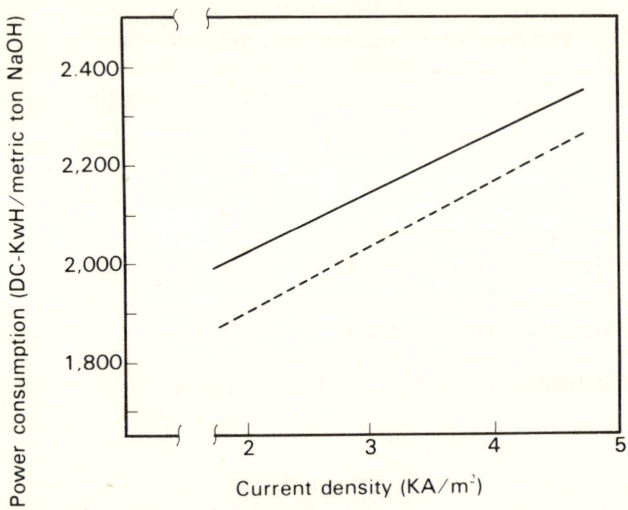

Fig. 6.8 – Power consumption versus current density in commercial electrolyzer (1982).
——— Steel cathode
------- Activated cathode

Figure 6.9 shows the view of the Asahi Chemical electrolyzer composed of cells 1.5 m in height and 3.6 m in width.

6.11 NEW PLANT FOR MANUFACTURE OF ION-EXCHANGE MEMBRANES FOR CHLOR-ALKALI PRODUCTION

1. Asahi Chemical started up the world's most advanced ion-exchange membrane manufacturing plant in 1981 and commenced production of new high performance perfluorocarboxylic and sulfonic acid type membranes at this plant. The plant facilities include monomer synthesis, polymerization, and membrane fabrication for the company's high performance membranes. The facilities for monomer synthesis and polymerization are located in Nobeoka Japan, and membrane fabrication is in Kawasaki.

Fig. 6.9 – Asahi Chemical's electrolyzer.
Unit cell: 1.5 m height × 3.6 m width.
Annual caustic soda capacity: 20 000 metric ton/year per electrolyzer
Number of cell per electrolyzer: 82 cells.

2. The capacity of the plant is large enough to provide membranes for chlorine production totalling 1.2 million tons per year, and can be further expanded to meet the future increases in demand.
3. With the startup of membrane production plant, Asahi Chemical has become the only company in the world to develop and offer complete capability in membranes, electrolyzers, electrodes, and process know-how, the four basic technologies necessary for ion-exchange membrane chlor-alkali electrolysis.

6.12 REFERENCES

[1] Seko, M. (to Asahi Chemical Ind. Co., Ltd.), Japanese Patent Publication 55-1 351 (Jan. 12, 1980).

[2] Seko, M. (to Asahi Chemical Ind. Co., Ltd.), Japanese Patent Publication 55-14 148 (Apr. 14, 1980).

[3] Seko, M. (to Asahi Chemical Ind. Co., Ltd.), French Patent 2 263 312 (May 11, 1979).

[4] Seko, M. (to Asahi Chemical Ind. Co., Ltd.), US Patent 4 178 218 (Dec. 11, 1979).

[5] Seko, M. (to Asahi Chemical Ind. Co., Ltd.), German Patent Publication 2 510 071 (Nov. 27, 1980).

[6] Seko, M. (to Asahi Chemical Ind. Co., Ltd.), German Patent Publication 2 560 151 (Nov. 27, 1980).

[7] Seko, M. (to Asahi Chemical Ind. Co., Ltd.), UK Patent 1 497 748 (Jan. 12, 1978).

[8] Seko, M. (to Asahi Chemical Ind. Co., Ltd.), UK Patent 1 497 749 (Jan. 12, 1978).

[9] Seko, M., *Industrial and Engineering Chemistry, Product Research and Development*, 1976, **15**, 286–292.

[10] Seko, M., *New development of the Asahi Chemical membrane chlor-alkali process,* presented at the Chlorine Institute Inc., 22nd Chlorine Plant Managers Seminar, Atlanta, Georgia, USA, Feb. 7, 1979.

[11] Seko, M., *New development of the Asahi Chemical membrane chlor-alkali process*, presented at Oronzio De Nora Symposium, Venice, Italy, May 15–18, 1979.

[12] Seko, M., *New development of the Asahi Chemical membrane chlor-alkali process*, presented at Electro. Chem. Tech. Group, London, UK, June 13–15, 1979.

[13] Ogawa, S., *Asahi Chemical membrane chlor-alkali process*, presented at Seminar on Developments in chlor-alkali industry, New Delhi, India, Mar. 7–8, 1980.

[14] Grot, W. G. F., Munn, G. E., and Walmsley, P. N., *Perfluorinated ion exchange membranes*, presented at 141st National Meeting of Electrochem. Soc., Houston, Texas, USA, May 7–11, 1972.

[15] Seko, M., Yamakoshi, Y., Miyauchi, H., Fukumoto, M., Kimoto, K., Watanabe, I., Hane, T., and Tsushima, S. (to Asahi Chemical Ind. Co., Ltd.), US Patent 4151053 (Apr. 24, 1979).

[16] Kimoto, K., Miyauchi, H., Ohmura, J., Ebisawa, M., and Hane, T. (to Asahi Chemical Ind. Co., Ltd.), Laid Open Japanese Patent Application 55-160 029 (Dec. 12, 1980).

[17] Grot, W. G., Molnar, C. J., and Resnick, P. R. (to E. I. Du Pont de Nemours & Co.), Belgian Patent 866 122 (Oct. 19, 1978).

[18] Seko, M., Yamakoshi, Y., Miyauchi, H., Fukumoto, M., Kimoto, K., Hane, T., and Hamada, M., (to Asahi Chemical Ind. Co., Ind.), Laid Open Japanese Patent Application 53-125 986 (Nov. 2, 1978).

[19] Sata, T., Nakahara, A., and Ito, J. (to Tokuyama Soda Co., Ltd.), Laid Open Japanese Patent Application 53-137 888 (Dec. 1, 1978).

[20] Onoue, K., Sata, T., Nakahara, A., and Ito, J. (to Tokuyama Soda Co., Ltd.), US Patent 4 200 711 (Apr. 29, 1980).

[21] Molnar, C. J., Price, E. H., and Resnick, P. R. (to E. I. Du Pont de Nemours & Co.), US Patent 4176 215 (Nov. 27, 1979).

[22] Asawa, T., Oda, Y., and Gunjima, T. (to Asahi Glass Co., Ltd.), Laid Open Japanese Patent Application 52-36 589 (Mar. 19, 1977).

[23] Seko, M., Yamakoshi, Y., Miyauchi, H., Kimoto, K., and Masuda, Y. (to Asahi Chemical Co., Ltd.), US Patent 4123 336 (Oct. 31, 1978).

[24] Seko, M., Yamakoshi, Y., Miyauchi, H., Fukumoto, M., and Kimoto, K. (to Asahi Chemical Ind. Co., Ltd.), Laid Open Japanese Patent Application 54-6 887 (Jan. 16, 1979).

[25] Putnam, P. E. (to E. I. Du Pont de Nemours & Co.), US Patent 3 301 893 (Jan. 31, 1967).

[26] Connally, D. J. (to E. I. Du Pont de Nemours & Co.), US Patent 3 282 875 (Nov. 1, 1966).

[27] Resnick, P. R. (to E. I. Du Pont de Nemours & Co.), US Patent 3 560 568 (Feb. 2, 1971).

[28] Kimoto, K., Miyauchi, H., Ohmura, J., Ebisawa, M., and Hane, T. (to Asahi Chemical Ind. Co., Ltd.), Laid Open Japanese Patent Application 55-160 007 (Dec. 12, 1980).

[29] Kimoto, K., Miyauchi, H., Ohmura, J., Ebisawa, M., and Hane, T. (to Asahi Chemical Ind. Co., Ltd.), Laid Open Japanese Patent Application 55-160 008 (Dec. 12, 1980).

[30] Kimoto, K., Miyauchi, H., Ohmura, J., Ebisawa, M., and Hane, T. (to Asahi Chemical Ind. Co., Ltd.), Laid Open Japanese Patent Application 55-160 030 (Dec. 12, 1980).

[31] Kimoto, K., Miyauchi, H., Ohmura, J., Ebisawa, M., and Hane, T. (to Asahi Chemical Ind., Co., Ltd.), Laid Open Japanese Patent Application 56-12 362 (Feb. 6, 1981).

[32] Kimoto, K., Miyauchi, H., Ohmura, J., Ebisawa, M., and Hane, T. (to Asahi Chemical Ind. Co., Ltd.), Laid Open Japanese Patent Application 56-15 260 (Feb. 14, 1981).

[33] Kimoto, K., Miyauchi, H., Ohmura, J., Ebisawa, M., and Hane, T. (to Asahi Chemical Ind. Co., Ltd.), Laid Open Japanese Patent Application 56-16 460 (Feb. 17, 1981).

[34] Kimoto, K., Miyauchi, H., Ohmura, J., Ebisawa, M., and Hane, T. (to Asahi Chemical Ind. Co., Ltd.), Laid Open Japanese Patent Application 56-90 054 (July 21, 1981).

[35] Kimoto, K., Miyauchi, H., Ohmura, J., Ebisawa, M., and Hane, T. (to Asahi Chemical Ind. Co., Ltd.), Laid Open Japanese Patent Application 56-92 263 (July 25, 1981).

[36] Yoshida, M., Masuda, Y., and Kashiwada, A. (to Asahi Chemical Ind. Co., Ltd.), Laid Open Japanese Patent Application 55-110 786, 56-116 891 and 56-145 927 and US Patent 4 323 434.

CHAPTER 7

Development of Asahi Chemical Chlor-alkali Technology

MAOMI SEKO, AKIRA YOMIYAMA, SHINSAKU OGAWA, HIROSHI ONO,
Asahi Chemical Industry Co., Ltd., Tokyo, Japan

Asahi Chemical Industry Co., Ltd. has succeeded in the development of unique technology for chlor-alkali production using a bipolar cell of metal construction, which allows high productivity of the individual cell and low energy consumption even at high current density.

In March of this year, Asahi Chemical expanded its own ion exchange membrane chlor-alkali plant with an annual production capacity of 80 000 metric tons of caustic soda to 120 000 metric tons, by the addition of two large electrolyzers composed of cells 1.5m in height and 3.6m in width. These large electrolyzers also incorporate several recent improvements relating to membrane, electrodes and cell, and have achieved a performance of 2100 d.c.-kWh per metric ton of caustic soda at $4.0\,kA/m^2$. With these electrolyzers, the area of the cell room has been reduced by 20% and the plant cost by 10%, compared with the electrolyzers composed of cells 1.2m × 2.4m.

7.1 INTRODUCTION

Asahi Chemical began the world's first commercial operation of the ion exchange membrane chlor-alkali process in 1975 [1, 2], and has developed its own complete system for the design and manufacture of membranes, electrodes, cells, heat recovery evaporators and overall plant design. The membrane chlor-alkali process is marked by:

(a) Low energy consumption.
(b) Low investment cost.
(c) Easy operation.
(d) Low operating cost.
(e) High product purity.

Of particular importance is the perfluorocarboxylic acid membrane invented by Asahi Chemical, which has proved essential to the economic operation of any membrane chlor-alkali process [3–10]. Asahi Chemical has recently begun the sale of this membrane to customers around the world.

The bipolar cell of metal construction which has been developed by Asahi Chemical allows a high current density and a large membrane area in the individual cell, and thus makes low-cost plant construction possible. Since March of this year, Asahi Chemical has conducted stable commercial operation of cells in which the size of the individual membrane is 1.5 m × 3.6 m (approx. 5 ft × 12 ft).

Asahi Chemical has also developed a heat recovery evaporator which employs the heat generated in the electrolyzer to concentrate the product caustic soda to 35–40% with no steam consumption [11–14].

Recent progress in membrane, electrode and cell have resulted in decreased unit consumption of electricity, from the 2700 d.c.-kWh per metric ton of caustic soda in 1976 at the time of announcement of the first membrane chlor-alkali process, to the present 2100 d.c.-kWh with an activated cathode and 2200 d.c.-kWh with a steel cathode, at a current density of 4.0 kA/m^2.

Since 1976, Asahi Chemical has built or licenced a complete package of membrane, cell, and design know-how to the companies indicated in Table 7.1. The total capacity of these and Asahi Chemical's own facilities is about 520 000 metric tons of caustic soda per year, or about 70% of the total worldwide capacity of membrane chlor-alkali plants.

This chapter describes Asahi Chemical's process design, and in particular the cell design concepts necessary for a successful membrane process.

Table 7.1

Chlor-alkali plant capacity employing Asahi Chemical process

Company	Location	Capacity (metric tons of NaOH per year)	Startup
Asahi Chemical Ind. Co., Ltd.	Nobeoka, Japan	40 000	1975
Denki Kagaku Kogyo K.K.	Ohmi, Japan	60 000	1976
Asahi Chemical Ind. Co., Ltd.	Nobeoka, Japan	40 000	1977
Prince Albert Pulp Co., Ltd.	Saskatoon, Canada	30 000	1978
St. Anne Chemical Co., Ltd.	Nackawic, Canada	10 000	1979
N. Z. Forest Products Ltd.	Kinleith, New Zealand	10 000	1981
Asahi Chemical Ind. Co., Ltd.	Nobeoka, Japan	40 000	1982
Tasman Pulp & Paper Co., Ltd.	Kawerau, New Zealand	10 000	1982
Akzo Zout Chemie Nederland b.v.	Rotterdam, Netherlands	280 000	1982

7.2 CHARACTERISTICS OF THE MEMBRANE PROCESS

The principles of the membrane process differ from those of the diaphragm process in several important respects. Since the membrane is permselective for Na^+ ion only and is impermeable to brine, the membrane process offers the following advantages over the diaphragm process:

(a) Products of high purity; product caustic soda containing less than 50 p.p.m. of salt, and product chlorine containing less than 0.5 of oxygen.
(b) Capability for operation at high electric current levels, resulting in lower investment cost.
(c) Capability for operation at high temperature and pressure, resulting in higher energy efficiency and simplification of the process.
(d) Capability for utilizing interrupted power, because of wide turndown ratio of electrolyzers.

On the other hand, the following factors must be carefully considered in design of the membrane process:

(a) Limiting current density and cell design.
(b) Complete neutralization of anolyte with hydrochloric acid.
(c) Optimum concentration of product caustic soda.
(d) Operating cell pressure.

7.2.1 Limiting current density

Because the membrane is selectively permeable to Na^+ ion, the transport number of the Na^+ ion in the membrane is higher than that of bulk solution of anolyte, resulting in the formation of a boundary layer depleted of salt on the anode side of the membrane, as shown in Fig. 7.1. The material balance of the depleted layer is described by the following equation:

$$\frac{I}{F}(T_{Na}-t_{Na}) = \frac{D}{d}(C-C_0) \qquad (7.1)$$

where

I = current density (ampere/cm^2)
T_{Na} = transport number of Na^+ ion in the membrane (current efficiency)
t_{Na} = transport number of Na^+ ion in the anolyte
F = Faraday's constant (96 500 ampere sec/eq)
D = NaCl diffusion constant in the depleted layer (cm^2/sec)
C = concentration of NaCl in bulk solution of anolyte (eq/cm^2)
C_0 = concentration of NaCl in depleted layer at surface of membrane (eq/cm^3)
d = thickness of depleted layer (cm)

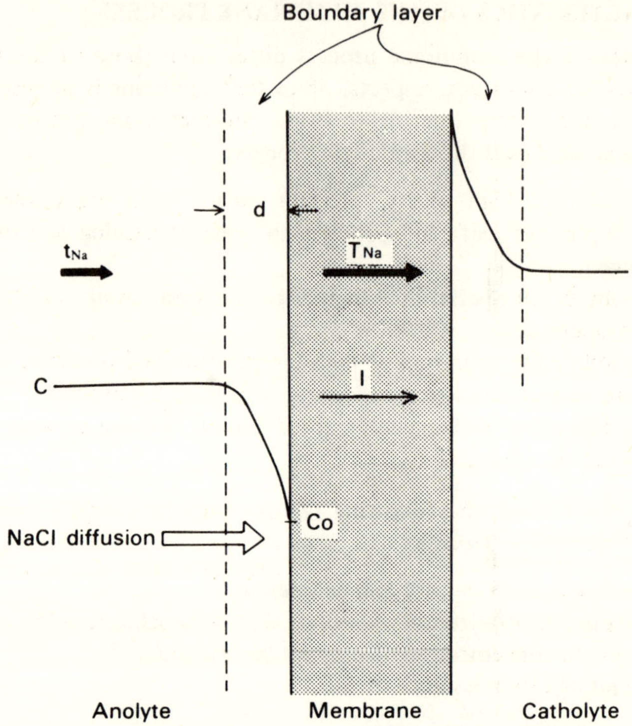

Fig. 7.1 – Concentration profile on the surface of membrane.

The left-hand side of this equation represents the rate at which salt is removed by the electric current from the surface of the membrane due to the difference of Na^+ ion transport numbers in the solution and the membrane. The right-hand side represents the rate of supply of salt to the membrane surface by diffusion from the bulk phase of solution. At the limiting current density (I_0), C_0 becomes zero and the following equation applies:

$$\frac{I_0 d}{C} = \frac{DF}{T_{Na} - t_{Na}} \tag{7.2}$$

The value of the right-hand side of this equation is determined by the membrane characteristics and the electrolysis temperature, and d is determined by the anolyte flow conditions and the cell structure. As the supply of salt to the membrane surface becomes impossible at this limiting current density, the water in the depleted layer is decomposed, and the current flow results partly from the migration of generated hydrogen ion. This phenomenon, which is called polarization causes lower current efficiency and higher electrolysis voltage [15].

Stable operation cannot be attained with a membrane cell based on the design concepts utilized for the diaphragm process, in which no such local polarization occurs as the asbestos is not permselective and a pressure difference is applied to effect brine migration from anolyte to catholyte. For this reason, the required flow conditions of electrolyte and cell structure in the membrane process are different from those of the diaphragm process, and must be designed to achieve the maximum possible limiting current density. For this purpose, the following measures in particular are essential:

(a) Minimization of d, by adopting a cell structure which yields sufficient turbulence at the surface of the membrane, utilizing a high rate of flow and locating the membrane near enough to the anode to allow agitation of the boundary layer by the evolving chlorine gas [16].
(b) Appropriate selection of anolyte effluent concentration. Although the limiting current density can be increased by increasing the concentration of the anolyte effluent, this results in the increased cost of brine purification by chelate resin. Therefore the optimum concentration of anolyte effluent should be obtained by appropriate design of cell.
(c) Maximization of D, by increasing the operating temperature.
(d) Uniform current density throughout the membrane and uniform concentration throughout the cell, through optimum cell design.

7.2.2 Dimensional stability of membrane, electrode and cell

Since turbulence is necessary to allow material transport of ion to the membrane surface across the boundary layer, a flat and planar membrane surface must be maintained during electrolysis. Any curvature in the membrane surface will lead to local stagnation of the electrolyte and polarization, which will in turn cause greater distortion of the membrane surface and reduce the electric current, with a mutually accelerating effect.

To obtain the rigid planar membrane surface, a dimensionally stable Teflon® woven reinforcing cloth is used to reinforce the membrane [17]. In no case should any film which shrinks and swells depending on the condition of external solution be utilized in the membrane process.

The cell and the gaskets which hold the membrane in place must also be of high dimensional stability. The cell must be metal, and the gaskets must have even greater dimensional stability than the membrane. If the gaskets expand or contract under the forces applied during cell assembly, the membranes may be damaged or torn and stable operation will become impossible.

7.2.3 Neutralization of brine by hydrochloric acid

As indicated by Fig. 7.2 the anolyte and the catholyte are separated by the liquid-impermeable membrane. Accordingly, it is necessary to neutralize the hydroxide ion at the membrane surface which migrates through the membrane

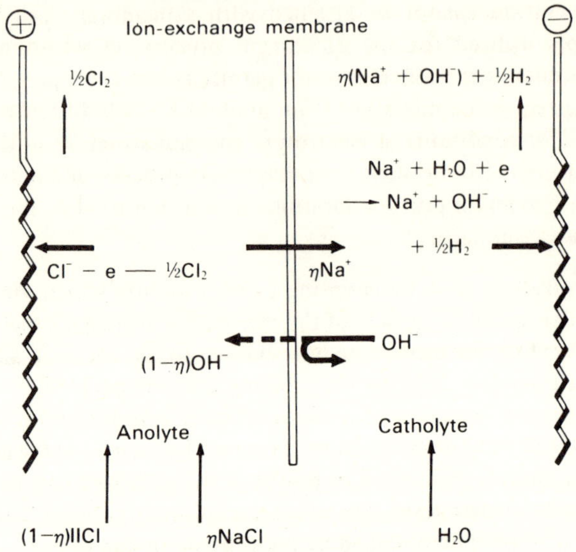

Fig. 7.2 – Material balance in electrolysis.

at a rate of $(1-\eta)$ equivalents per Faraday (η = current efficiency). The hydrochloric acid used for this neutralization is injected into the cell through the brine stream inlet. Even with the 96% current efficiency of the perfluorocarboxylic acid membrane, a large amount of hydrochloric acid is necessary for this purpose. As an example, taking the utilization ratio of salt in one pass to be 50% and assuming the use of saturated salt solution ($\approx 5\text{N}$) as feed brine, the anolyte influent passing into the cell through the anolyte inlet must contain:

$$5\ \text{mol/l} \times 0.5 \times \frac{4}{96} = 0.104\ \text{mol/l of HCl}\ .$$

With a perfluorocarboxylic acid membrane containing only carboxylic acid groups, the membrane would be made non-conductive by contact with this concentrated hydrochloric acid and it would therefore be impossible to neutralize completely the hydroxide ion in the anolyte. With a perfluoro membrane containing both carboxylic and sulfonic acid groups, a sufficient dissociation of the functional groups is obtained due to the low pK_a of the sulfonic acid group in the membrane facing the anolyte, thus effectively preventing non-conductivity in the membrane.

The lack of a means to effect complete neutralization with a membrane containing only carboxylic acid groups would result in a higher oxygen content in the product chlorine, a shorter anode life and the accumulation of ClO_3^- in the anolyte. This accumulation of ClO_3^- would make chemical purification economically impractical, and would result in reduced concentration of the salt in the brine and thus necessitate bleeding of the brine.

The economic impracticality of the removal of ClO_3^- is a compelling reason for the choice of a membrane containing both sulfonic and carboxylic acid groups, as the only way to allow complete neutralization in the cell by hydrochloric acid.

In the diaphragm process any ClO_3^- which may be formed is removed as part of the caustic soda stream, since all of the brine enters the catholyte.

7.2.4 Optimum concentration of caustic soda

The maximum electric conductivity of caustic soda solution occurs at a concentration of about 20% at the ordinary electrolysis temperature and the membrane conductivity tends to decline sharply with caustic soda concentration in the catholyte exceeding 20%, as discussed previously in this presentation. The advanced perfluorocarboxylic acid membrane now available has a current efficiency of over 90% in a broad range of caustic soda concentrations (20–40%), but minimum power consumption is achieved at caustic soda concentrations of approximately 20–30%, where the electrolysis voltage is lowest. Although caustic soda of any concentration up to 40% can be produced in the membrane process, the constraints of energy cost and investment cost must be considered in relation to any increase in electrolysis voltage.

7.2.5 Operating pressure

In the diaphragm process, a small difference in hydraulic pressure is applied between anolyte and catholyte to transport anolyte through the diaphragm to the catholyte compartment. It is therefore impractical to pressurize the anolyte, since this would cause chlorine gas to dissolve in the anolyte and thus mix with the catholyte.

In the membrane process, however, pressurized operation can be utilized, resulting in various advantages [18]:

(a) Reduced operating voltage due to reduction of gas volume in the cell and equalization of the current density vertically along the cell.
(b) Reduced membrane vibration and consequent reduction in damage to the membrane and electrode for long-term operation.
(c) Reduced water content in the generated gas, resulting in easier drying of chlorine with less heat loss.
(d) Reduced investment cost for product gas transportation, and elimination of need for compressor or blower is some cases.
(e) Possibility of utilizing higher operating temperature.

7.3 BASIC DESIGN CONCEPT OF ASAHI CHEMICAL'S ELECTROLYZER

While the membrane process has many important advantages over the diaphragm process, it also has the following disadvantages:

(a) Need for purification of brine with chelate resin, to reduce the concentration of multivalent ion to less than 0.1 p.p.m.
(b) Relatively high cost of the membrane.
(c) Need for sufficient turbulence near the membrane surface, to prevent polarization.

To overcome these disadvantages, a design concept quite different from that of the diaphragm process is necessary, with the following basic minimum requirements:

(a) High current density.
(b) Low energy consumption at high current density.
(c) Uniform current density in the cell.
(d) Large capacity of individual cell with large electrode and membrane area.
(e) Dimensional stability of cell structure and gaskets.
(f) Long service life of anode and membrane.

The following sections describe the design concepts of Asahi Chemical's cell.

7.3.1 High uniform current density

Based on considerations of ion transport at the membrane surface, the achievement of a high ratio of I/C (I = mean current density, C = anolyte effluent concentration) is a basic prerequisite of good cell design. In Asahi Chemical's cell, with an anolyte effluent concentration of 2.7 N (158 g/l) and a current density of 4 kA/m^2, an I/C ratio of 148 (I = mA/cm^2, C = meq/cm^3) has been attained. This is very high compared with other cells.

Since the quality of the cell design depends largely on this I/C, any cell design having a non-uniform or low I/C will tend to lead to difficulty in maintaining long-term stable operation due to the local polarization.

A good membrane cell must allow operation at a current density considerably higher than the 2.5 kA/m^2 which can be achieved in the diaphragm cell. Operational difficulties may be expected to occur in operation of a membrane cell designed for much the same current density as the diaphragm cell without paying particular attention to the characteristics of the membrane process and maintenance of stable voltage in long-term operation is considered most difficult. An operating current density of 4 to 6 kA/m^2 at a sufficiently low voltage must be maintained in the membrane cell.

To obtain the high, uniform current density, it is necessary to maintain (a) an identical voltage difference between the entire areas of anode and cathode,

(b) identical voltage and current in each of the many cells, and (c) as little variation as possible in the current density vertically along the membrane surface. Since it is difficult for the monopolar cell to meet requirements (a) and (b) as discussed in section 7.3.4 below, it is of limited suitability in the membrane process. The bipolar cell is inherently superior for this purpose, as it eliminates variation in voltage and current among the large number of cells in one electrolyzer and provides a constant voltage difference between the entire areas of anode and cathode, independently of the vertical height in the cell.

7.3.2 Low energy consumption at high current density

The design and performance of the membrane, cell and electrode are all important in attaining this objective.

The only membrane that satisfies the requirement for high current efficiency and low resistance is the double-layer membrane containing both sulfonic and carboxylic acid groups. The cell is designed to provide freedom from the influence of generated gas, a small ohmic drop across the electrolyte and a minimization of local polarization. The electrode is selected for a low overvoltage.

7.3.3 High capacity individual cell

The individual electrodes and cells are designed as large as practically possible, particularly in order to reduce investment costs. This is evident from the following approximate relations, in which the height and width of the cell are assumed to be L:

Capacity L^2
Cost of equipment:
  Electrode L^2
  Structural materials L
  Piping, nozzles, etc. $1/L^2$
  Membrane $L \sim L^2$

These relations clearly indicate that the cell cost per unit capacity decreases with increasing cell dimensions. Figure 7.3 shows the relation between the investment cost index for a chlor-alkali plant and the area of individual membranes and electrodes.

7.3.4 Bipolar and monopolar cells

In the monopolar configuration as shown in Fig. 7.4, individual cells are connected electrically in parallel in the electrolyzer and the electrolyzers are connected electrically in series. The voltage across each electrolyzer is low but the total current is relatively high. For the purpose of power supply, each electrode in the electrolyzer is connected by bus bar to the corresponding electrode in the adjacent cell unit.

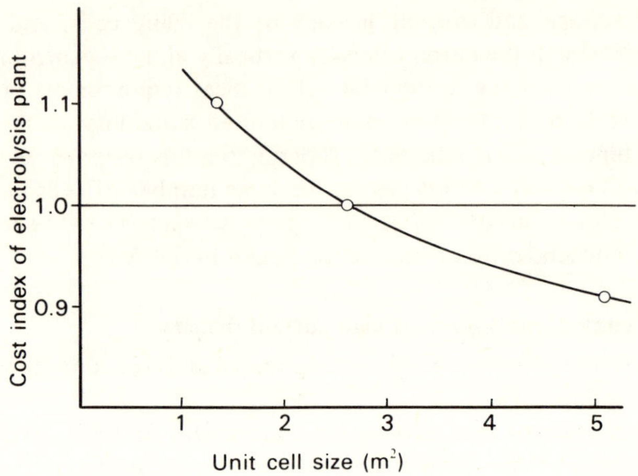

Fig. 7.3 – Construction cost index of electrolysis plant versus unit cell size (m^2).

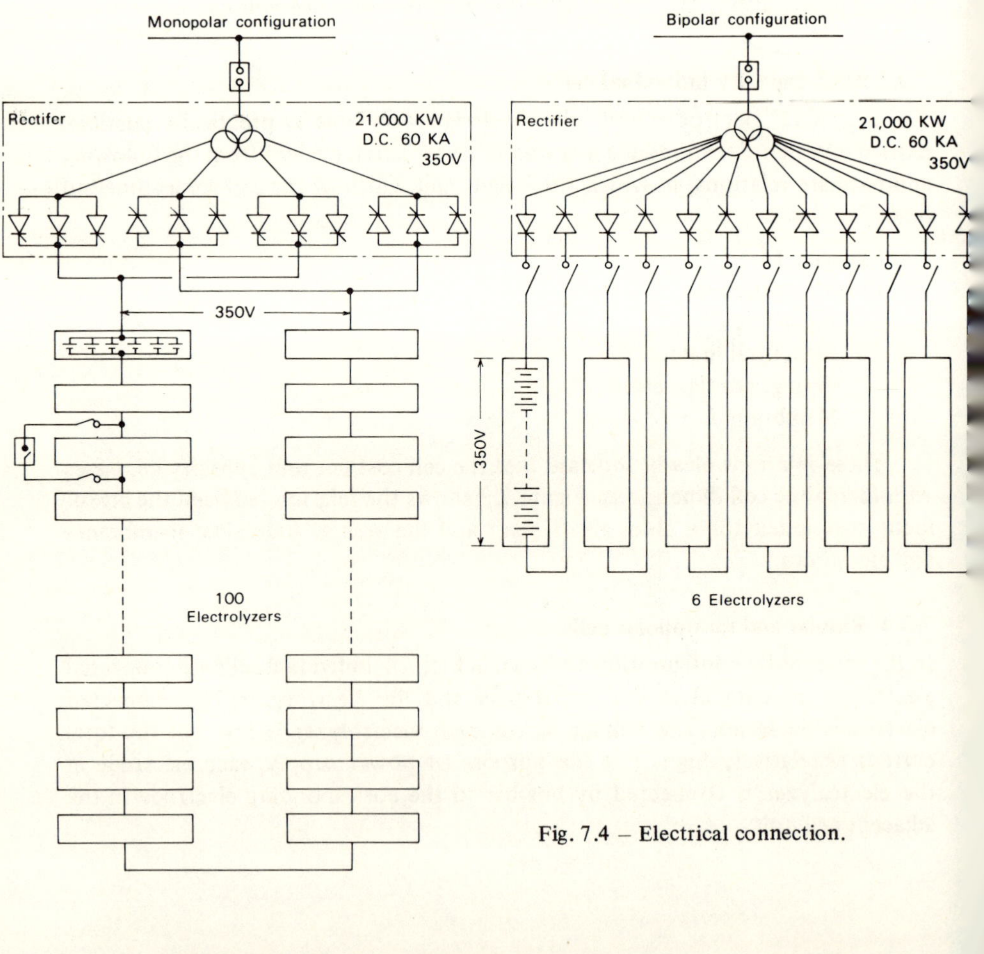

Fig. 7.4 – Electrical connection.

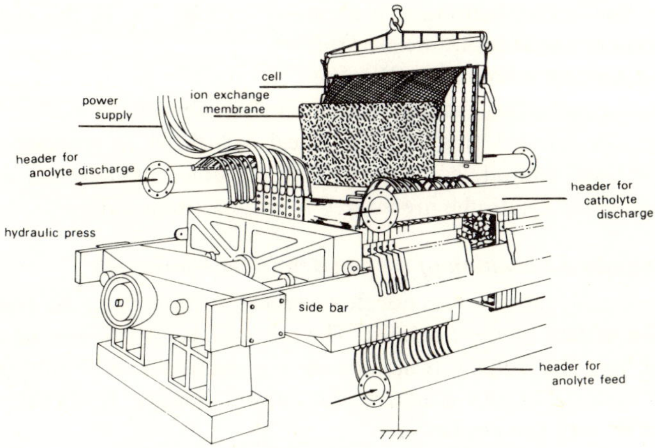

Fig. 7.5 – Asahi Chemical's bipolar electrolyzer.

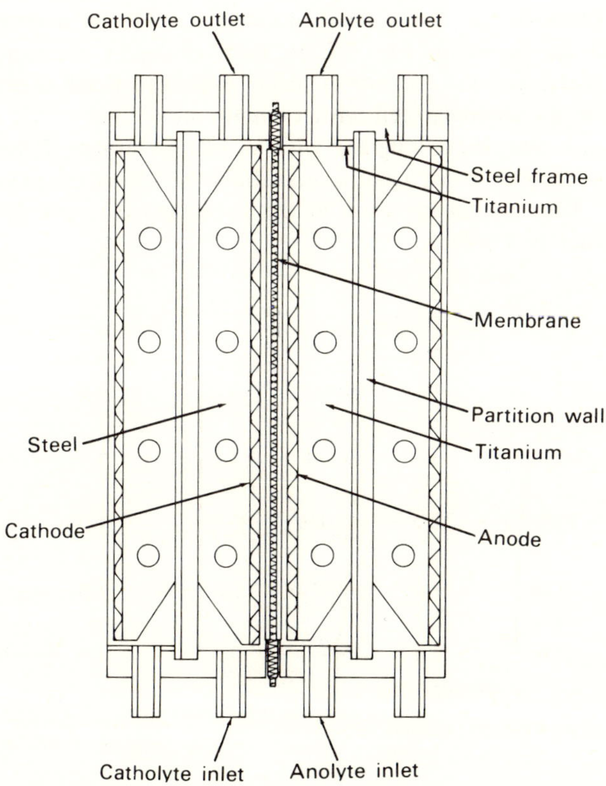

Fig. 7.6 – Structure of cell units.

In the bipolar configuration, also shown in Fig. 7.4, all the cells are connected electrically in series in the electrolyzer.

Asahi Chemical's bipolar cell in the filter press type electrolyzer is shown in Fig. 7.5. The individual cell is shown in Fig. 7.6. A partition wall of conversion rolled explosion-bonded titanium clad steel is utilized to separate the anode compartment and cathode compartments in the unit cell [19], and the components including the electrodes are welded to this partition wall.

7.3.4.1 *Uniform distribution of current density in bipolar cell*

In the bipolar cell, current does not flow vertically either along the partition wall or along the surface of the electrode. The potential is therefore equal at all parts of the wall and the voltage is applied by the electrodes uniformly throughout the cell compartment. As a result, the current density is virtually identical throughout the cell and electrolyzer.

In the monopolar cell, where vertical current flow occurs along the surface direction of each electrode, the voltage drop between anode and cathode varies vertically in the cell. Consequently the current density varies with the vertical position. Examples of the vertical potential profile of monopolar cell and bipolar cell are shown in Fig. 7.7. In these examples the anode is titanium expanded metal and the cathode is nickel expanded metal sheet. Calculations show that in a monopolar cell with titanium anode and nickel cathode, each 0.75mm thick and 36cm long and with an average current density of 3 kAm^2, the current density varies from a maximum of 4.1 kA/m^2 to a minimum of 2.6 kA/m^2. The total ohmic drop in the anode and the cathode was found to be approximately 0.26V.

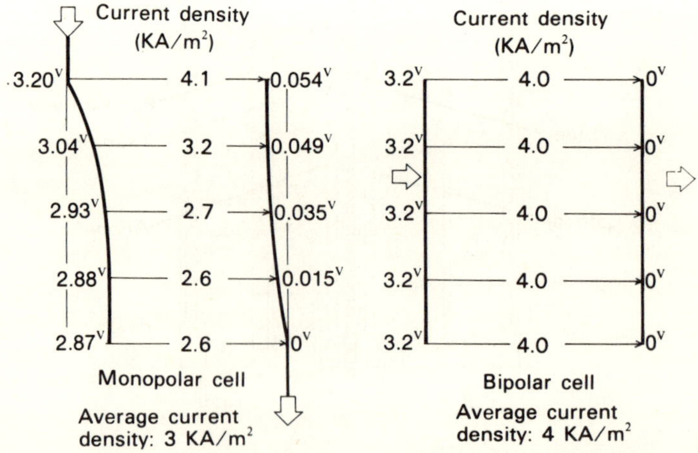

Fig. 7.7 – Examples of vertical potential profile and current density distribution.

This considerable potential drop along the height of the electrodes in the monopolar cell also makes it impractical to utilize electrodes of large surface area. The maximum size of electrode that can be used in the monopolar cell is 0.5 m^2.

The current density also varies among the monopolar cells, as the current varies with slight differences in cell resistance due to differences in dimension, ageing of membrane and electrodes, and resistance of current supply parts.

7.3.4.2 *Voltage loss in monopolar bus bar*

In the monopolar configuration, a voltage loss at bus bar connections is unavoidable. At an operating voltage of 3.5 V per cell, the loss per cell due to intercell bus and structural ohmic drop is 0.2–0.3 V [20], which is quite significant in comparison with the total ohmic drop of about 0.7–0.8 V in the modern cell.

In Asahi Chemical's bipolar cell, the voltage drop through the partition wall is only 3 mV at 5 kA/m^2.

7.3.4.3 *Operation and maintenance*

In the bipolar configuration, the current in each electrolyzer can be regulated independently with a thyrister, permitting easy starting and stopping of each electrolyzer. Automatic detection of abormal cells is also easy through measurement of voltage of the individual cells. If a short circuit occurs in any cell, the electrolyzer containing the defective cell can be automatically tripped off [21]. This makes possible the safe design of a cell with reduced distance between anode and cathode. In the monopolar configuration, such detection is rather difficult, and the risk of serious accident involving the entire electrolyzer is present due to the lack of a partition wall.

7.3.4.4 *Leakage current*

In both the monopolar and the bipolar configurations, the total applied voltage is about the same and leakage current may occur in both.

In the diaphragm process, a dropper cup is provided to interrupt the leakage current by dropwise supply and discharge of the electrolyte. Such a device is not appropriate for the membrane process, in which it is preferable to supply a large quantity of electrolyte in order to obtain uniform concentration throughout the compartment. In Asahi Chemical's bipolar process, however, leakage current is reduced to the negligible level of a few tenths of one per cent with no electrolytic corrosion, through the use of a specially developed system.

In the membrane process, electrolysis must be carried out at high current density and high temperature, with pressure applied to the electrolyte. The membrane process should permit the use of large capacity cells. It is now clear that to obtain uniform, high current density particularly in a large-scale plant, the use of metal cells in a bipolar configuration is essential. However, no company other than Asahi Chemical has succeeded in the development of long-term commercial operation with bipolar cells. A major reason for Asahi Chemical's success is the partition wall made of conversion roll explosion bonded clad titanium steel (and/or stainless) sheet [19], which is manufactured only by Asahi Chemical. Although explosion clad sheet is manufactured by other companies, its high price makes it economically impractical for use in the partition wall.

Titanium sheet and steel sheet can also be bonded by special electric welding, but extensive trials with partition walls of such material have shown that it is not suitable for partition wall because in long-term operation it tends to crack and leak in the vicinity of the welded portion, due to mechanical strain and differences in the thermal expansion of these metals.

Although there are rumours to the effect that titanium hydride tends to separate the partition wall layers, these remours have no basis in fact. We have been using these partition walls commercially in membrane cells since 1975, without any problem whatsoever.

Such a problem does not occur because the partition wall is located away from the cathode and the hydrogen overvoltage of the cathode is much smaller than that of the partition wall due to the deposition of iron on the cathode, so that very little hydrogen generation occurs on the partition wall. For the same reason, any hydrogen migration which might occur through the partition wall would be very small.

One important development goal of Asahi Chemical is a further substantial reduction of the cost of the partition wall.

Asahi Chemical has also conducted extensive research for the development of large-capacity, low-cost monopolar cells which are free from the conventional disadvantages inherent in the monopolar cells, but has found little evidence that the monopolar cell is economically advantageous.

7.3.5 Corrosion resistant cell materials

Titanium is the only structural material appropriate for the inner walls of the anode compartment. Although the use of plastic cells has been reported, their effectiveness is highly doubtful in view of the high temperatures involved (about 90°C) and the likelihood of creep due to the considerable stress involved in securing the cell assembly, which would in turn tend to cause membrane damage because of the non-uniform application of stress on the membrane. Long-term operation of a plastic cell would also tend to result in leakage and consequently in corrosion of equipment. It would, moreover, seem even more difficult to manufacture and use large-capacity individual cells with plastic as a structural material.

For similar reasons, the cathode compartment can be manufactured from steel or stainless steel, but not from plastic or rubber.

7.3.6 Dimensional stability

For stable long-term performance, the cell and its component parts must be of high dimensional stability. This also makes it essential to utilize metal materials. Deformation of the gasket may be permitted perpendicular but not parallel to the membrane surface and careful design of gaskets is also required.

7.3.7 Long anode service life

The anode should be highly resistant to alkali in the membrane process [2, 22], in contrast to that of the diaphragm process.

7.4 EXPERIENCE WITH OTHER TYPES OF ELECTROLYZERS

Asahi Chemical has also 30 years of industrial experience in the fields of membrane electrolysis and electrodialysis including the following applications:

(a) Salt production by seawater dialysis.
(b) Potable water production by desalination of seawater or brackish water.
(c) Electro-hydrodimerization of acrylontrile [23].

Long-term, extensive experience in industrial operation of such electrochemical processes has allowed conclusive evaluation of various types of electrolyzers and electrolyzer materials, including rubber or plastic cell bodies, and plate-type cell structures.

7.4.1 Rubber cell bodies

Electrodialysis for the concentration and desalination of seawater is generally carried out around ambient temperatures. In these applications the outer circumference of rubber gaskets forms the cell body. Such a cell can be used only with non-corrosive aqueous solutions at temperatures near the ambient, and in unit operations where there is no danger of corrosion or external leakage, since only limited forces may be applied in assembling the cell.

7.4.2 Plastic cell materials

Electro-hydrodimerization of acrylonitrile is carried out with bipolar cells containing partition walls of PVC, HDPE or PP.

Problems were encountered in the past with the use of partition walls made of these plastics without reinforcing material due to creep and deformation of the wall at opening temperatures as low as 60°C. To overcome such problems, large amounts of talc are now incorporated into the partition wall. Nevertheless the operating temperature is still limited to 60°C to avoid creep or deformation of the wall.

7.4.3 Plate-type cells

For a period of 6 years from 1968 to 1974, during the development stages of the membrane chlor-alkali process, Asahi Chemical conducted trail operations of plate-type chlor-alkali cells with internal porting, similar in structure to plate-type exchangers, in a pilot plant with an annual capacity of 4400 metric tons of caustic soda. Based on the results of these trial runs, such cells were judged unsatisfactory and were therefore not commercialized. Three major disadvantages were found in this type of cell. In long-term operation, plate-type chlor-alkali cells tend to result in leakage of anolyte or catholyte between the cell compartments and the porting and also between different porting lines. The sources of such leakage are rather difficult to determine even if the cell is disassembled. The second serious problem is the crevice corrosion of the plates by the electrolyte leakage, and the resulting short service life of the cell. The third problem is related to the slits needed to connect the cell compartments and the internal porting. The slits must serve to allow a uniform electrolyte flow rate to and from each compartment, by providing a considerable pressure drop over the short distance between the compartments and the porting. For this purpose, the slits must be quite narrow, but they are therefore prone to clogging by particles from the gaskets and other foreign substances. If any slit is clogged, a uniform electrolyte flow rate cannot be obtained to or from the compartments and this tends to result in a gradual increase in operating voltage.

If the plate-type cell is made of elastic bodies, the cell structure which allows the internal flow of both a concentrated caustic soda solution and brine containing free chlorine necessitates the use of different kinds of elastic bodies as spacers. The inclusion of numerous spacers of different composition and complex forms in the electrolyzer secured with bolts tends to cause electrolyte leakage and tearing of the membrane due to the non-uniform pressures on the sealing surfaces. It is also difficult to return to service any elastic spacer in which creepage has occurred.

7.5 RECENT IMPROVEMENTS IN COMMERCIAL ELECTROLYZERS

7.5.1 Membranes

The membrane characteristics are as described previously in our discussion and the related paper. It should be noted that Asahi Chemical is now preparing to begin sale of its membranes to customers throughout the world.

7.5.2 Rough surface membranes

An extreme narrowing of the gap between anode and cathode results in increased electrolysis voltage, as shown in Fig. 7.8. Careful observation of the membrane surface during such electrolysis shows countless hydrogen bubbles attached to the membrane surface facing the catholyte. It has been found that these fine bubbles increase the voltage drop through the membrane by restricting the current pathways and thus causing non-uniform distribution in the membrane.

As the gap between the anode and cathode is reduced, it becomes increasingly difficult to remove these bubbles because of obstruction by the cathode, and the increased voltage results. Studies have shown that the simplest and most effective means of eliminating this disadvantage is by giving a roughened surface to the membrane during its manufacture [24]. No hydrogen bubbles attach to this rough surface membrane, and no voltage rise occurs even if the gap between the anode and cathode is reduced to virtually zero, as shown in Fig. 7.8.

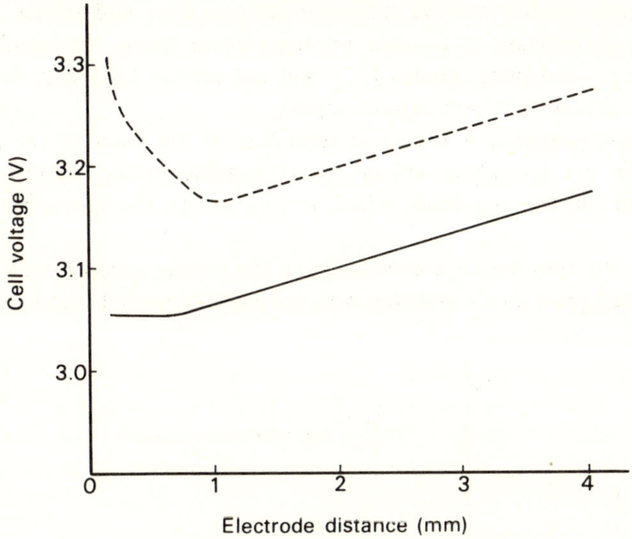

Fig. 7.8 – Electrode distance versus cell voltage
------ Conventional membrane
——— Rough surface membrane
1.2m × 2.4m cell
Current density: 4 kA/m²
Temperature: 90°C.

7.5.3 Electrodes

7.5.3.1 *Anode surface*

The pathway of ions in the membrane during electrolysis can be observed from the measurement of the concentration of ^{45}Ca isotope in the membrane if brine containing ^{45}Ca isotope is electrolyzed. This measurement has showed that the pathway of ^{45}Ca ions in the membrane is strongly influenced by the shape of the anode surface.

If the electrolysis is conducted with ordinary expanded metal anode, the ^{45}Ca pathways in the membrane are connected in portions of the membrane where projections on the anode face the membrane. This indicates that the shape of the anode surface facing the membrane has far greater influence on the current distribution in the membrane and the ohmic drop across the membrane than that of the cathode.

Since the local concentration of current pathways may cause a critical current density which is higher than the limiting current density and thus adversely affect cell performance, the shape of the anode is an important factor in cell design.

A search for an anode shape which is favourable to the uniform current distribution in the membrane has shown that the most effective anode has a structure which readily permits sufficient anolyte flow and release of gas and has as many projections as possible on the surface facing the membrane. For example, perforated plate anodes [25] and rod anodes [26] were found to be highly favourable for uniform current density.

The anode potential is the same regardless of the shape of the anode and the reduction of electrolysis voltage is attributable to the uniformity of the current density in the membrane which is obtained by the appropriate shape of anode.

Figure 7.9 shows the difference between the voltage drop across the member with perforated plate anode and that with ordinary expanded metal anode.

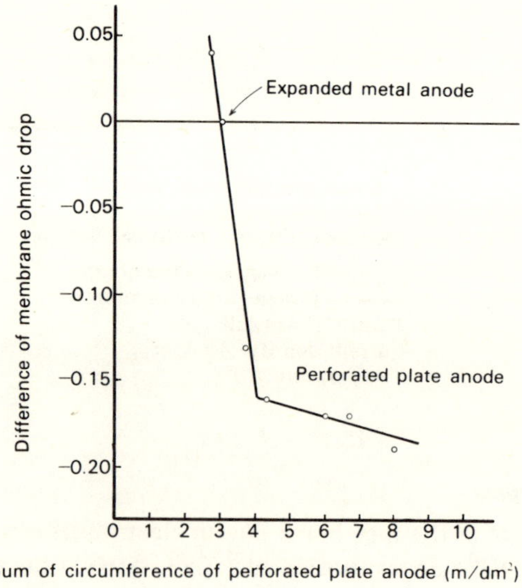

Fig. 7.9 – Comparison of ohmic drop between expanded metal anode and perforated plate anode
Membrane: Asahi Chemical's membrane (EW = 1350)
Current density: 5 kA/m²
Operating temperature: 90° C.

7.5.3.2 *Activated cathode*

The cathode overvoltage can be reduced by activating the cathode with nickel or other catalytic material or by increasing the cathode surface area.

Because of the high cost of activated nickel and other catalysts, corresponding savings must be obtained through reduced electric consumption and sufficient service life of the cathode. The service life is particularly important in industrial applications and Asahi Chemical is now conducting a several-year test program at its commercial plant in order to confirm the service life. Figure 7.10 shows some of the performance results for activated cathodes.

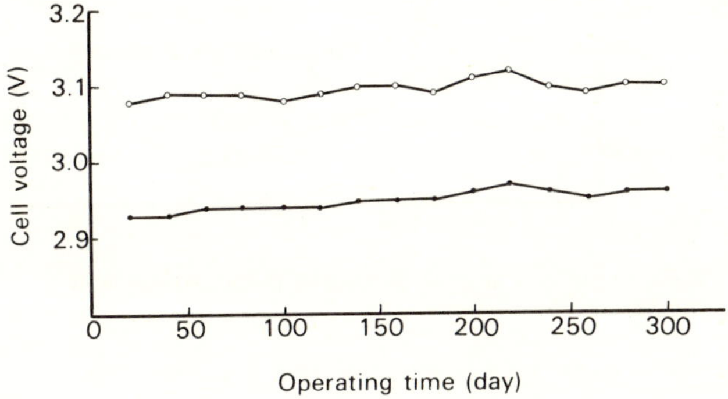

Fig. 7.10 – Durability of Asahi Chemical's activated cathode in commercial cell
-○- Cell voltage without activated cathode
-●- Cell voltage with activated cathode
1.2 m × 2.4 m cell
Current density: 4 kA/m^2
Temperature: 90°C.

7.5.3.3 *Cathode surface*

The increase of cathode surface is also important to the reduction of electrolysis voltage, as shown in Figure 7.11. This figure shows the difference between the cell voltage with ordinary expanded metal cathode and that with perforated plate cathodes having various surface areas on the side facing the membrane. The total surface area of perforated cathode is the sum of the surface area facing the membrane and the inner wall area of the perforation. As shown in Fig. 7.11, it is clear that the overvoltage of the cathode can be reduced simply by increasing the ratio of the cathode surface area to the conducting surface area of the membrane without changing the material.

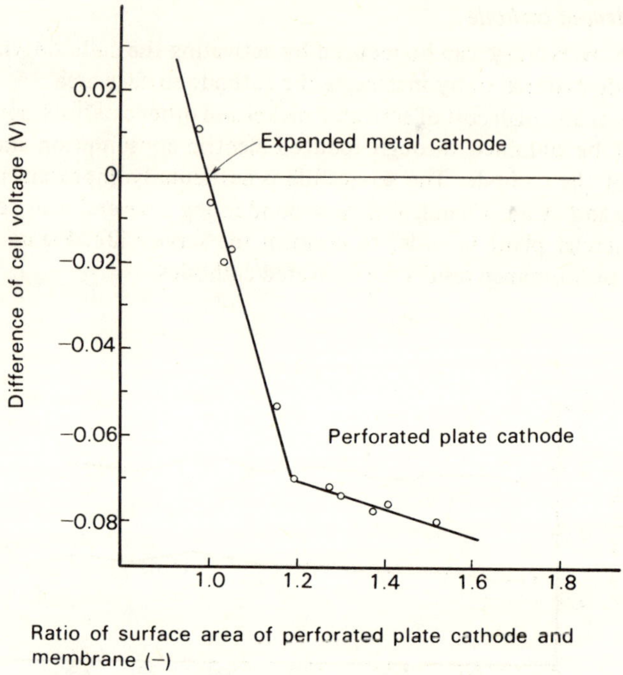

Fig. 7.11 – Difference of cell voltage between expanded metal cathode and perforated plate cathode.
Membrane: Asahi Chemical's membrane (EW = 1350)
Current density: 4 kA/m^2
Operating temperature: 90°C.

7.5.4 Electrolyzers

In March of this year, Asahi Chemical increased the capacity of its ion exchange membrane chlor-alakli plant to 120 000 metric tons of caustic soda per year by installing two of the electrolyzers shown in Fig. 7.12, each with an annual capacity of 20 000 metric tons of caustic soda.

This great incease in electrolyzer capacity was made possible by the application of reliable technology based on the proven performance of the metal cell of bipolar configuration.

These large electrolyzers also incorporate the recent improvements described in this paper, and have achieved a reduction in power consumption to 2100 d.c.-kWh per metric ton of caustic soda.

Fig. 7.12 – Asahi Chemical's electrolyzer. Unit cell: 1.5m height × 3.6m width. Annual caustic soda capacity: 20 000 metric ton/year per one electrolyzer. Number of cells per electrolyzer: 82 cells.

7.5.5 Power consumption

Continued efforts to further reduce the energy consumption have resulted in important savings. This is clearly illustrated in Fig. 7.13, which shows the power consumption of recently installed commercial cells with steel and activated cathodes. The marked reduction in power consumption over the past seven years was made possible by important improvements in technology.

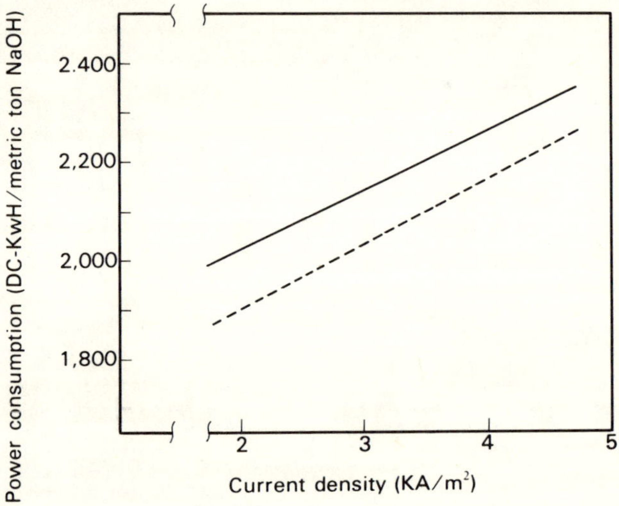

Fig. 7.13 – Power comsumption versus current density in commerical electrolyzer (1982)
——— Steel cathode
------ Activated cathode.

7.6 REFERENCES

[1] Seko, M., *Industrial and Engineering Chemistry, Product Research and Development*, 1976, **15**, 286–292.

[2] Seko, M., *Commerical Operation of the Ion Exchange Membrane Chlor-alkali Process*, presented to Amer. Chem. Soc. Centennial Meeting, New York, April 4–9, 1976.

[3] Seko., M. (to Asahi Chemical Ind. Co., Ltd.), Japanese Patent Publication 55-1351 (Jan. 12, 1980).

[4] Seko, M. (to Asahi Chemical Ind. Co., Ltd.), Japanese Patent Publication 55-14 148 (Apr. 14, 1980).

[5] Seko, M. (to Asahi Chemical Ind. Co., Ltd.), French Patent 2 263 312 (May 11, 1979).

[6] Seko, M. (to Asahi Chemical Ind. Co., Ltd.), US Patent 4 178 218 (Dec. 11, 1979).

[7] Seko, M. (to Asahi Chemical Ind. Co., Ltd.), German Patent Publication 2 510 071 (Nov. 27, 1980).

[8] Seko, M. (to Asahi Chemical Ind. Co., Ltd.), German Patent Publication 2 560 151 (Nov. 27, 1980).

[9] Seko, M. (to Asahi Chemical Ind. Co., Ltd.), UK Patent 1 497 748 (Jan. 12, 1978).

[10] Seko, M. (to Asahi Chemical Ind. Co., Ltd.), UK Patent 1 497 749 (Jan. 12, 1978).

[11] Seko, M., *The Asahi Chemical Membrance Chlor-alkali Process*, presented at the Chlorine Institute Inc., 20th Chlorine Plant Managers Seminar, New Orleans, La., USA, Feb. 9, 1977.

[12] Ogawa, S., *Asahi Chemical Membrane Chlor-alkali Process*, presented to the Seminar of Developments in Chlor-alkali Industry, Indian Institutue of Chemical Engineers, New Delhi, India, March 7–8, 1980.

[13] Kajihara, A., Ogawa, S., Kobayashi, T., and Seko, M. (to Asahi Chemical Ind. Co., Ltd.), Laid Open Japanese Patent Application 52-52 897, UK Patent 1 513 109 and US Patent 4 090 932.

[14] Ogawa, S. (to Asahi Chemical Ind. Co., Ltd.), Japanese Patent Publication 5-26 882, UK Patent 1 576 432 and US Patent 4 132 588.

[15] Seko, M., Ogawa, S., and Takemura, R. (to Asahi Chemical Ind. Co., Ltd.), Laid Open Japanese Patent Application 52-9 700, UK Patent 1 543 249 and US Patent 4 276 130.

[16] Ogawa, S., and Yoshida, M. (to Asahi Chemical Ind. Co., Ltd.), Japanese Patent Publication 52-32 865, UK Patent 1 487 284 and US Patent 4 108 742.

[17] Yamakoshi, Y., Miyauchi, H., Watanabe, I., Fukumoto, M., and Tsushima, S. (to Asahi Chemical Ind. Co., Ltd.), Published Examined Japanese Utility Model Application 55-36 510 and 56-8 901, and US Patent 4 072 793.

[18] Ogawa, S., and Yoshida, M. (to Asahi Chemical Ind. Co., Ltd.), Japanese Patent Publication 56-43 116, UK Patent 1 547 062 and US Patent 4 105 515.

[19] Seko, M., Ogawa, S., Ajiki, N., and Yoshida, M. (to Asahi Chemical Ind. Co., Ltd.), Japanese Patent Publication 52-32 866, UK Patent 1 503 799 and US Patent 4 111 779.

[20] Grother, M. P., and Harke, C. J., *The Development of Hooker's H-2A and H-4 Cells*, presented to Chlorine Bicentennial Symposium, 1974 (Electrochemical Society, Inc.).

[21] Ogawa, S., Isu, T., and Kuroiwa, T. (to Asahi Chemical Ind. Co., Ltd.), Laid Open Japanese Patent Application 51-110 481.

[22] Seko, M., Ogawa, S., Yoshida, M., and Nakamura, A. (to Asahi Chemical Ind. Co., Ltd.), Japanese Patent Publication 52-20 440, UK Patent 1 481 767 and US Patent 4 005 004.

[23] Seko, M., A new process for electro-hydro dimerization of acrylonitrile, presented to Chemical Eng. Congress (AIChE), Kyoto, Japan, Oct. 10–14, 1972.

[24] Yoshida, M., Masuda, Y., and Kashiwada, A. (to Asahi Chemical Ind. Co., Ltd.), Laid Open Japanese Patent Application 55-110 786, 56-116 891 and 56-145 927, and US Patent 4 323 434.

[25] Yoshida, M., and Matsuoka, H. (to Asahi Chemical Ind. Co., Ltd.), Laid Open Japanese Patent Application 56-146 884.

[26] Yoshida, M., Matsuoka, H., Munekuni, H., and Korenaga, T. (to Asahi Chemical Ind. Co., Ltd.), Laid Open Japanese Patent Application 57-63 684.

7.7 DISCUSSION

1. Prof. Kahn of India asked about the structure of the Asahi Chemical bilayer membrane. Was the polymer cross linked to incorporate both sulphuric and phosphoric groups, what was the ratio of the two functional groups and was an organometallic reaction used for the copolymerization step?

   Dr Seko stressed that the membrane is never cross linked. Generally, it was copolymerized with TFE (not using an organometallic reaction) and due to the crystallinity of the TFE polymer component the membrane becomes insoluble. The thickness of the carboxylic layer facing the anolyte is about $7\mu m$ the remainder of the membrane is about $100\mu m$ so this gives a guide to the ratio of the functional groups.

CHAPTER 8

Olin Membrane Cell Technology

PETER A. DANNA, KEN E. WOODARD, Olin Corporation, Stamford, Connecticut, USA

Olin Corporation has been a major factor in the US chlor-alkali industry for over 80 years. Through its predecessor companies, Castner Electrolytic Company, Mathieson Alkali Works, and Olin Mathieson Chemical Corporation, Olin has been among the leaders in developing and applying new technology in the production of chlorine and caustic.

In view of increased power costs associated with chlor-alkali production, and recognizing the significant technology advances made in membranes for use in a chlor-alkali environment, Olin began research on new membrane cell technology in the mid-1970's at its Regional Technical Center at Charleston, Tennessee. This successful effort has led to the development of an innovative, highly reliable system.

The program spans the entire range of technical knowhow required to develop equipment, evaluate its operation and assemble complete engineering and hardware packages for chlor-alkali plants. This chapter describes a few key elements of this program.

8.1 PROGRAM SCOPE

The development of Olin membrane cell technology has been an integrated program of system components evaluated at multiple scales with the development of the support systems necessary to provide a commercial system. This is illustrated in Fig. 8.1.

Olin's membrane cell technology program development effort is composed of experience gained in the operation of membrane cells at five levels of scale. In laboratory and bench scale, new membranes are evaluated for operational characteristics, impurity tolerance, and life tests of membranes, anode coatings and cathode surfaces.

In pilot scale, engineering evaluation of the cell hardware concepts and configurational testing of electrode geometrics are conducted. Here, the test

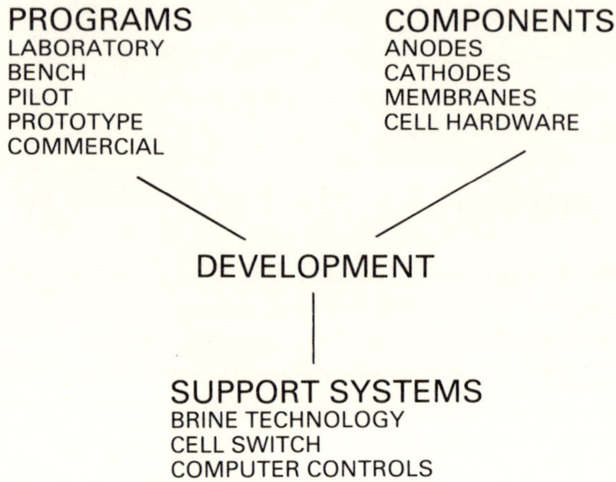

Fig. 8.1 – The development of Olin membrane cell technology.

hardware is also large enough to examine such things as hydrodynamic effects in cell operation.

In prototype scale, the cell design is evaluated in commercial size unit cells for overall performance of the membrane, anode, cathode, and operational characteristics of the engineered system.

Finally, in commercial scale, the full size electrolyzer is evaluated under routine to day operational situations at current loads up to 160 kA in a commercial mercury cell circuit and production environment. An OM-PAC^{tm} 70 commercial cell was operated successfully over a 2-year period in our cell circuit at Charleston, Tennessee evaluating the performance of both Flemion® and Nafion® membranes. This cell contains 12 anode elements that are 1.5 m wide × 2.2 m high, designed for an operational current density of 2.25 kA/m^2. As will be described later, commercial current densities since this cell was run have now escalated to two to four times that value.

8.2 COMPONENT DEVELOPMENT AND EVALUATION

In would now like to discuss another phase of Olin's effort, namely component development and evaluation.

8.2.1 Anode coatings

Experience has been gained in anode coatings produced by Electrode Corporation and by Conradty; the latter are available in the US through Grelcon. These are used in all scales of the program. The performance of anode coatings for long life in Olin's OM-PAC^{tm} Membrane Cell has been demonstrated.

8.2.2 Membranes

In the area of membranes, at the last meeting, Dr Howard Yeager presented a paper on the use of radiotracers to measure the sodium ion transport of membranes in the chlor-alkali cell environment [1]. Over the past 3 years, Olin, working with Dr Yeager, has evaluated and reported at The Electrochemical Society meetings on the transport properties of the perfluorinated carboxylate and sulfonate ion-exchange polymers in the concentrated solution environments of membrane chlor-alkali cells. This work has led to the development of a sophisticated system that can automatically, under computer control, develop the sodium ion transport properties of membranes. And, of course, sodium ion transport is directly related to the caustic current efficiency of the membrane cell.

A specially designed chlor-alkali membrane cell is used to determine the sodium ion transport by radiotracer techniques at the conditions of temperature, current density, electrolyte concentration that are of interest in today's high performance membranes. Data obtained by this system is utilized, via modern computer tools, to develop what we call membrane response surfaces.

These surfaces provide us with a road map through the complex current efficiency surface created by (1) the temperature, (2) the current density, (3) the caustic concentration and (4) the brine concentration. You can see from the comparison of the response surface constant value contour plots of Nafion 901® membranes at 2 and 4 kA/m^2 at 84°C shown in Figs. 8.2 and 8.3 that current efficiencies of 96% are achievable at different combinations of electrolyte concentrations.

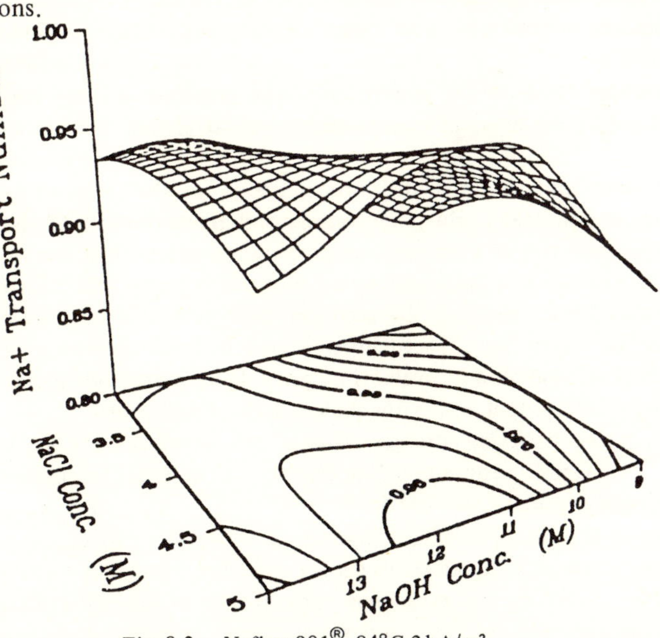

Fig. 8.2 – Nafion 901®, 84°C 2 kA/m^2.

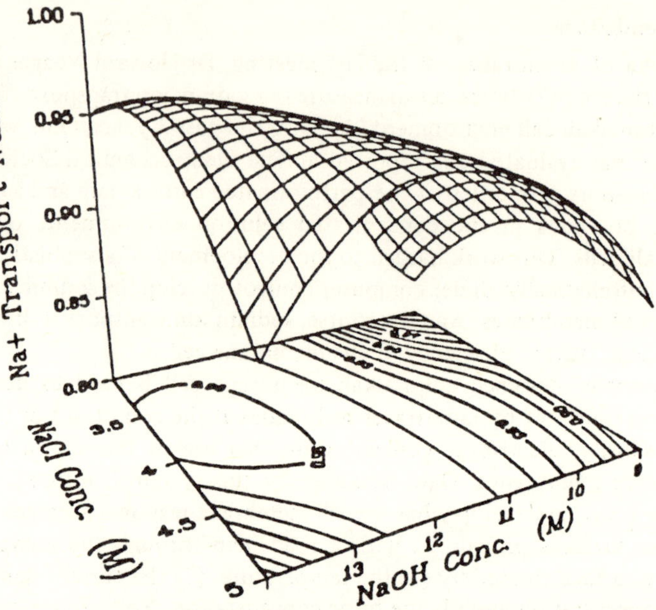

Fig. 8.3 – Nafion 901®, 84°C 4 kA/m².

As shown in these figures, the effect of increased current density is to shift the maximum in the surface to higher caustic concentrations and lower brine concentrations.

When the temperature is increased, the response surface also changes in shape, changing the areas of maximum current efficiency as shown in Figs. 8.4 and 8.5.

You may tend to rank this procedure as another interesting laboratory technique among many. But the complete characterization of the membrane response surface you've seen here, was completed in less than one week's system operation, whereas it would take 6 to 9 months to gather this information using more conventional methods. In addition, test system results have accurately described the performance actually observed in large operating cells. We feel this method of membrane evaluation gives Olin a real advantage in rate of technology development and is unequaled in the world today.

Now, going back to commercial size cell development, the later version OM-PAC^{tm} 50 Cell is a commercial monopolar cell designed to operate at 3 kA/m^2 current density, 160 kA current load, producing a nominal 5 metric tons/day of caustic soda. The cell utilizes titanium and nickel construction with natural 'gaslift' circulation of the electrolytes for uniform consistent membrane performance. The caustic current efficiency performance of this cell utilizing the latest high performance Nafion 900® series membranes at brine salt depletion of 70%,

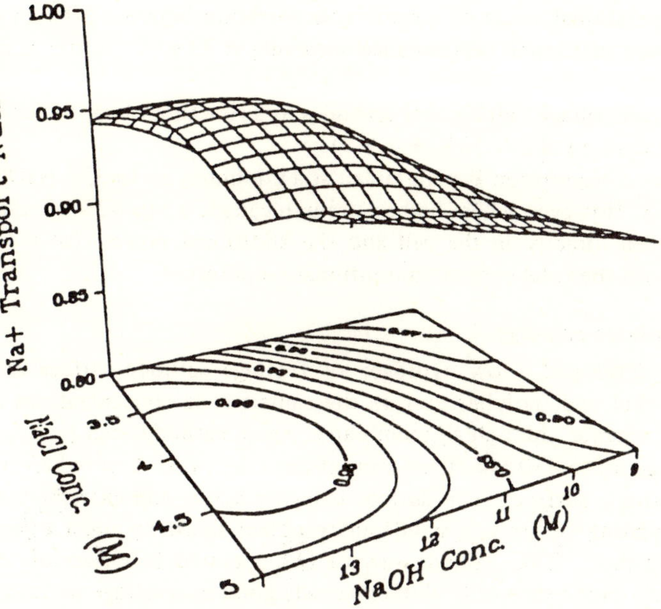

Fig. 8.4 – Nafion 901®, 90°C 2 kA/m².

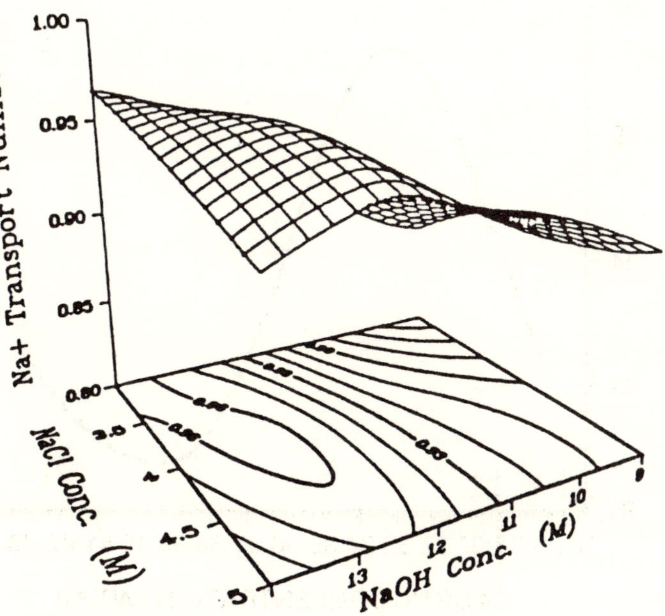

Fig. 8.5 – Nafion 901®, 90°C 4 kA/m².

has an operational range of caustic concentration between 30% and 42% with peak current efficiency performance occurring at 33 wt-% caustic as indicated in Fig. 8.6.

The cell voltage, with nickel cathodes, is a function of caustic concentration and is a minimum at 29–30% (Fig. 8.7).

Power consumption is maintained at a minimum up to 33% NaOH as shown in Fig. 8.8. However, where higher strength caustic up to 42% is required, it can be produced directly in the cell and the additional power cost is highly competitive with the total cost of conventional evaporation.

8.2.3 Cathode coatings

Olin has developed a low hydrogen overvoltage cathode surface consisting of Raney nickel and molybdenum. In the caustic soda concentrations of the new high-performance chlor-alkali membranes, these cathodes can reduce the overall cell voltage up to 350 mV at cell current densities of 3 kA/m^2 when compared to nickel cathode surfaces. This surface is highly active and extremely resistant to the overplating of iron, a significant problem faced by most other low-overvoltage surfaces. This cathode surface has operated in excess of 550 days in bench scale membrane cells with extremely good overvoltage performance. Pilot cells operating with finite gap using this cathode have achieved 3.08 V at 3 kA/m^2,

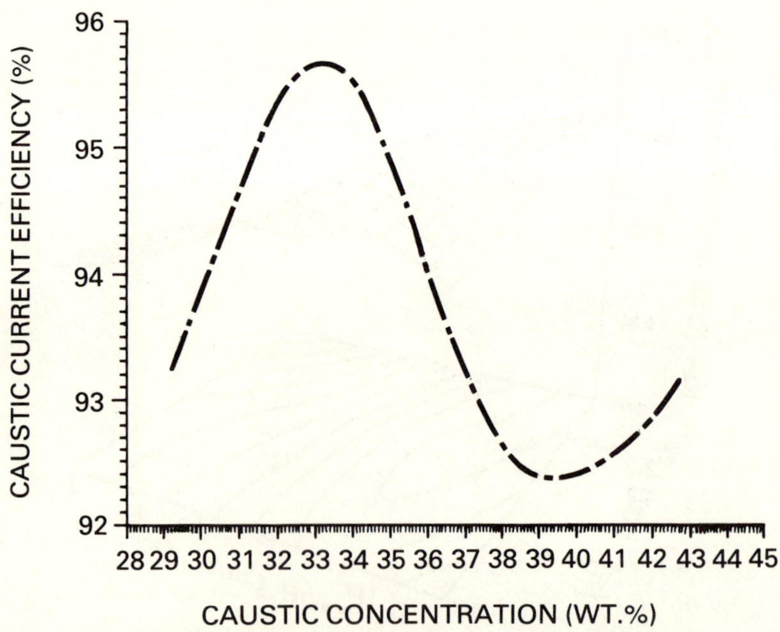

Fig. 8.6 – Current efficiency profile.

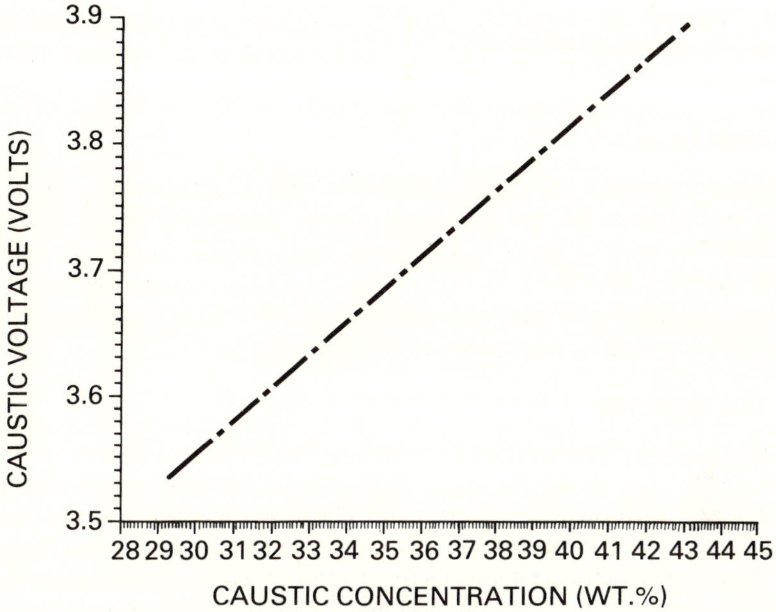

Fig. 8.7 – Voltage performance.

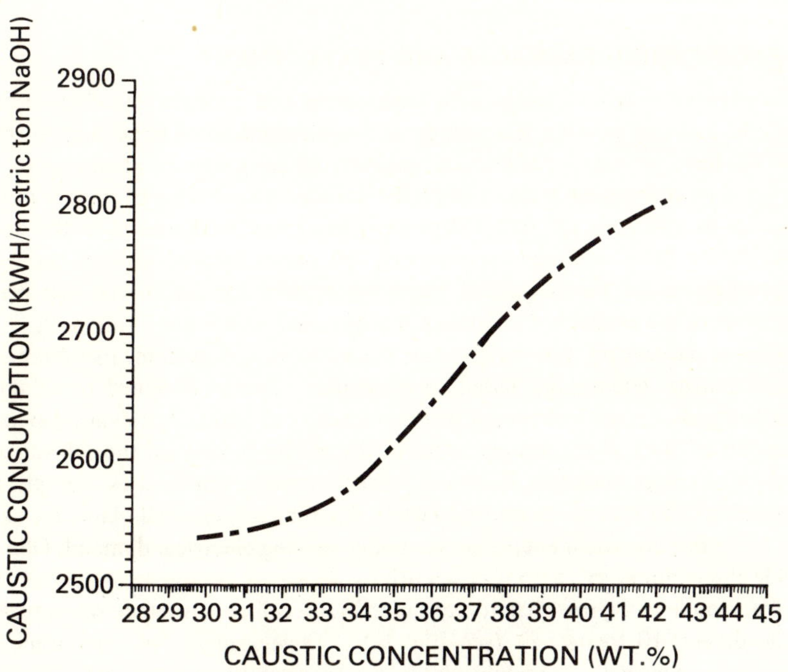

Fig. 8.8 – Power consumption.

2150 d.c.-kWh/metric ton NaOH, using the latest high performance membranes. Commercial scale production of this system for a commercial scale call has been achieved.

8.3 BRINE QUALITY

The high performance of today's membrane cells is achieveable only if the correct environment for the membrane can be maintained. Brine treatment technology to control alkali metal earths, trace metals, and other harmful cations has been developed at Olin. It has been demonstrated in the highly successful operation of a prototype cell in excess of 2 years using Nafion 901® membranes, maintaining high levels of current efficiency.

8.4 CELL STARTUP

Olin, jointly with Westinghouse Corporation, has developed a membrane cell jumper switch that allows the initial application of low current on startup of these new high performance membranes with sequential increase to achieve finally full circuit load. This allows the membrane to be started up at a current that does not 'damage' the membrane during its first period of transport, resulting in higher average current efficiency over the membrane's life and allows the cell circuit to be maintained at normal loads during individual cell changeouts.

8.5 ENGINEERING PACKAGES AND DEVELOPMENT

Olin continues to work closely with engineering and construction firms around the world and can provide technology and equipment for a wide range of plant sizes. We have provided preliminary engineering packages on plants up to 700 tons per day and we have also completed detailed engineering of membrane cell plants up to 150 tons per day. Olin's experience and background in the design of circulating brine systems for mercury cell plants located all over the world has complemented the design of the brine system for membrane cell plants. Brine systems are available to handle salt from solar, mines and brine wells.

Olin is pioneering, just as it did in the mercury cell system, membrane cell process control technology based on computer systems designed to minimize the utilization of electrical energy in a membrane cell plant. Achieving maximum utilization of electrolytic energy requires the ability to control the cell variables at conditions that minimize electrical power consumption on these complicated response surfaces that we presented earlier. When membrane cell plants are called upon to utilize electrical energy better under varying electrical demand, Olin can provide the technology to do it efficiently.

We should recognize that today's membrane cell performance was a technological dream 10 years ago (see Fig. 8.9). Developments in membrane cell concepts and especially in membranes have now established this as the technology of the chlor-alkali industry.

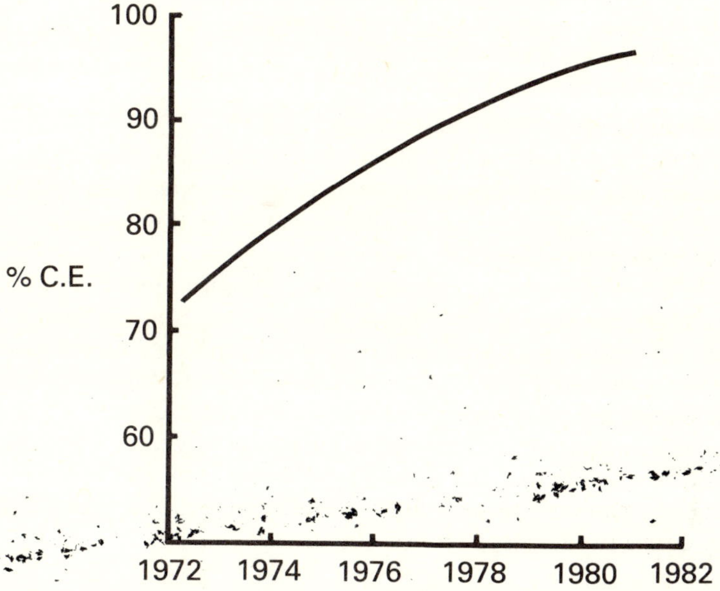

Fig. 8.9 – Membrane cell: caustic current efficiency versus time.

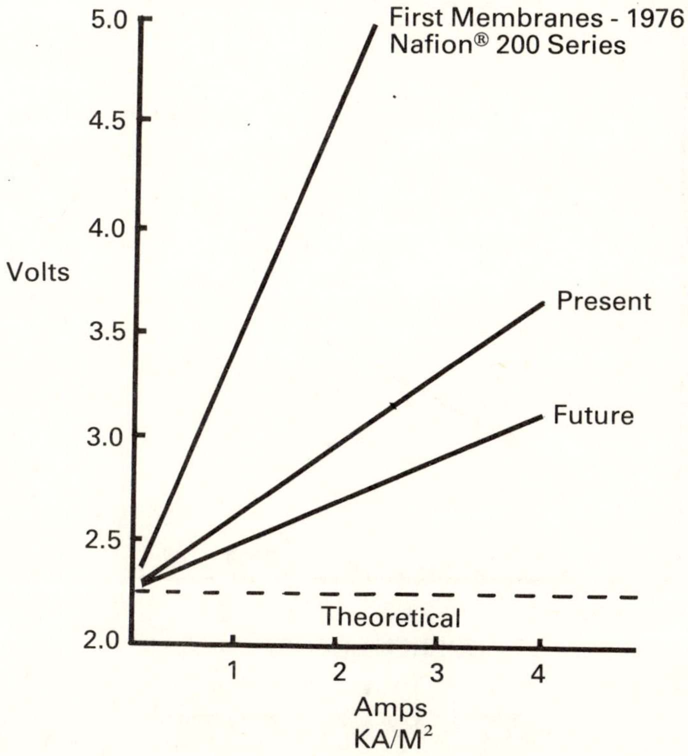

Fig. 8.10 – Olin membrane cell.

As you are aware, a wide variation of cell design concepts is evident among the various cell licensors. From Olin's viewpoint, what we are seeing are voltage–current relationships that are now equivalent to mercury cell performance in terms of cell coefficients. This is illustrated in Fig. 8.10.

Olin's design philosophy is to build substantial cells, with long life materials of construction and adequacy of components to enable commercial operation in the 5 to 10 kA/m^2 range in the not too distant future. In this continuing effort, we have developed the Electrode Membrane Composite (or EMC) Cell that combines the latest improvements in membranes, cathodes, and cell hardware.

Cell voltages that translate to commercial d.c. power consumptions of 1950 kWh/metric ton NaOH at 2 kA/m^2 and 2175 at 4 kA/m^2 are now being demonstrated in pilot scale cells, Fig. 8.11. A commercial size cell utilizing this technology will be operational later this year at a current density of 5 kA/m^2.

The advancements in membrane cell technology clearly have established this as the most energy efficient and desirable for future chlor-alkali plants of the 1980s. Olin will continue to be a leader in developing and applying this new technology during the decades to come.

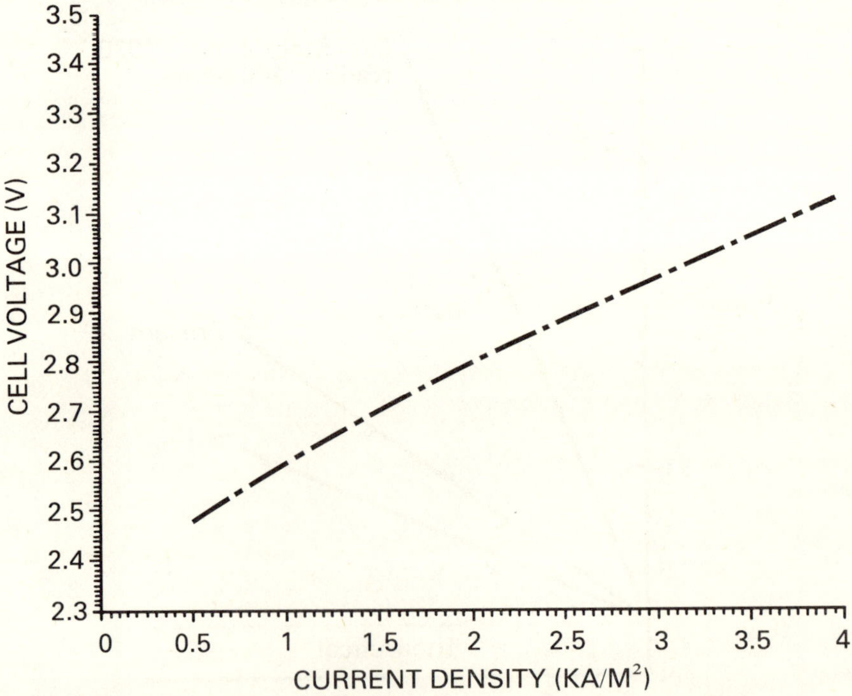

Fig. 8.11 – Olin EMC cell performance.

8.6 REFERENCE

[1] Dotson, R. L., and Yeager, M. L., 'Fundamentals of Transport and Diffusion through Chlor-Alkali Cell Membranes', in *Modern Chlor-Alkali Technology*, (ed. M. O. Coulter), Ellis Horwood, Chichester, 1980.

CHAPTER 9

Uhde's Membrane Cell Technology

H. ISFORT, Uhde GmbH, Dortmund, West Germany

9.1 INTRODUCTION

While the mercury-cell and the diaphragm-cell processes for the electrolysis of alkaline chlorides are still part of a traditional field of activity, Uhde has been engaged in the research and development of the membrane-cell process since the early seventies. The membrane cell has, indeed, passed through various stages of design which may broadly be defined as three cell generations. These cell generations are characterized in particular by the gradual reduction of the energy requirements.

Figure 9.1 highlights the rapid reduction of energy requirements for membrane cells over a period of about 5 years. Up to 1979, the d.c. power requirements for membrane cells of the first generation amounted to about 2800 kWh per ton of caustic. For the second generation, built between 1979 and 1981, the figure was 2600 kWh per ton of caustic while the power requirements

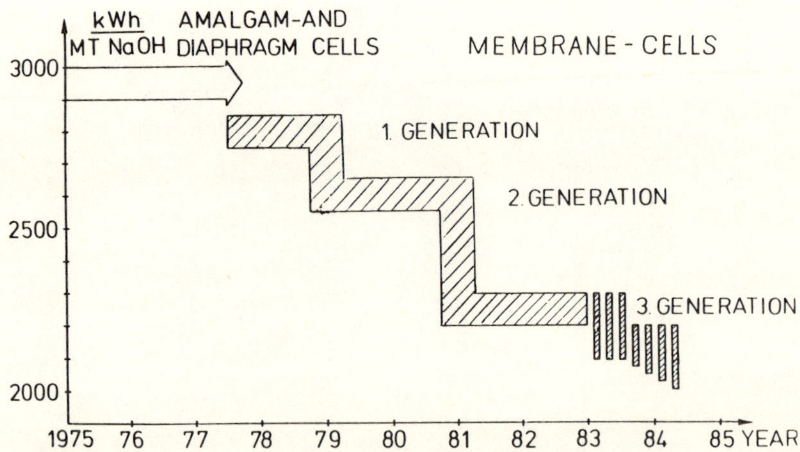

Fig. 9.1 – Development of energy consumption for alkalichloride electrolysis.

could be lowered to 2200 to 2300 kWh per ton of caustic with the advent of the third generation in 1981. A further reduction of electric power consumption – although only at a smaller rate – may be expected from future improvements relating to the membranes, for example. It is too early to predict further progress in the economy of the membrane-cell process by the application of air or oxygen cathodes, pressurized cells or other innovations. What may be expected, however, is a longer service life of electrode coatings and membranes and a higher specific load of membrane cells.

Figure 9.1 reflects only the d.c. power consumption of the membrane cells while energy requirements of the diaphragm cells include the steam consumption for concentrating the caustic soda solution to 50%.

The characteristic features of the three generations of membrane cells are outlined in this chapter.

9.2 FIRST MEMBRANE-CELL GENERATION (1975 to 1979)

This generation of cells was initiated by the availability of membranes manufactured by Du Pont for commercial-scale application. While service life and energy requirements of these membranes were satisfactory, the cells produced caustic soda solution at a low concentration only. For the application of these membranes, Uhde cooperated with Hooker in the development of electrolysis cells for producing chlorine and dilute caustic soda solution for specific consumers, such as the pulp and paper industry.

The cooperation betweeen Hooker and Uhde in the development of membrane cells continued their previous cooperation in the field of diaphragm cells.

Figure 9.2 summarizes the major design characteristics of the cells. The cells were equipped with membrane type Nafion® 324 and 390 of Du Pont, suitable for a caustic concentration ranging from 8 to 15%. They contained expanded-metal type titanium anodes with DSA* coating.

	1. GENERATION – MEMBRANE CELLS 1975–1979
MEMBRANE:	NAFION 324, 390 FOR NaOH 8–15%
ANODES:	MESH-TYPE, TITANIUM WITH ACTIVATION
CATHODES:	MESH-TYPE, STEEL
ELECTRODE – GAP:	3–6mm
CELL – CONFIGURATION:	MONOPOLAR AND BIPOLAR, FILTERPRESS
CELL – ELEMENT DESIGN:	FRAME WITH INTERNAL BUS BARS FRAME MATERIAL: PLASTIC

Fig. 9.2 – Cell-design parameters.

The cathodes feature the same structural shape as the anodes and were fabricated from ordinary carbon steel. The electrode gap ranged from 3 to 6 mm. Development work was focused on a monopolar cell for pilot plant operation while commercial plants were equipped with Hooker's bipolar cells only. Both cells were of the filter-press type. The cell elements consisted of frames with internal bus bars for the power supply and distribution of anodes and cathodes. Plastic material was used for fabricating the frames.

The Hooker cell was earmarked by types MX-1 and MX-2. Cells of the MX-1 type were used for the plant of Reed Paper Co. at Dryden, Canada, and commissioned as early as 1975. The plant capacity was 20 000 tons of NaOH per year. Cells of the MX-2 type were installed in two plants started up in 1980 and constructed for Fort Howard Paper in Oklahoma with a capacity of 4000 tons NaOH per year, and for Tofte Cellulosefabrik A/S in Norway with a capacity of 14 000 tons NaOH per year.

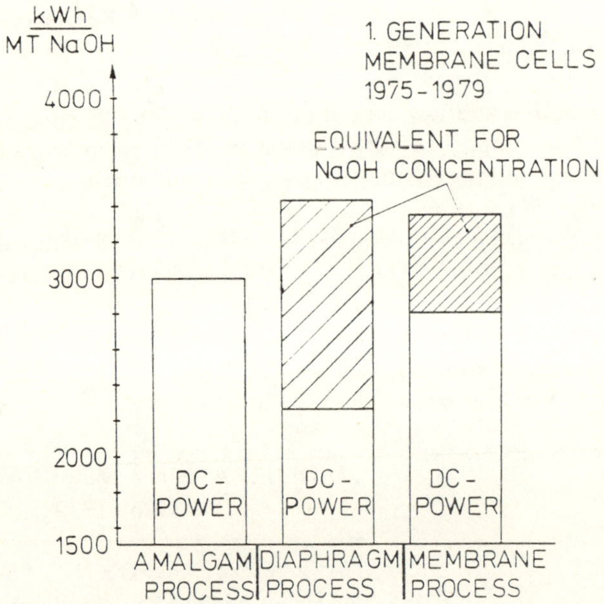

Fig. 9.3 – Energy requirement basis NaOH 50%.

Figure 9.3 reflects the total energy requirements of a membrane cell of this generation – referred to the production of caustic soda solution at 50% – compared with the mercury-cell and the diaphragm-cell processes. Energy requirements for membrane cells are obviously higher than for mercury cells because of the equivalent energy input for concentrating the cell liquor from 15 to 50%. Typical first generation installations are shown in Figs. 9.4 to 9.7.

Fig. 9.4 – Uhde/Hooker monopolar membrane cell with plastic frames.

Fig. 9.5 – Hooker MX-1 membrane cell, Dryden/Canada.

Fig. 9.6 – Hooker MX-2 membrane cell, Tofte/Norway.

Fig. 9.7 Membrane cell plant, Tofte/Norway

9.3 SECOND MEMBRANE-CELL GENERATION (1979 TO 1981)

The second membrane-cell generation is characterized by:

(1) Improved membranes.
(2) Metallic cell frame.

The improved membranes were at the root of the real breakthrough of membrane-cell technology for a broader spectrum of commercial applications. They made it possible to obtain caustic soda solution of a higher concentration at reduced power consumption.

A modification of the cell design and the use of metal for cell construction contributed jointly to lowering the energy requirements and raising the service life of the membranes and membrane cells. Figure 9.8 summarizes the design features of the second membrane-cell generation. Apart from the Nafion® 390

	2. GENERATION – MEMBRANE CELLS 1979–1981
MEMBRANE:	NAFION 390 FOR NaOH 8–15% 901 FOR NaOH 35% FLEMION 230 FOR NaOH 35%
ANODES:	MESH-TYPE, TITANIUM WITH ACTIVATION
CATHODES:	MESH-TYPE, STEEL AND STAINLESS STEEL WITH ACTIVATION
ELECTRODE – GAP:	3mm
CELL – CONFIGURATION:	MONOPOLAR – FILTERPRESS BIPOLAR – SINGLE ELEMENT TYPE
CELL – ELEMENT DESIGN:	FRAME WITH INTERNAL BUS BARS FRAME MATERIAL: METAL

Fig. 9.8 – Cell-design parameters.

membrane, the membranes available at that time were Nafion® 901 of Du Pont and Flemion® 230 of Asahi Glass, these two being capable of producing caustic soda solution at about 35%. The mechanical design of the anodes remained unchanged while stainless steel was used in addition to carbon steel for the cathodes. The application of cathode activating coatings permitted a reduction of the cell voltage, too. The electrode gap could be reduced to and held at exactly 3 mm because of the improved mechanical design of the cell. The advantages provided by the innovations in the frame design and by the use of metal (carbon steel/stainless steel and titanium, respectively) may be outlined as follows:

- Maintenance of closer fabrication tolerances which permit an exact spacing of electrodes, as said before, and ensure less cell leakage.
- Life expectancy of metal is significantly higher than that of plastic material.
- Cells can be operated at higher working temperatures, which contributes to reducing energy requirements.

Development work extended to both monopolar and bipolar cells. While the monopolar version was again of the filter-press type, the design of the bipolar cell is based on the principle of individual elements. The typical features of these monopolar and bipolar cell elements will be better understood by referring to Fig. 9.9.

The monopolar element consists of a rectangular metal frame to which electrodes of expanded metal are welded on both sides. Electric power is supplied through jacketed copper rods which, in turn, are welded to the expanded metal with current distributors between copper rods and expanded metal. The anode element is made of titanium throughout while carbon steel or stainless steel is used for the cathode element. The membrane is inserted between the anode element and the adjacent cathode element and is sealed by means of a rubber gasket to prevent external leakage. A large number of these cathode and anode elements are assembled to form a cell block in which the cell elements are held together in the form of a filter-press.

In the bipolar cell element, anode and cathode are each accommodated in a rectangular half-shell. The membrane is inserted between these two half-shells, and the element is sealed by means of a gasket and an external flange. The current conductors penetrate the half-shell walls and are welded to the anode and cathode, respectively. Again, a number of elements are lined up and tied together firmly in an effort to obtain the necessary electrical element-to-element contact, but not in order to prevent gas or liquid leakage which is aimed at by this operation for the monopolar version of the filter-press type.

The first metallic monopolar cell was a small-size unit designed for trial operation in the pilot plant of Hoechst. The second step was the assembly of a 70 kA block which was installed and successfully operated in the pilot plant at Leverkusen. This pilot plant also incorporated a cell of Asahi Glass running in series with the Uhde cell. The Asahi cell was operated in compliance with a license agreement signed by Asahi and Uhde, by virtue of which Uhde has the right to offer both the complete Asahi cell and the Flemion® membranes for use in Uhde cells.

The Uhde monopolar cell is also used in the membrane-cell electrolysis plant of Hoffmann La Roche in Scotland. The plant is under construction and will

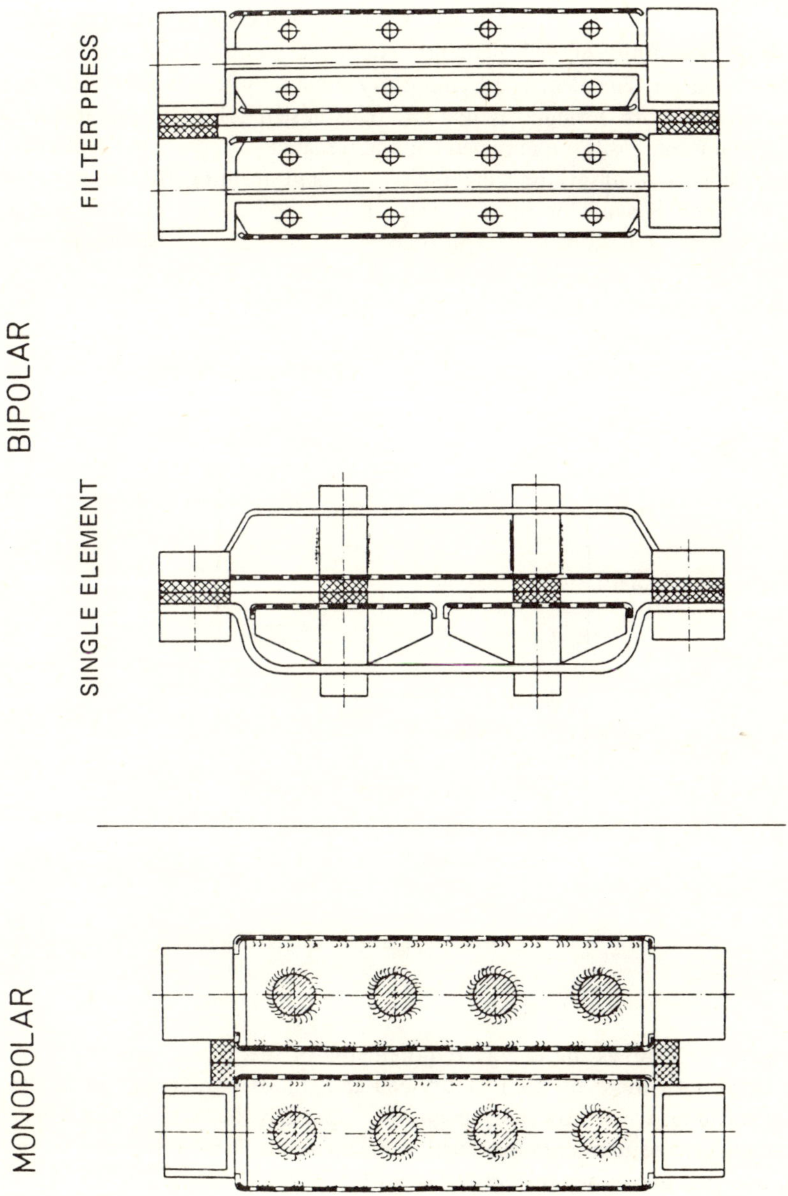

Fig. 9.9 – Typical features of monopolar and bipolar cell elements.

comprise 28 cells designed for a load of 80 kA to produce 23 000 tons of caustic soda solution per year.

The bipolar cell configuration was developed and built by Hoechst. The cell block is an assembly of cell elements which are composed of the half-shells referred to before. The electrical element-to-element contact is achieved through copper contacts with spring-loaded stems. This design offers the advantage that a cell element can easily be removed for maintenance of membrane replacement without time-consuming emptying operations and refilling of the entire cell block. This cell is available in two sizes of 1 m^2 and 2.7 m^2 electrode surface. Both types have undergone long-term trials in the pilot plant of Hoechst.

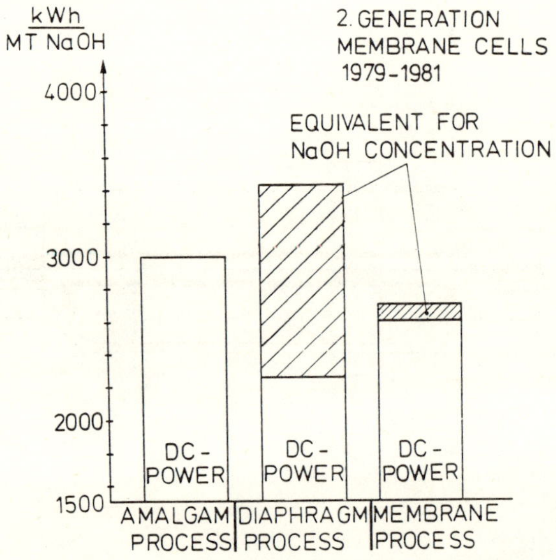

Fig. 9.10 – Energy requirement basis NaOH 50%.

Figure 9.10 shows the energy requirements of this generation of cells compared again with those of the mercury-cell and the diaphragm-cell processes. A distinct advantage can now be recognized, especially because of the markedly reduced energy requirements for caustic concentration. Typical second generation membrane cells are shown in Figs. 9.11–9.18.

Fig. 9.11 – Hooker/Uhde monopolar membrane cell with metal frames, pilot cell.

Fig. 9.12 – Hooker/Uhde monopolar membrane cell, 70 kA.

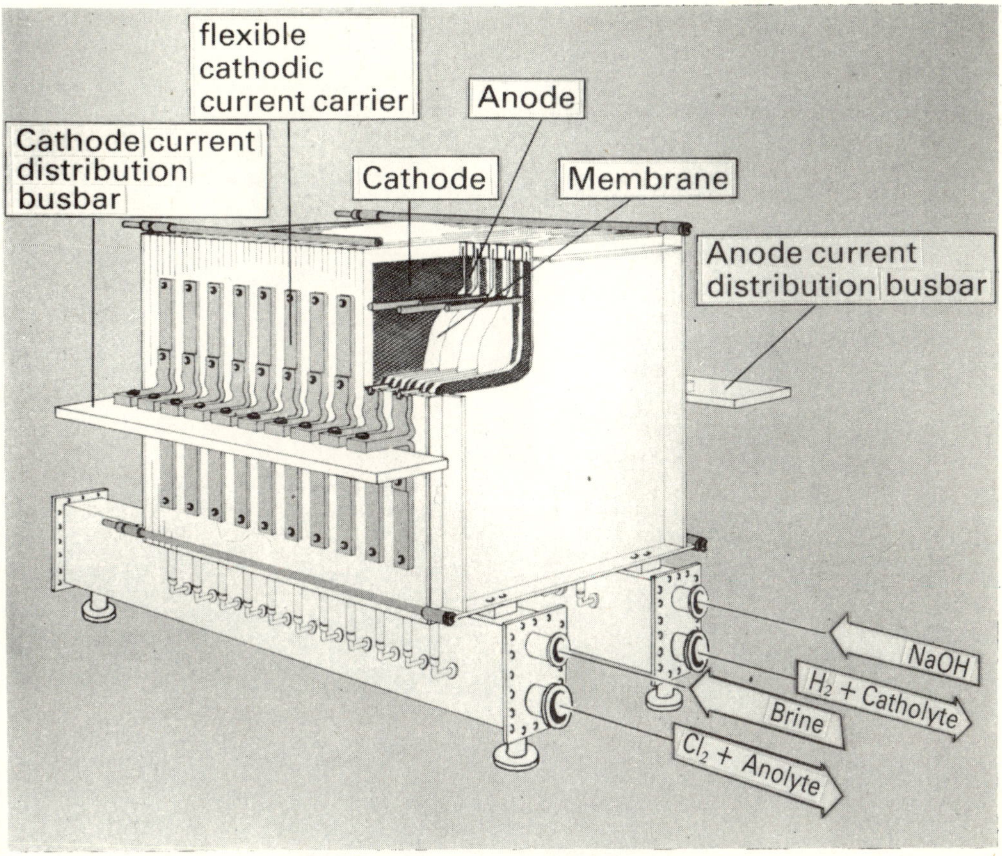

Fig. 9.13 – Hooker/Uhde monopolar membrane cell for Hoffmann La Roche, Scotland.

Fig. 9.14 – Model of cell room of Hoffmann La Roche, Scotland.

Fig. 9.15 – Asahi Glass and Uhde monopolar membrance cells at Bayer AG, Leverkusen.

Fig. 9.16 – Hooker/Uhde monopolar membrane cell element.

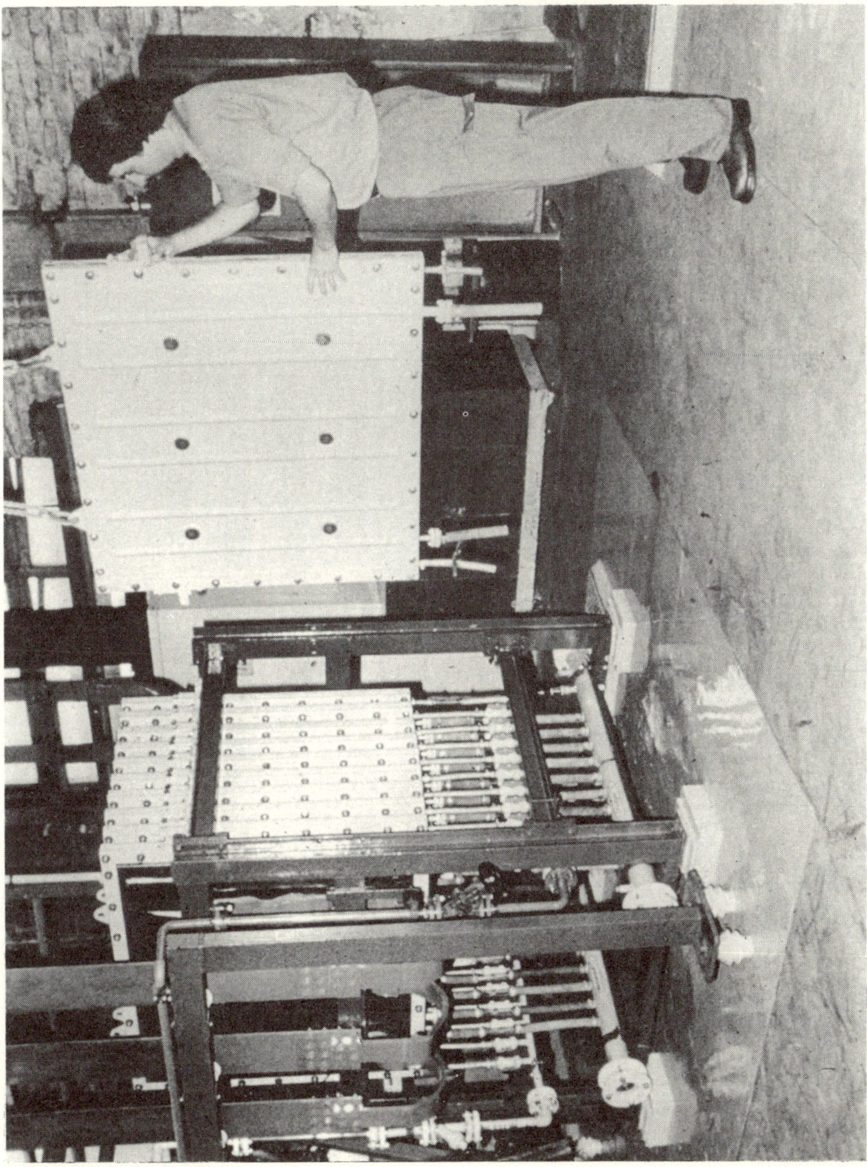

Fig. 9.17 – Hoechst bipolar membrane cell, 1 m^2 electrode surface.

Fig. 9.18 – Hoechst bipolar membrane cell, 2.7 m² electrode surface.

9.4 THIRD MEMBRANE-CELL GENERATION (SINCE 1981)

The development of the third membrane-cell generation was initiated again by the availability of membranes with improved characteristics. Better surface quality and reduced voltage drop are the principle improvements. Membranes of the preceding generation featured a surface with hydrophobic properties which favoured the adhesion of gas bubbles. The ultimate aim to reduce the electrode gap to a minimum ('zero gap') could not be achieved because of this property of the membranes. On the contrary, it was found that the cell voltage increased again when the gap was reduced below a certain definite minimum.

This undesirable effect could be eliminated by enhancing the hydrophilicity of the surface of the membrane, which paved the way for designing this third generation cell. As far as we are aware, this new type of membrane can presently be procured from Asahi Glass only.

Figure 9.19 shows a summary of design features of the third membrane-cell generation. The Nafion® membranes have been included although they do not yet possess the hydrophilic properties of the Flemion® membrane. They can also be used in this new type of cell without at present bringing about the desirable minimum voltage.

The mechanical design of anodes and cathodes was significantly modified to ensure a satisfactory discharge of gas bubbles and simultaneously a good electric current distribution across the membrane at the lowest possible investment cost. Anode coating remained unchanged while cathode coatings with improved activated effect are available. This activating coating can be deposited on a steel or nickel structure.

	3. GENERATION – MEMBRANE CELLS FROM 1981
MEMBRANE:	NAFION 901 FOR NaOH 35%
	NX 902 FOR NaOH 35%
	FLEMION 723 FOR NaOH 35%
	753 FOR NaOH 35%
ANODES:	IMPROVED STRUCTURE, TITANIUM WITH ACTIVATION
CATHODES:	IMPROVED STRUCTURE, STEEL/ NICKEL WITH ACTIVATION
ELECTRODE – GAP:	0mm
CELL – CONFIGURATION:	MONOPOLAR, FILTERPRESS, MODULE
CELL – ELEMENT DESIGN:	PLATE

Fig. 9.19 – Cell-design parameters.

The design of the individual cell element was markedly simplified as compared with the previously described cell element of the second cell generation which was relatively complex and expensive. The welded frame was replaced by electrode plates which can be fabricated from a single metal in one shaping operation. The width of the electrode in the direction of current flow was significantly reduced to hold ohmic losses across the electrode at a minimum. Two plates with an interposed gasket form an anode and a cathode, respectively. This gasket is sized and shaped to provide the anode space and cathode space, respectively, including the channels for the input and discharge of electrolyte and for gas discharge. Again, the membrane is clamped and held in place between anode and cathode with the aid of a gasket. A cell block is assembled by simply stacking a number of electrode plates and gaskets in a way similar to a plate heat exchanger.

Attention was drawn earlier to the narrow shape of the electrode plates which reduces the permissible load of a cell block composed of these plates. For a cell block comprising about 100 elements, the permissible load is in the order of about 100 kA which can be accepted only for electrolysis plants of small production capacity. Efforts are being made by Uhde to overcome this disadvantage.

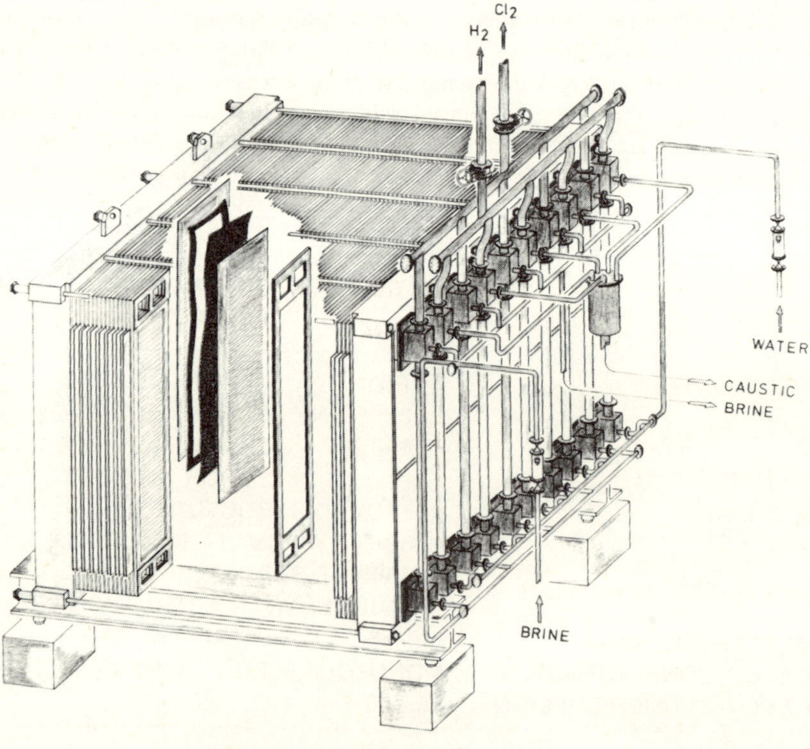

Fig. 9.20– Typical Uhde cell module.

The key to solving this problem is the modular design of membrane cells of the new generation.

Figure 9.20 shows a typical cell module.

A module consists of a number of cell blocks arranged so that they can be tied together by a single frame while the electric current flows from cell block to cell block through directly connected anode plates and cathode plates.

For the inlet and outlet of the electrolyte and for the discharge of gaseous products, a single pipe system is provided to serve all cell blocks within one module, said pipe connection being equipped with the necessary instruments and valves. Restriction orifice plates are installed for distributing the electrolyte to the various cell blocks.

Figure 9.21 shows the interconnection of the cell modules by copper bus bars. Copper connectors are provided between bus bar and head end of the module for a bypass switch. After breaking the bus bar and pipe connections, a complete cell module can be removed with the aid of a hydraulic-lift truck and replaced by a new module.

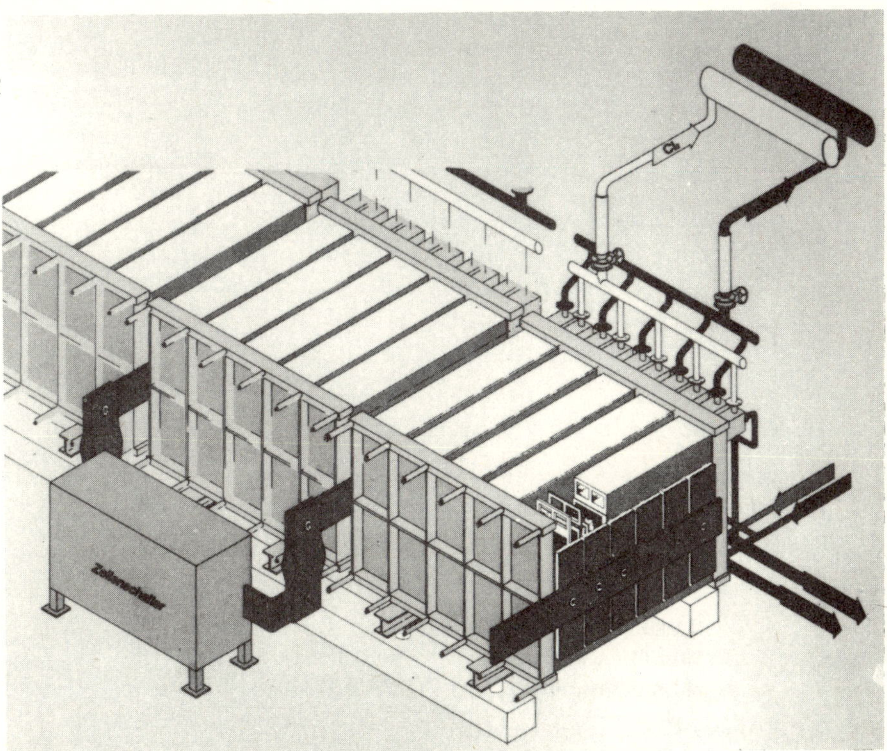

Fig. 9.21 – Interconnection of the cell modules.

Figure 9.22 shows the arrangement of cell modules in a fully equipped cell room of a plant for the production of 200 tons caustic soda solution per day. Space requirements are reduced to about two-thirds as compared with the arrangement of separate cell blocks.

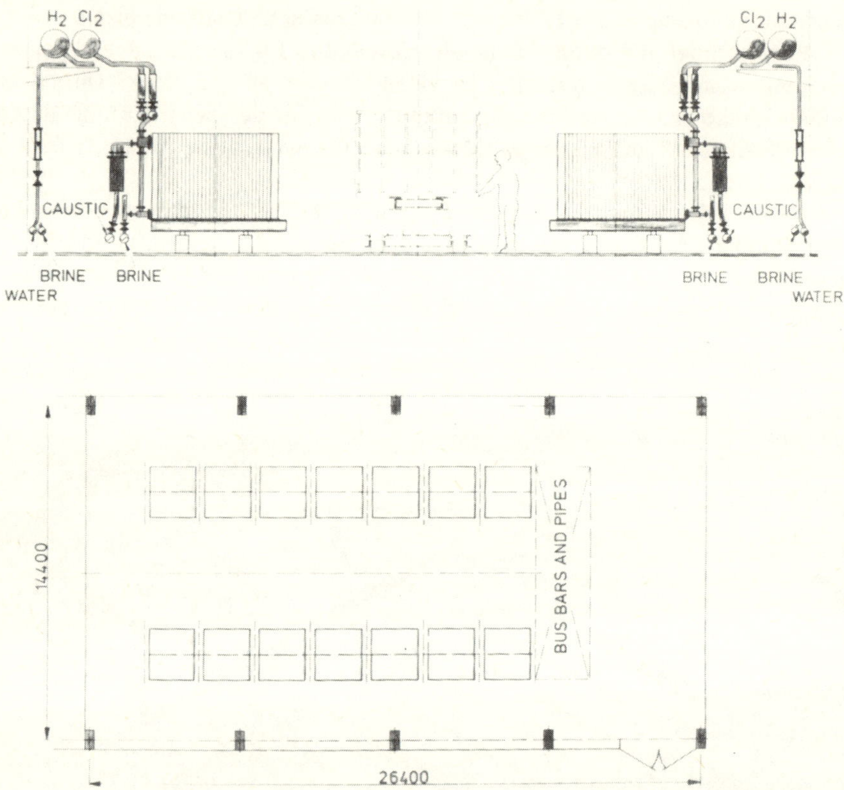

Fig. 9.22 – Arrangement of cell modules for a 200 tons per day caustic soda plant.

The principle advantages achieved with the modular design of this new type of cell may be summarized as follows:

(1) Reduced capital costs for the cells and the cell room, due to
   – uncomplicated electrode design
   – simplification of end plates and clamping device
   – simplification of piping and valves in the cell room.

(2) Reduced space requirements.

(3) Easy handling of cell modules. Labour costs are cut by reducing the number of cells per production unit by a factor of 3 to 5.

A cell module of this design has been running for some weeks in the pilot plant of Hoechst in Frankfurt. Operating results are satisfactory.

At the same time, an AZEC cell (Asahi Glass Zero gap Electrolytic Cell) and the latest ICI-FM21 cell have been running for some weeks in the pilot plant operated jointly by Bayer AG and Uhde at Leverkusen. These cells are connected in series with the AGC and the Uhde monopolar cells of the second generation and are rated for a load of 70 kA.

Figure 9.23 reflects the typical energy requirements for the membrane cells of the third generation in comparison with the mercury cell and the diaphragm cell. The significant advantage on energy requirements in conjunction with the marked lowering of capital costs outlined before is likely to gain increasing importance also in plant owners' considerations for converting their existing mercury-cell or diaphragm-cell plants to membrane-cell technology.

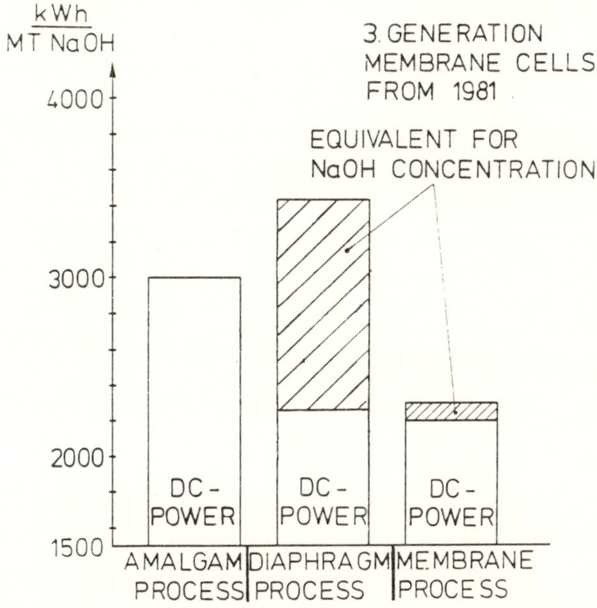

Fig. 9.23 – Energy requirement basis NaOH 50%.

Figure 9.24 is a summary of the technical data referring to the various membrane-cell versions that Uhde can offer to interested parties. This broad spectrum of electrolytic cells enables Uhde to provide the optimum solution to any requirement with regard to plant capacity, plant location, local conditions, and the re-equipment of mercury-cell or diaphragm-cell facilities with membrane cells.

Figure 9.25 is a review of the partners with whom Uhde is cooperating in the field of membrane-cell technology in particular and in the field of alkaline

Performance data
(examples)

	Bipolar Hoechst		Monopolar HUMM	MM	MODUL	AZEC
No. of elements	24	15	14	78	390	124
Amperage (kA)	3	8	70	70	70	70
Current density (kA/m^2)	3	3	3	3	3	3
Active area (m^2)	24	40	24	24	117	24
Active area per element (m^2)	1	2.7	1.7	0.3		0.2
Power comsumption (kWh/t NaOH)	2600		2600	2200†		2200
Cell dimensions:						
Length (m)	2.8	1.8	2.1	2.1	2.1	2.4
Width (m)	1.5	2.6	1.9	0.6	2.2	1.2
Height (m)	2.0	2.3	2.2	2.0	2.0	2.0
Capacity (tpd NaOH) approx.	2.5	4	2.5	2.5	12	2.5

The energy consumption depends on the operating conditions and the type of membrane used and varies accordingly.

† expected

Fig. 9.24 – Summary of technical data.

HOECHST:†	CELLS, PILOT OPERATIONS (CELLS, BRINE PURIFICATION)
BAYER:†	PILOT OPERATIONS (CELLS, BRINE, PURIFICATION)
HOOKER:	CELLS
ASAHI GLASS:	MEMBRANES, CELLS
DU PONT:	MEMBRANES

† LICENSORS FOR CHLORINE CAUSTIC PLANT OPERATIONS

Fig. 9.25 – Cooperation UHDE Partners.

chloride electrolysis technology in general. All of these renowned parties have been engaged in development work in this field for many years; some of them are major producers of chlorine and caustic. The benefits derived from this cooperation add to the experience gathered from the operation of pilot plants and the construction and commissioning of membrane cell plants put Uhde in an excellent position as an engineering company for the design and construction of alkaline chloride electrolysis facilities of the new era.

Typical third generation membrane cells are shown in Figs. 9.26 and 9.27.

Fig. 9.26 – Uhde membrane cell module.

Fig. 9.27 – Membrane cells at Bayer AG, Leverkusen: ICI-FM21, AGC monopolar cell, Hooker/Uhde monopolar cell, AGC AZEC.

CHAPTER 10

Energy Saving Advances in the Chlor-alkali Industry

V. H. THOMAS, E. J. RUDD, Eltech Systems Corporation, Center St, Chardon, Ohio, 44024, USA

10.1 INTRODUCTION

Energy and its impact upon the chlor-alkali industry has provided the major theme for this 1982 Conference from the standpoint of:

- the cost of electricity today and tommorrow,
- the availability of energy,
- the need for conservation of our energy resources, and conservation through more efficient utilization.

With respect to the latter, several technical developments which effectively reduce the power usage in the production of chlorine, sodium hydroxide and sodium chlorate have been discussed at this meeting.

The concern is very real. For example, as is shown in Table 10.1 there has been a rapid escalation in the cost of electrical power to US industry since 1977.

Table 10.1
Average industrial electrical power rate in the United States (cents per kWh)

Year	Rate	Increase (%)
1977	2.33	
1978	2.59	11
1979	2.85	10
1980	3.41	19.6
1981	3.96	16
1985	6.69	
1990	12.9	

Should the average rate of increase be maintained, the chlor-alkali industry in America would be using power at 6.7 cents per d.c.-kWh in 1985 and 13 cents per d.c.-kWh by the end of 1990.

Eltech Systems Corporation, as a licensor of electrochemical technology, has maintained a commitment to research and development programs directed towards improved technology particularly from the standpoint of energy consumption. Over the past decade, this commitment has led to commercially available technology for the production of chlorine and sodium hydroxide, using either the diaphragm process or membrane process, with lower and lower power usage (Fig. 10.1 and Fig. 10.2).

10.2 ENERGY REQUIREMENTS IN CHLOR-ALKALI PROCESSES

The total energy required to produce one metric ton of sodium hydroxide (50 w-%), by the electrolysis of brine using the three major technologies available today, is shown as Fig. 10.3. These data are believed to be 'state-of-the-art' for 1979–1980 and are regarded as typical for the particular technology.

It is immediately obvious that the major consumer of energy is the electrolysis, the operation of the electrolysis cell, be it a diaphragm, mercury or membrane cell. However, it should be noted that in the diaphragm cell process the energy required to evaporate and purify the cell liquor is larger than that used in the electrolytic cell, being 43% of the total energy requirement. In this regard, Diamond Shamrock will bring on line this summer a cogeneration system at the Battleground Plant (Houston, Texas). The system includes two gas-fired turbines and one steam turbine, the hot gases being used to produce high pressure steam, part of which is then diverted to the evaporation and purification units.

It is well established that the power consumed in an electrolysis is a function of the cell voltage and current efficiency, i.e.

$$\text{Power consumption} \propto \frac{\text{Cell voltage}}{\text{Current efficiency}} .$$

It is also clear that the minimum power consumption is realized with the minimum cell voltage and maximum current efficiency.

The current efficiencies obtained in today's diaphragm cell technology typically range from 93 to 95%. Furthermore, recent developments in ion exchange membrane cell technology have led to cathodic current efficiencies of 95–96% consistently being obtained in commercial membrane electrolyzers. Inprovement would seem unlikely and certainly will be difficult to achieve. However, improvements in current efficiency can certainly be cost effective. For example, an increase of 2% in the current efficiency is equivalent to a reduction of 50 d.c.-kWh/metric ton NaOH in the power consumption. Considering a

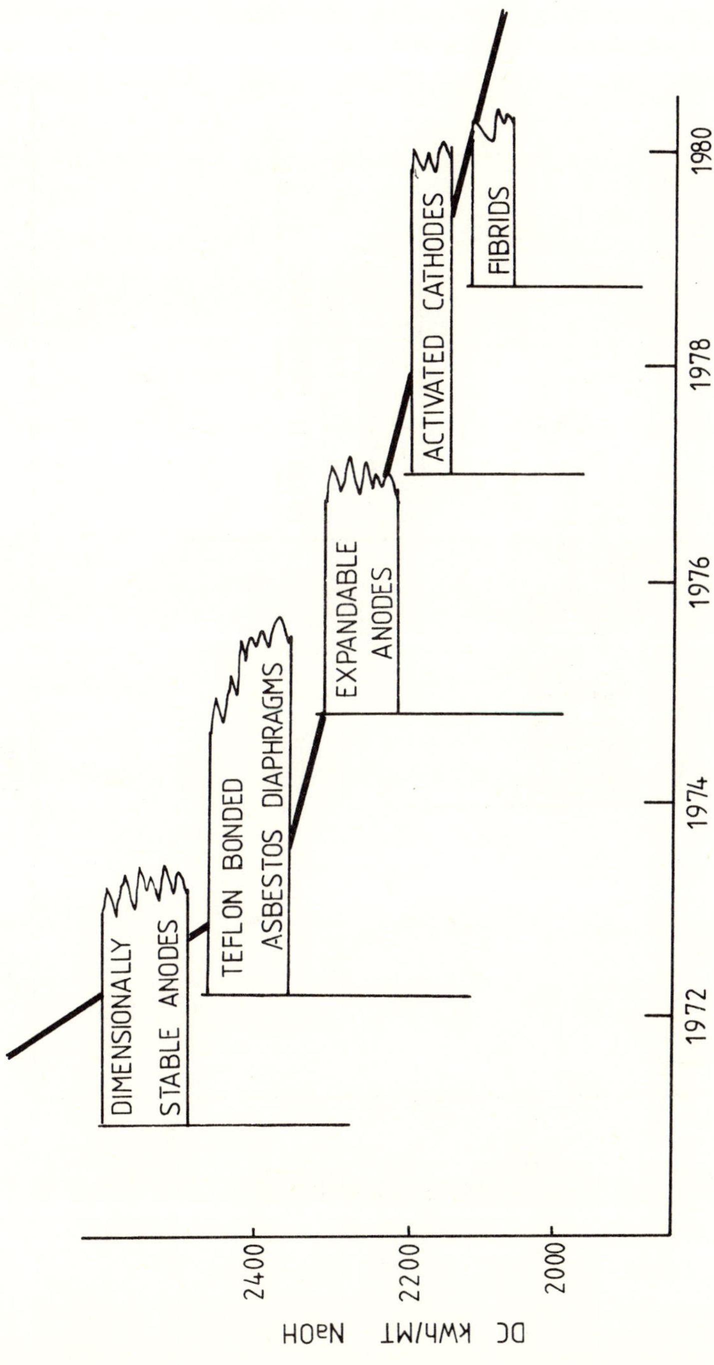

Fig. 10.1 – Development of Eltech Systems technology for production of chlorine and sodium hydroxide using the diaphragm process.

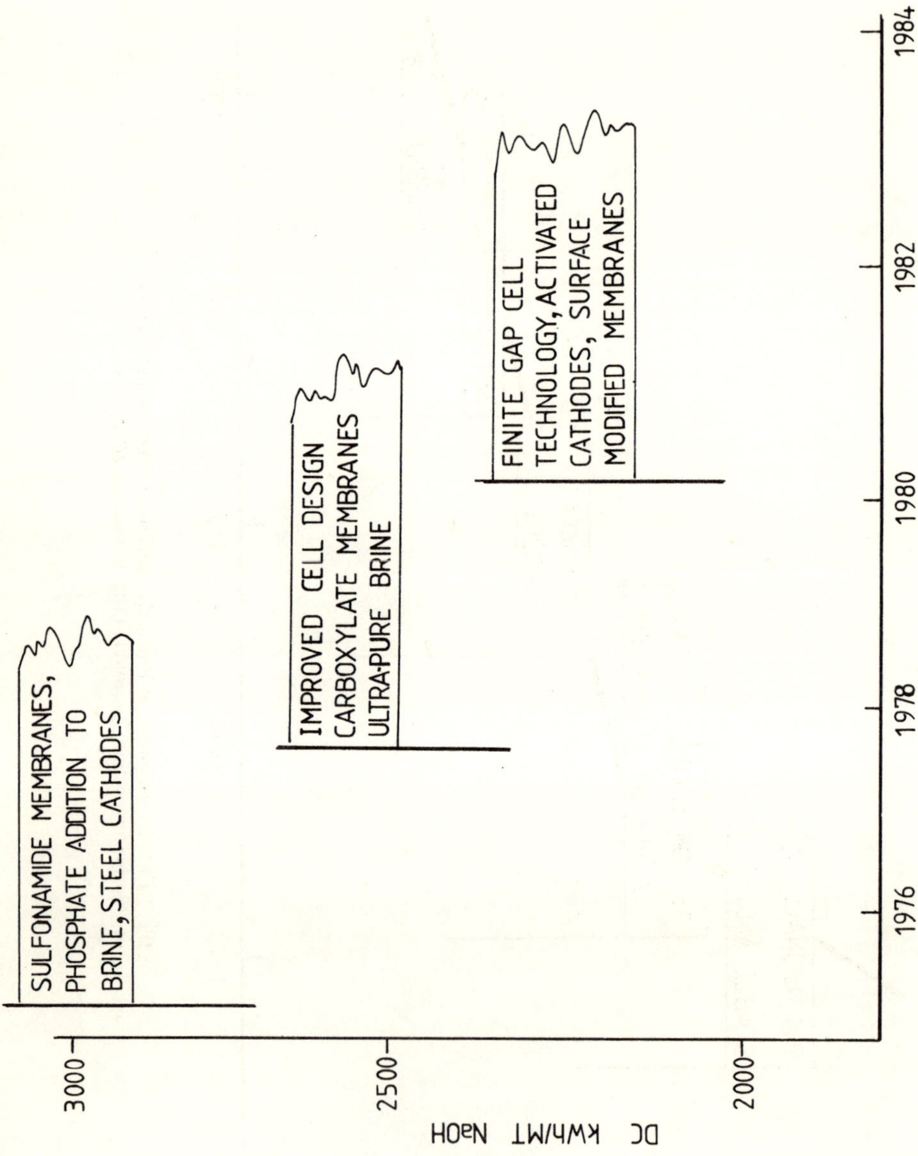

Fig. 10.2 – Development of Eltech Systems technology for production of chlorine and sodium hydroxide using the ion-exchange membrane process.

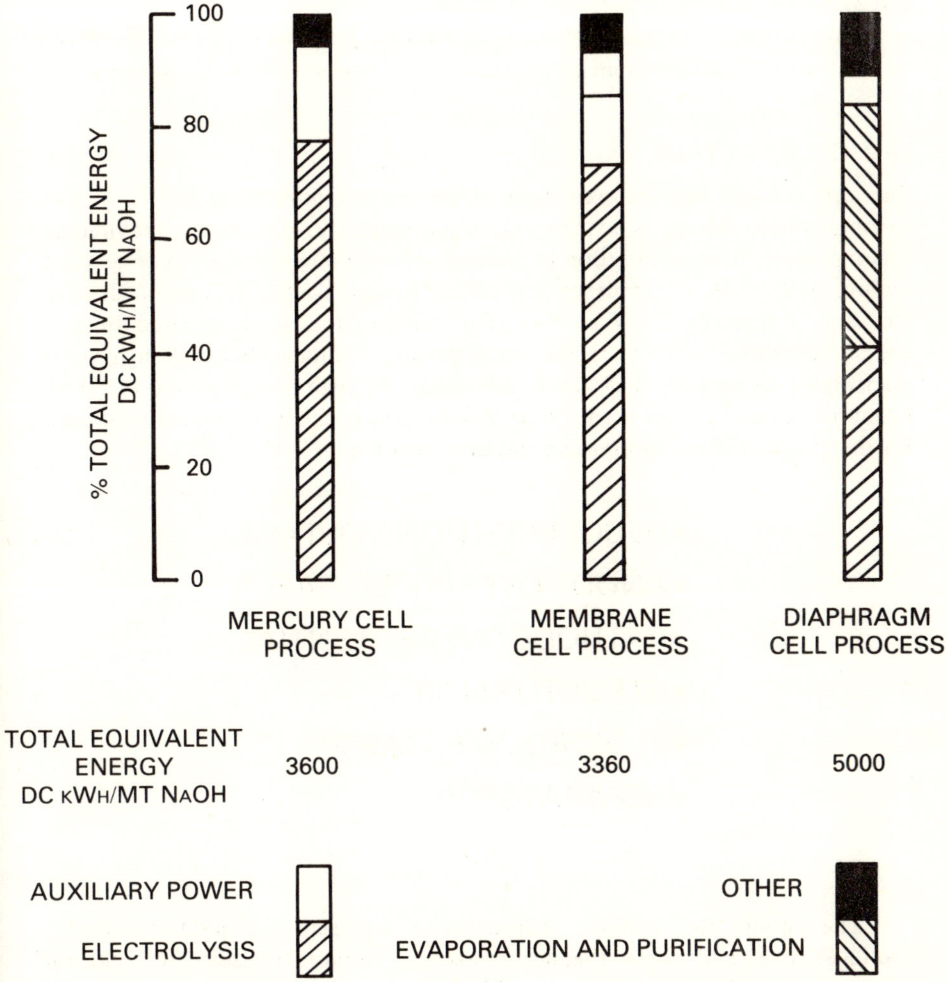
% TOTAL EQUIVALENT ENERGY
DC kWh/MT NaOH
100
80
60
40
20
0
MERCURY CELL PROCESS
MEMBRANE CELL PROCESS
DIAPHRAGM CELL PROCESS
TOTAL EQUIVALENT ENERGY DC kWh/MT NaOH
3600
3360
5000
AUXILIARY POWER
ELECTROLYSIS
OTHER
EVAPORATION AND PURIFICATION

Fig. 10.3

100 metric ton/day chlor-alkali plant, this reduction in the power consumption translates into savings of \$61 000 annually in terms of today's operating costs and \$250 000 in 1990 (power rates from Table 10.1).

But the major savings in power consumption are most likely to be realized by the research and development programs that focus upon the cell voltage.

10.3 CELL VOLTAGE

The cell voltage may be expressed as several components (Fig. 10.4). The decomposition voltage is a thermodynamic quantity, the energy required to decompose an aqueous solution of sodium chloride into chlorine, hydrogen and sodium hydroxide. It is independent of cell design, electrodes, current density, etc., and is typically 2.2 V at 90°C. Operation of the cell at elevated pressure and temperature will reduce the decomposition voltage, but although water electrolysis technology has taken advantage of high temperatures and high pressures, presently there seems to be little interest in similar operating conditions for the production of chlorine and sodium hydroxide.

- DECOMPOSITION VOLTAGE
- ANODE OVERPOTENTIAL
- CATHODE OVERPOTENTIAL
- STRUCTURAL IR
- ELECTROLYTE, GAS IR
- SEPARATOR IR

Fig. 10.4

The anodic and cathodic overpotentials represent the additional energy (voltage), over and above the decomposition voltage, necessary to make the reaction proceed at a practical rate. The cell voltage also includes a number of so-called ohmic or resistive losses, due to the passage of electricity through (1) the metal electrodes, (2) the gas-filled electrolytes, and (3) the separator.

The contribution of these several components to the total cell voltage is shown diagrammatically as Fig. 10.5 for the diaphragm cell process and for the membrane cell process. Recognizing that the overpotential of the anode is as low as it practically will ever be, it can be seen that the other components (structural IR, electrolyte IR, separator IR, cathodic overpotential) offer possibilities of significant savings in energy. The possibility of reducing the decomposition voltage will be discussed separately.

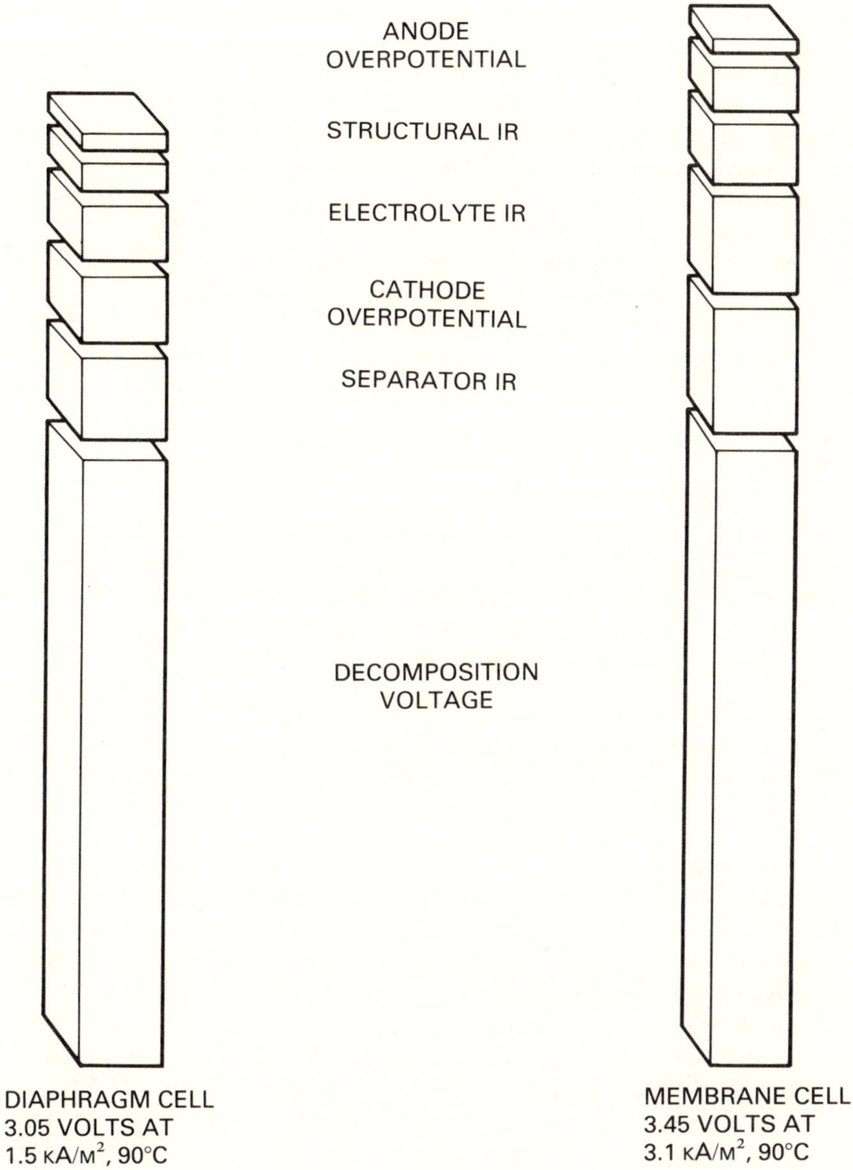
ANODE
OVERPOTENTIAL
STRUCTURAL IR
ELECTROLYTE IR
CATHODE
OVERPOTENTIAL
SEPARATOR IR
DECOMPOSITION
VOLTAGE
DIAPHRAGM CELL
3.05 VOLTS AT
1.5 KA/M², 90°C
MEMBRANE CELL
3.45 VOLTS AT
3.1 KA/M², 90°C

Fig. 10.5

10.3.1 Diaphragm cell process

The polymer-bonded asbestos separator represented a major advance in diaphragm cell technology when it was introduced in the mid-1970s, but the chlor-alkali industry today is concerned that future environmental constraints may restrict and even forbid the use of asbestos. At Eltech Systems, a new generation of dimensionally stable separators has been developed at least to the laboratory stage today, separators that provide superior performance due to lower resistance. Furthermore, the physical stability of the separator may allow the development of a 'zero gap' diaphragm cell. With respect to the cathodic overpotential, Eltech Systems presently licences an activated or coated cathode for use in a diaphragm cell, but further developments in activated cathodes look most promising.

Hence, at this point in time, it is believed that perhaps by 1986–1987 the power consumption of a commercial diaphragm cell will be reduced to 1900 d.c.-kWh/metric ton NaOH. As is shown in Fig. 10.6 this may well be the 'bottom line' for this technology.

10.3.2 Membrane cell process

The same considerations can be applied to the membrane cell and it is feasible that energy savings will be achieved by addressing the cathode overpotential, the membrane voltage, electrolyte IR losses, and structural IR losses. To be competitive in this technology, the cell must incorporate an activated cathode. Furthermore, we believe that the cathode must:

(1) reduce the overpotential at 3.1 kA/m^2 at least 200 mV relative to a mild steel cathode, and
(2) maintain this performance level for at least 1 year and more probably 2 years.

It is important to stress that the activated cathode used in the diaphragm cell is not *necessarily* suitable for the membrane cell. Presently, Eltech Systems has identified several candidate cathodes that meet these criteria, and it is anticipated that such an activated cathode will be commercially available for use in a membrane cell in the near future.

Under the normal operating conditions, the voltage drop across today's commercial membrane may be 350–450 mV at 3.1 kA/m^2. This voltage is a function of the membrane composition, the thickness, the type of mechanical reinforcement that may be used, and in the case of the asymmetric or bi-membranes, the thickness of the barrier layer. It must be remembered that changes in the membrane that may reduce the voltage, may also lower the current efficiency. Thus, new developments must focus upon optimisation of the power consumption, particularly in the face of the escalating costs of energy. A further development of interest to the chlor-alkali industry would be an ion exchange membrane capable of producing 46–50 wt-% sodium hydroxide.

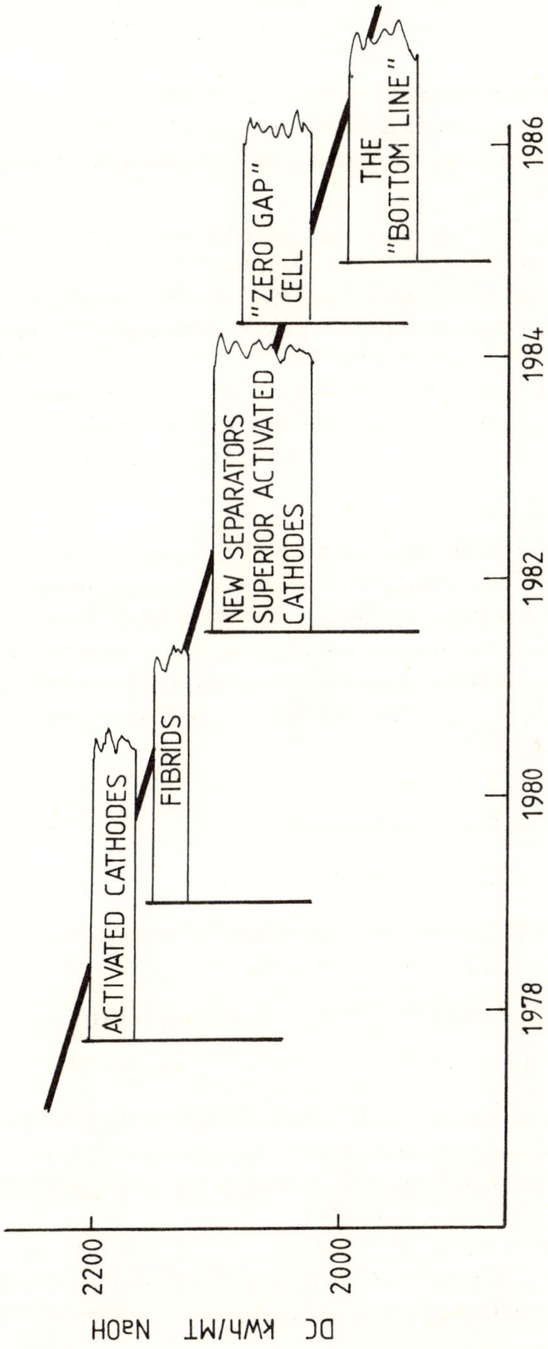

Fig. 10.6 – Future developments in diaphragm cell technology.

The development of possibly novel electrode-membrane configurations has led to demonstration of a membrane gap cell, using a cell design that lends itself to the production of a *low cost, high performance electrolyzer.* The membrane-gap cell is shown diagrammatically as Fig. 10.7. In this cell, the only electrolyte between the anode and cathode is the membrane itself. Data that have now been obtained consistently with both laboratory test cells and with pilot units support the following conclusions:

(1) In this mode of operation, using the pilot cells, the power consumption realized is 2000 d.c.-kWh/metric ton NaOH.
(2) It is expected that a commercial MGC unit will operate at 2.95–3.00 V at 3.1 kA/m^2 and demonstrate power consumptions of approximately 2100 d.c.-kWh/metric ton NaOH.
(3) The dependence of the cell voltage upon current density is lower than that usually observed, thus offering less of a penalty in power consumption when operating at current densities higher than 3.1 kA/m^2.

10.3.3 Oxygen cathodes

The major component of the cell voltage in both the diaphragm and the membrane cell is the decomposition voltage. This is a thermodynamic quantity and not a kinetic quantity and as stated previously only operation at higher temperatures (and necessarily, therefore, higher pressures) can reduce its contribution to the cell voltage. A more attractive approach is to use an alternative cathodic reaction to produce sodium hydroxide. The reduction of oxygen is just such a reaction,

$$O_2 + 2H_2O \xrightarrow{4e^-} 4OH^-$$

and therefore the overall reaction becomes

$$4NaCl + 2H_2O + 2O_2 \longrightarrow 4NaOH + 2Cl_2$$

The cathodic half-cell reaction, the reduction of oxygen, occurs at significantly lower potentials than the decomposition of water,

$$O_2 + 2H_2O \xrightarrow{4e^-} 4OH^- \qquad E_0 = 0.4\,\text{V versus NHE}$$

$$4H_2O \longrightarrow 2H_2 + 4OH^- \qquad E_0 = -0.83\,\text{V versus NHE}$$

so that in theory a savings of 1.23 V may be realized in the electrolysis of brine to produce chlorine and sodium hydroxide. Using a membrane with a cathodic current efficiency of 95% (today's technology) the reduction in power usage would be 870 d.c.-kWh/metric ton NaOH!

Eltech Systems described the application of oxygen reduction electrodes to chlor-alkali cells as early as 1945–1946. But, the technology is far from simple and has yet to be applied to a large industrial electrolytic process, such as the electrolysis of brine.

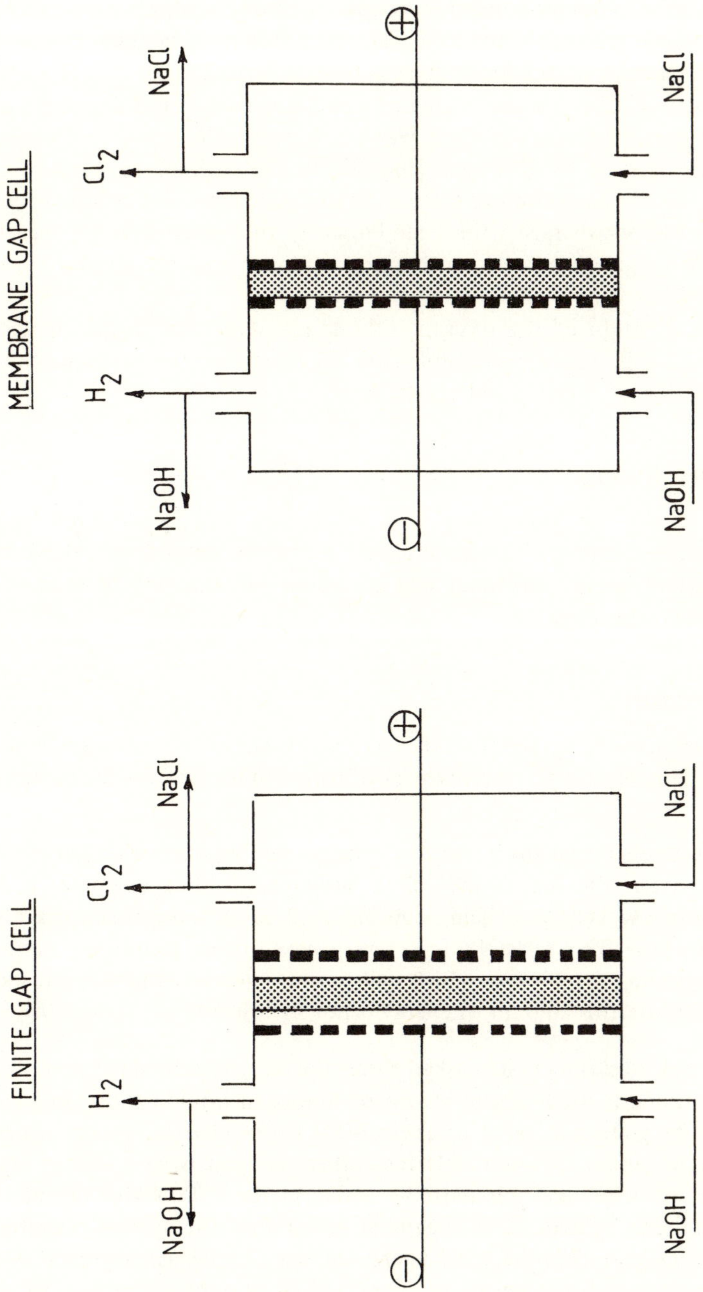

Fig. 10.7 – Diagrammatic representations of (a) the finite gap cell, and (b) the membrane gap cell.

Considerable expertise was required to overcome several problems associated with the use of a polymer-bonded, catalytic electrode with electrolyte at one face and a gaseous reactant at the other. The electrode must be microporous and the reduction reaction occurs inside the electrode structure.

The efforts of Eltech's electrochemists and electrochemical engineers were amplified by parallel studies at Case Western Reserve University in Cleveland, under the direction of Dr E. Yeager (Fig. 10.8). The technology has now progressed from successful laboratory scale units to a prototype electrolyzer brought on line at Muscle Shoals earlier this year. Presently, that electrolyzer is operating at 2.3–2.35V to give a power consumption of approximately 1650 d.c.-kWh/metric ton NaOH.

Commercialization of the oxygen cathode membrane cell is anticipated in 1985–1986 and will, without question, offer the lowest ever power consumption for the production of chlorine and caustic soda.

10.4 FUTURE GOALS

Figure 10.9 shows the 'milestones' that we believe will be achieved by the membrane cell technology. These are clearly attractive milestones indeed from the standpoint of energy efficiency and are milestones that will be realized by Eltech Systems technology.

10.5 DISCUSSION

1. Prof. Fleischmann remarked from the chair that in the energy costings discussed there had been no energy cost for the purification of liquor from the membrane cell.

   Mr Thomas replied that the amount of energy required for further purification of caustic from the membrane cell is insignificant. For example, if 33% caustic from the cell were being concentrated to 50% the steam usage was only about 0.5–0.6 t/t NaOH. In the very near future, membrane suppliers will be supplying membranes which will produce greater than 40% caustic so in the the future this amount of energy is not significant.

2. An un-named questioner then asked about the oxygen cathode system. If the oxygen cathode is used, about 355 m^3 of hydrogen would be lost. In Europe this must be given the value of about 400 kWh of elctrical energy which is equivalent to about 50 DM/t chlorine, otherwise you have a loss of nearly 420 kWh/t chlorine and this must be added to the 1700 kWh/t claimed for the air cathode system. This means in effect that the system operates at about 2100 kWh/t chlorine total energy in the German environment where 1 m^3 of hydrogen is equivalent to about 15 DM as opposed to zero value in the United States.

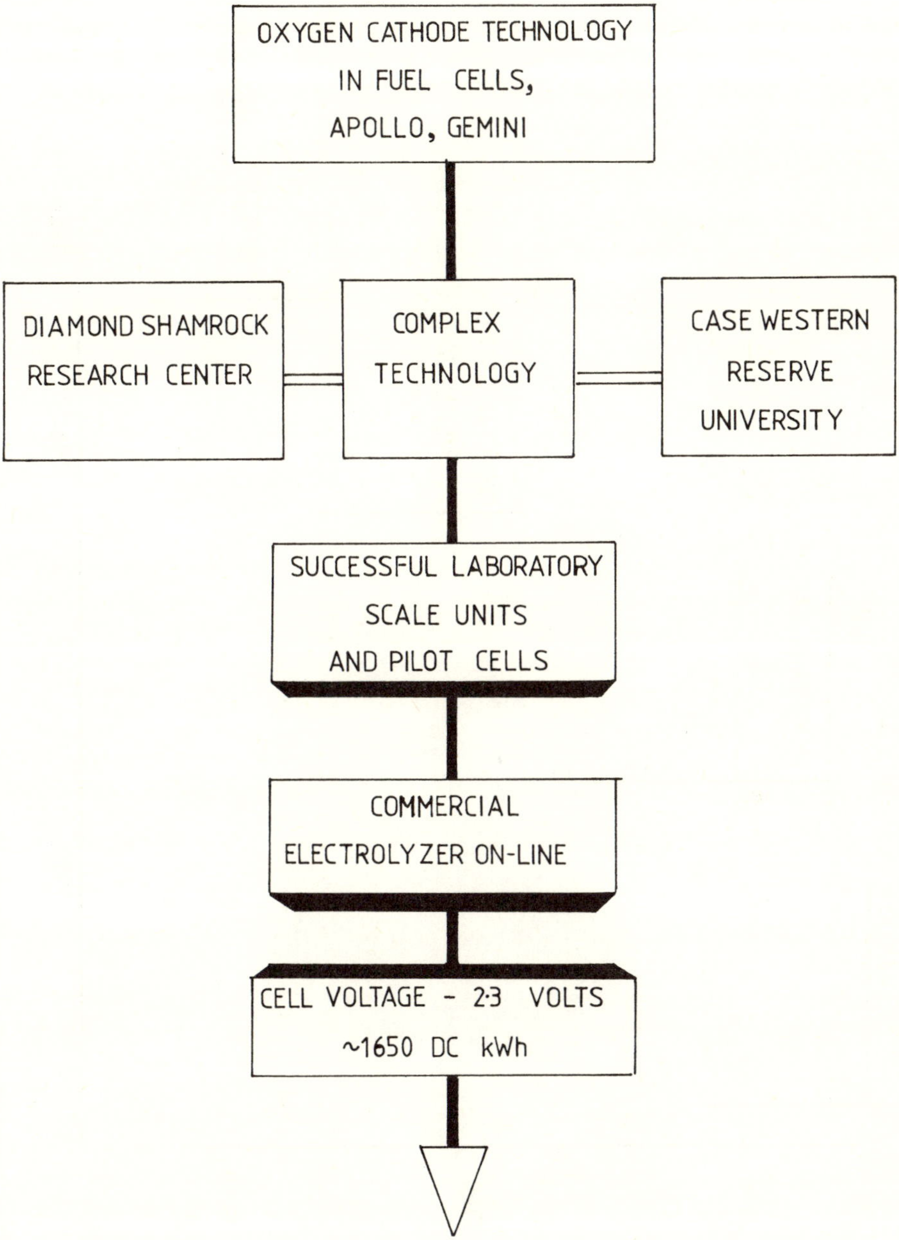

Fig. 10.8 – R & D programs that have led to a commercially sized membrane electrolyzer with an O_2 cathode.

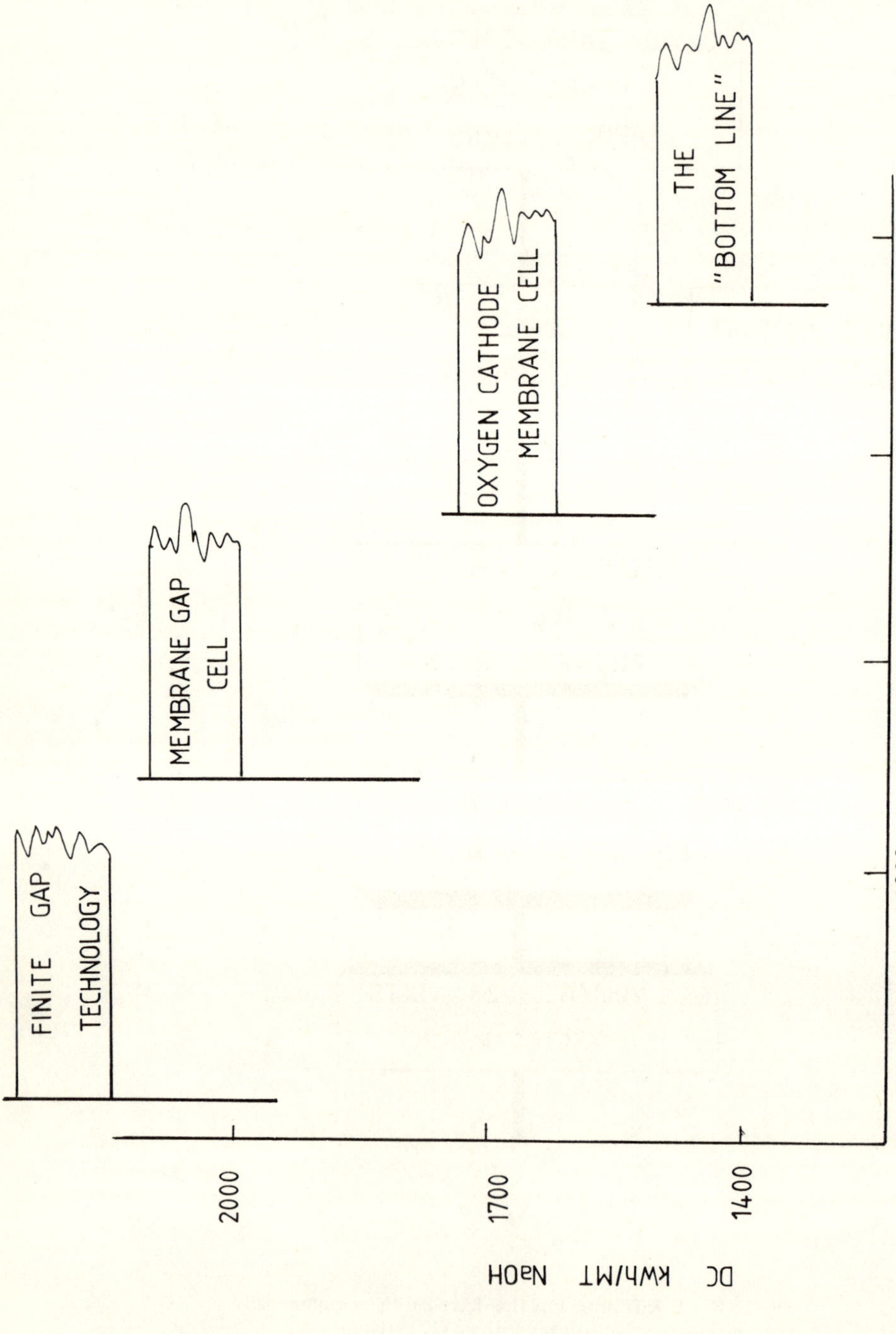

Fig. 10.9 – Future developments in membrane cell technology.

Mr Thomas agreed with the above conclusion and reiterated what he had said in his presentation, that an oxygen cathode will not be economic for every situation. In doing average economics we in Eltech Systems would penalize the total savings derived from the use of the oxygen cathode by the value of hydrogen used as fuel that would not be produced. On average, in the US we would penalize such that the net result of what can be attributed to the oxygen cathode is about 400 kWh/t, that is, the theoretical up to 800 kWh/t reduced back by whatever value you put on the hydrogen. If the hydrogen from an installation is being sold as a chemical with higher value, then the oxygen cathode cannot be justified at the present time.

3. In reply to a question about the Eltech Systems oxygen depolarized cathode cell operating in Albama.

Mr Thomas said that it was a 1 m^2 cell producing about 0.1 t/day of chlorine at a current density of about 2.5 kA/m^2. A number of catalysts had been tried but the most successful one was platinum based. Considerable work was going on to reduce the amount of precious metal to very small quantities.

Mr A. K. Rao of DCM Chemical Works, India, entered into the discussion by saying that in his experience the Indian Government feels that the technology is still developing and therefore Indian producers must wait for a couple of years or so for the technology to be finalized before they can take a decision as to whether to go for membrane technology.

What would be the speaker's view? Is it necessary to wait until the final power savings are achieved?

As a chlorine producer and in view of current market prices, Diamond Shamrock cannot afford to rebuild or replace existing facilities or build new ones at the present time. To answer your question honestly you must consider what condition your present plants are in, what is their energy consumption and what is the escalation of energy likely to be. In other words what are your individual economics relative to producing chlorine and caustic in your market place. Normally, in any technology we try as much as possible to ensure that it does not become obsolete in 2, 3 or 5 years. In other words it is technology that if we sell today, it can be adapted in the future to take the latest results of the R & D programme.

So in your own situation you have to see where you are today and where you want to be in 5 years based on your own projections, then you can make a decision today with the complete understanding that certain developments coming along (not only from Eltech Systems but for others) would be readily adaptable to update the technology you basically have. This is the only prudent and economic way to view technology. If you wait, there is never going to be a finite ending when there will be no further developments.

If we review back our industry it is only by our continued searching and developing that we have been able to survive. This process will continue, developments are always going to come, we must be intelligent enough to see that whatever we do today or 2 years from now does not become completely obsolete 5 or 10 years hence.

SECTION C

Application of Membrane Cell Technology

CHAPTER 11

Criteria for the Selection of Membrane Cell Technology

E. PEARSON, Robert Spence Associates Ltd., Saskatoon, Canada

11.1 SUMMARY

The number of companies offering membrane cell technology has grown rapidly. Because of the wide variation in cell design concept, a system approach is required for proper evaluation. Important technical and commercial criteria for detailed impact assessment are presented in a format suitable for further analysis. The capital and operating costs for each alternative must be optimized using local site specific information prior to making economic comparisons.

11.2 BACKGROUND

Membrane cells were first commercialized in the mid 1970s. At that time they were considered a novel approach to the resolution of the mercury regulatory constraints that were plaguing the chlor-alkali industry throughout the world. The first commercial installations were in countries with the most serious concerns in this area. These first designs were somewhat of a technical curiosity to the industry offering potential advantages over both the amalgam and the diaphragm cells. The application of the concept was made possible by the introduction of perfluoro compounds suitable as the basic membrane precursor. Only a few companies had the background and resources to develop this technology to the commercial scale. In order to attract interest it became necessary to make extensive comparisons to the familiar technologies. The standard presentation in one form or another, states that the membrane cell offered the following advantages:

(1) reduced total energy consumption
(2) high purity products
(3) elimination of environmental problems
(4) easier operations
(5) more compact cell room design.

The approaches have continued. Today, however, there is a different knowledge base – there is extensive commercial experience along with a much better appreciation of the concept and its potential [1]. The advantages used in the late 1970s to attract attention are now universally accepted. The technology has evolved rapidly in this period with the number of companies offering equipment increasing dramatically. The relative efficiency and energy saving claims have become much more significant. The membranes themselves are now available commercially and independently from the cell hardware promoting widespread development activity.

Unfortunately the tone of the presentations by the cell suppliers has not changed. Each is still basing technical literature on the premise that a membrane cell represents a viable alternative to the more conventional approaches [2, 3, 4]. This lack of unique definition is further complicated by the fact that each supplier has appoached the cell design from its own bias. The result is widespread variation in system concept as well as hardware detail. Therefore, the selection of appropriate membrane cells for a particular site or application, whether for an expansion, a retrofit or a grass roots situation, can not be based on the cell alone – it must be determined on a systems approach, which considers the impact of a particular cell choice on the total plant installation. The object of this presentation is to develop the criteria, and approach required to evaluate the various cells on a system basis and to suggest a workable technique for the ultimate selection.

11.3 SYSTEM DEFINITION

Figure 11.1 illustrates a typical example of a membrane cell schematic as it would appear in a supplier publication. Information is provided on the basis of the

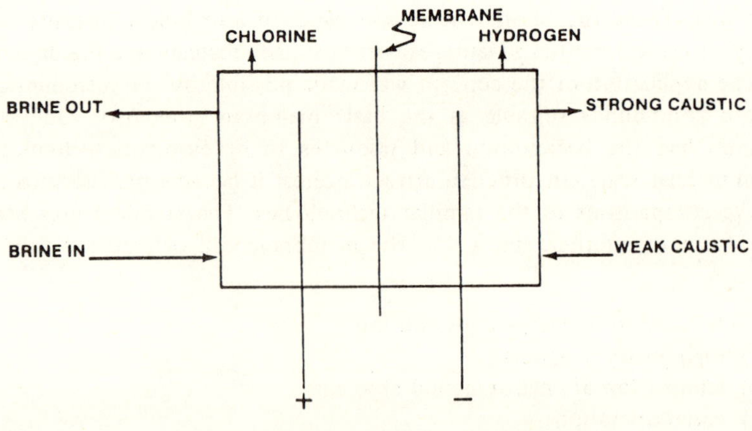

Fig. 11.1 – Typical membrane cell representation.

streams shown entering and exiting this cell. Figure 11.2 gives a better interpretation of the actual system with an indication of all of the various key plant sectors influenced by the cell. This system is far more complex, but the real considerations must be for the actual system inputs and outputs which are effectively the nets of the various internal block flows. The thorough understanding of these interrelations is paramount in the final selection process because of the wide variation in conceptual approach used by the designers. Therefore, in order to study and develop the system related impacts, into comparative mode for assessment application, a chart such as shown in Fig. 11.3

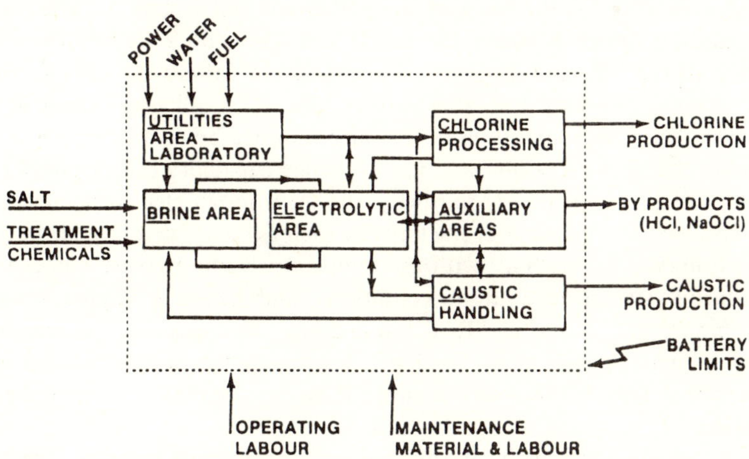

Fig. 11.2 – Membrane cell system: schematic.

SECTOR ITEM	OPERATING COST IMPACT	CAPITAL COST IMPACT	TECHNOLOGY (ABSOLUTE VALUES)	
			A	B etc.

Fig. 11.3 – Technology comparison chart.

is used. Columns are provided for indication of which areas are impacted and whether the impact is capital or operational in nature. At this point the various supplier information can be formatted in terms of the major technical and commercial parameters. As much as possible, the actual data should be presented on a unit output basis.

11.4 TECHNICAL CRITERIA

11.4.1 Membranes

The membrane itself is the heart of the membrane cell system. Its effectiveness as a separation device between the anode and the cathode defines the current efficiency of the cell, and its chemical composition defines the caustic concentration that corresponds with this current efficiency. All membranes in commercial use today are adaptations of the same fluorocarbon matrices; the physical and electrochemical properties and ion exchange capability can vary widely [5]. Two-year life expectancies are routinely quoted from an electrochemical standpoint; however, the physical characteristics such as compressibility at the gasket surfaces, dimensional stability with the temperature and pressure ranges experienced during startup and upset conditions, and handling requirements, are equally important in useful life determination. The quantity of membranes reserved as spare, needs to be specified along with the conditions and special treatment that may be necessary to store these membranes or prepare them for installation.

Most cell suppliers have selected a specific membrane supplier. The reason for this choice has to be understood along with any future limitations this might present. All cell suppliers like to state that membranes are interchangeable in their hardware; this may be physically true but impractical from a service, contractual or patent standpoint.

11.4.2 Raw materials

(a) *Salt and brine*

The consumption of salt is limited to the theoretical consideration of sodium chloride decomposition. Although the degree of depletion of sodium chloride from the brine varies from cell to cell, there is little variation in the final effective unit consumption. It should be noted that the hydrochloric acid and caustic soda used in the brine treatment and dechlorination steps produce sodium chloride. These amounts are discounted using a system approach because these equivalent amounts of caustic and chlorine do not appear as outputs.

All membrane cells require a high purity brine in order to produce stable long term performance. A chelate type ion exchange resin is used to reduce the levels of multivalent cations to near detection levels. The cost of the installation and operation of this resin system has to be evaluated. Some suppliers have included proprietary resins and bed designs as part of the technology package. There is also an effluent created in this step which must be considered.

The widest variations in this area are in the actual specifications for the feed and depleted brine from the cells. Temperatures, inlet and outlet flow rates, pressure, acidity (pH), and the approach for functional control of these elements, are all unique to a singular design. Some systems have a high level of recycle with makeup and withdrawal, others are once through. The power costs for cell fluid movement have to be included in the operating cost impacts. The control of pH or acidity in the cell must be understood. Some suppliers state that use of acidified brine is optional, but there are implications to the cell gas composition, current efficiency, and voltage [6].

These parameters have major impacts on the process complexity and materials of construction in the brine and cell room areas.

(b) *Power*

Stated power consumptions vary significantly. With power consumption representing 25–30% of the operating costs, a thorough examination of the factors involved in arriving at the stated power usage is important. To apply the system approach properly the cell power consumption should be considered on an a.c. basis including all bus losses and rectifier efficiency. The cell power consumption is the product of the applied voltage and current per unit of output. Inherent design features which affect the total voltage, cause voltage drops, or affect the efficiency of conversion of the applied current, are unique to the particular cell and cell room design. Current efficiency considerations are almost completely related to the membrane performance. In order to insure true representations of actual operating scenarios, the value employed for calculations should be the average anticipated over the life of the membrane. Voltage considerations are hardware related; such items as the type and number of connections, either intercell or intracell, anode/cathode and membrane spacing, and materials of construction, influence the electrical conductance ability. The electrical resistance and therefore the voltage drop associated with the fluid path is dependent on temperature, pressure, and concentrations of the fluids. The total voltage drop associated with the sum of these resistances is a function of the applied current. For any cell design a curve of total voltage versus current can be created. A more useful comparison curve is that of power consumption versus current density. Typical examples of these curves are shown on Fig. 11.4. The slope of the curve is an indication of the sensitivity of that particular cell design to capital cost. A flat curve indicates that spending additional capital money to purchase more cells and reduce the current density does not significantly reduce the unit operating cost.

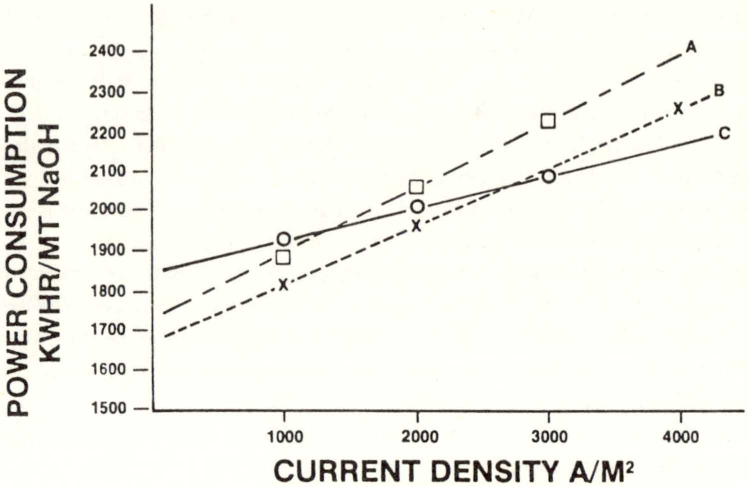

Fig. 11.4 – Current density A/m².

11.4.3 Products and byproducts

(a) *Chlorine*

Any inert materials in the chlorine gas (oxygen, air) become concentrated during the liquefaction process limiting the potential chlorine recovery by conventional means. In cases where the chlorine gas is dried and compressed in order to be suitable for liquefaction and storage, the temperatue and pressure of the gas evolved from the cells dictates the degree of compression and cooling required. In retrofit applications it is imperative that these conditions be stated in terms of the existing equipment and design so that the necessary changes to effect compatibility are recognized. There can be substantial benefits available in such applications. For example, if the chlorine gas stream into an existing system can be altered from 96% purity at 25 mm hg vacuum to 98% purity at 0.1 kg/cm^2, a 15% increase on drying and liquefaction capacity is possible.

(b) *Hydrogen*

Membrane cell hydrogen is pure and chlorine free, making it an ideal fuel source. The degree of handling and processing can be further simplified if the hydrogen is available under pressure. Some hydrogen is required internally for hydrochloric acid synthesis and must be properly recognized in the operating cost area.

(c) *Caustic soda*

The major important feature of the caustic soda is the exit concentration from the cells. In diaphragm conversion applications where existing evaporation equipment may be modified, lower concentrations may be of less importance. In other instances the higher the concentration the less equipment and operational consideration is required to produce merchant material. Some technology suppliers have available heat recovery systems designed to use the low grade heat available from the cells for caustic concentration. In any situation the higher the exit concentration, the lower the storage volume required to maintain process inventories. Caustic soda purity may be of interest but because of the uniformity of specification it is not likely to be an important evaluation criteria.

11.4.4 Operating characteristics

The significance of most of the operating variables has already been covered in the discussions on the brine and cell product areas. These are, however, absolute in nature and reflect the cell design philosophy only. When considering systems approach, it is necessary to understand the overall operational design concepts. Visits to actual operating plant and demonstration facilities are necessary to determine the impacts of these concepts on the system dynamics. Membrane cells in general are easier to operate. None the less it is important to determine if the cell suppliers' ideas about cell control, analytical control, and response dynamics are compatible and consistent with those desired. This assessment is particularly critical when the technology is to be introduced in situations where existing operators and supervision have established their own methods and work procedures. This fact is also true at any installation visitied, and must be understood in order to extrapolate and interpret the system under consideration. The analytical requirements should be determined in terms of frequency, skill requirements, and equipment preprequisites. The number of operations personnel and supervision, for the total system and not just the cell area, develops from this assessment.

11.4.5 Maintenance

The maintenance evaluation needs to be subdivided into two major areas: an appreciation of the time and effort expended on the maintenance function on an annual basis over the consideration period, and an appreciation of the impact of this requirement on the basic facilities design.

In the time and effort sector, the proper selection of materials of construction for the service conditions is important. This includes not just cell hardware and piping but electrical insulation and isolation devices and instrumentation. Component design for the maximum amount of interchangeability of parts is desirable from the standpoint of the management of spare parts inventory and a difficulty of assembly. The size and weight of individual pieces determines the complexity of material handling systems. The degree of sophistication of components may limit the ability of local skills and equipment to provide the necessary services. Anode costs, both initial and ongoing, are considered as maintenance costs with the supplier's guarantee used for the life determination. The methods and costs associated with perpetual anode supply form an important factor beyond this initial period.

The design impact considerations cover the requirement of spare units, or for installed spare capacity to insure that the actual output of the plant compensates for the maintenance downtime anticipated. The determination of the type and magnitude of space required to perform the maintenance function includes building design for access aisleways, overhead cranes, and storage areas. The average manning component, and the maximum crew required to perform a function, determine the cost and the ability to perform the work within a time frame.

11.4.6 Technology trajectory

The final selection can not be made solely on the basis of the existing position, but must have due regard for the potential of the technology.

The membrane cell technology field is relatively new. It is therefore reasonable to assume that there will be rapid evolution in the design and application over the next few years. Any technology choice must consider the applicability to modification and accomodation of improvements as they become commercially available. It is particularly important to consider developments known to be in the pilot testing stage at the point of the decision. A thorough examination of the supplier's research and development activities, and its ability to obtain the relevant work of others is required.

At the present time advanced development work is being carried out in the areas of high pressure operation, cathode coatings, solid polymer electrolytes, oxygen or air cathodes and coupled fuel cells [7, 8]. These types of developments have the potential of reducing the overall power input by 40% from the current levels.

11.5 COMMERCIAL CONSIDERATIONS

11.5.1 Engineering capability

An understanding of the extent and type of engineering provided as part of the technology package is necessary in order properly to anticipate the total project engineering requirement. This package must be evaluated as to the applicability of this basic supply in terms of project standards, local regulations, and construction practices. The relevant experience of the technology supplier is an important facet. Because of the need to integrate the cell technology into the remainder of the plant, the technology supplier should have relevant practical background in the total chlor-alkali plant engineering and design. Such experience would insure compatibility, as well as provide the owner with the opportunity to expand the engineering and design responsibility, with perhaps some overall cost savings or risk sharing benefit.

11.5.2 License terms and conditions

The license agreement, an important part of the technology package, dictates the interfaced responsibilities of the supplier and the owner. It is important that this document contain adequate protection for the owner for such items as patent infringement, and technological changes without being overly restrictive as to the operating requirements, flexibilities and reporting procedures. Any performance guarantees included have to be real, such that attainment is not conditional on achieving and maintaining a set of ideal operating conditions. The financial considerations associated with these agreements vary from a one time fee, to a running royalty basis. Such costs must be properly allowed in the capital and operating projections.

11.5.3 Operational support capability

No project or operation runs smoothly 100% of the time. Unforeseen circumstances materialize throughout the operating life of the facility. The plant gradually develops its own experience and related capabilities to assess and react to these situations properly. However during the initial operation phase it is important to be able easily to discuss and clarify anomalies with the technology supplier. The availability and the cost of this type of service should be established as part of the technology assessment. Even beyond the initial learning period it is valuable to have access to design concepts and the practical experiences of the supplier. In other words, plan for a lasting relationship with the selected supplier.

11.5.4 Licensor background

The company offering the technology must be visibly understood in order to determine the importance of this particular business area to its longer term goals and objectives. The key employees should be identified with the acutal roles that they would play in a project, whether in the construction or operational phases. Examination of previously completed projects, from the owners perspective, determines if there are any positive or negative aspects that could impact on the new project.

11.6 EVALUATION

The work program outlined presents the investigation areas and the formatted supplier information. The critical task remaining is the translation of this material into a recognizable and workable comparison mode.

One potential technology is selected as the base case. This selection should be done on the basis of the technology with which the most information is available. Using this base case the capital and operating costs should be detailed using formats as shown in Fig. 11.5 and 11.6. Capital costs should include all blocks identified in Fig. 11.2 so that the total impact is costed. Operating costs are based on the capital cost selected, again, considering all blocks. At this point the financial analysis is conducted on a maximum discounted cash flow basis with this one alternative to insure that the information provided or the proposal made by the technology supplier adequately reflects the economic and tax considerations that apply to the site specific example. This is particularly important in the current density versus power consumption area, where local trends and conditions can have a major influence. A simple calculation can be carried out with a programmable calculator [9]. If inflationary, currency considerations and debt servicing are included, more sophisticated calculation methods are required. Once this DCFRR has been maximized a valid data base has been created for comparison with the alternative technologies.

AREA	BASE CASE CAPITAL COST	ALTERNATIVE VARIATION +/(-)
ELECTROLYTIC		
CELLS		
MEMBRANES		
ANODES		
CIRCULATION EQUIP.		
CELL MAINTENANCE EQUIP.		
INSTRUMENTATION		
SPARES		
BUILDING		
LICENSE		
ENGINEERING		
BRINE		
ION EXCHANGE RESIN		
ION EXCHANGE EQUIP.		
DECHLORINATION EQUIP.		
BUILDING		
ENGINEERING		
CHLORINE PROCESSING		
COOLING EQUIP.		
DRYING EQUIP.		
COMPRESSION EQUIP.		
LIQUEFACTION EQUIP.		
ENGINEERING		
CAUSTIC HANDLING		
EVAPORATION EQUIP.		
STORAGE		
ENGINEERING		
AUXILIARIES		
HCl SYNTHESIS EQUIP.		
STORAGE		
VENT SCRUBBER SYSTEM		
EMERGENCY SYSTEM		
ENGINEERING		
UTILITIES		
MAIN SUBSTATION EQUIP.		
STEAM PRODUCTION EQUIP.		
COOLING WATER EQUIP.		
INERT GAS EQUIP.		
LABORATORY FACILITIES		
BUILDINGS		
ENGINEERING		

Fig. 11.5 – Capital cost determination (summary).

	BASE CASE		ALTERNATIVE VARIANCE +/(-)
POWER & UTILITIES	CONSUMPTION	COST	
A.C. POWER — EL.			
— (MOTIVE)			
FUEL — PURCHASED			
WATER — PROCESS			
— COOLING			
INERT GAS			
MATERIALS			
SALT			
BRINE TREATMENT			
SULFURIC ACID			
RESIN			
CELL MAINTENANCE MTL'S			
CELL ANODES			
CELL MEMBRANES			
GENERAL MAINTENANCE MTL'S			
LABOR			
OPERATIONS			
MAINTENANCE			
LABORATORY			
SUPERVISION			
OTHER			
ROYALTIES			
SERVICE			
DEPRECIATION			
TOTAL			

Fig. 11.6 – Operating cost determination (summary – unit plant output basis).

Each should be considered as a comparison to the base case, with differences showing as plusses or minuses to the capital or operating costs. Careful base case selection and information gathering minimizes the number of assumptions that are made at this point. Each alternative requires the application of the same optimization technique because of the varying sensitivities of technologies to capital and operating costs.

This procedure results in an accurate assessment of the relative financial situations, providing operating and capital valuations for input into the final selection decision.

In summary, the choice of a membrane cell technology is a site and situation specific one. There are no automatic or easy answers. A complete and thorough investigation using the criteria outlined in the presentation should provide the background required to make that decision.

11.7 REFERENCES

[1] Pearson, E., and Miyazaki, T., *Commercial Operation of Ion Exchange Membrane Chlor-alkali Process*, presented at 2nd World Congress of Chemical Engineering, Montreal, Canada, Oct. 1980.

[2] Nagamura, M., Ukihashi, O., and Shiragami, O., *The Progress of Flemion Membrane Chlor-alkali Process*, presented at the 25th Chlorine Plant Managers Meeting, Chlorine Institute, Atlanta, USA, Feb., 1982.
[3] Diamond Shamrock, DM-14 Membrane Electrolyzer, ES-ECL-4A, Diamond Shamrock, Electrolytic Sytems Div., Chardon, USA.
[4] O'Brian, T., and Gilliat, B., *The FM-21, A Novel Approach to Membrane Cell Design*, presented at the 25th Chlorine Plant Managers Meeting, Chlorine Institute, Atlanta, USA, Feb. 1982.
[5] Stinson, S., Electrolytic Membrane Cell Development Surges, *Chemical and Engineering News*, Mar. 1982, pp. 22–25.
[6] Seko, M., Ogawa, S., Yoshida, M., Yomiyama, A., and Takemura, R., (Asahi Chemical Industry Company Ltd), Canadian Patent 1088465, Oct. 28, 1980.
[7] Coker, T., Laconti, A., Balko, E., and McGray, G., (General Electric) United States Patent 4212714, July 15, 1980.
[8] Thomas, V., Air Cathodes for the Chlor-alkali Industry, presented at the 25th Chlorine Plant Managers Meeting, Chlorine Institute, Atlanta, USA, Feb. 1982.
[9] Abam Engineers, Process Engineering and Economic Evaluations of Diaphragm and Membrane Chlorine Cell Technologies, ANL/OEPM-80-9, Argonne National Laboratories, Argonne, USA, Dec. 1980.

11.8 DISCUSSION

1. Mr Y. C. Yen of SRI International opened the discussion by asking Mr Pearson to clarify a statement made about rectifier efficiencies. He was of the opinion that the efficiencies of present day rectifiers were much higher than the 90% mentioned in the paper.

   Mr Pearson replied that the figure stated was derived by considering the difference between a stated cell d.c. power consumption and the a.c. power consumption of the complete cell circuit, including bus bar losses, cell-to-cell losses and rectifier efficiencies. This is the total a.c. input and must be recognized as such in any economic study.

2. A question about the implication of membrane cost and lifetime was asked by Mr A. K. Rao of DCM Chemicals, India.

   Mr Pearson said that although he could make some comments he could not speak in absolute terms since membrane costs depend on the membrane supplier, the application in individual cells and the way in which the cell supplier handles the membrane cost in his technology package. It is possible that this may be on a lease basis which includes service costs and of course the financing costs are carried by the supplier or they could be purchased outright and it depends on the way in which the individual feels it is best economically for him to approach this financially. As far as the cost/unit of production was concerned, this was very difficult to state. Unit production costs of membrane in real dollar terms are reducing as large production facilities are being brought on stream by a number of membrane suppliers although because of inflation impacts this may not appear to be so. Membrane lifetime has increased from 1 year to over 2 years at the present time.

Mr Rao interjected to agree with the statement on lifetime but said he felt he could not agree that membrane costs were decreasing. In his experience membrane costs were now 30% higher than at the same time last year which was roughly equivalent to an increase of about $30/t caustic.

Mr Pearson replied by saying that the cost of membrane to the customer was artificial in that suppliers had very high development costs to recoup and they tended to change what the market will bear.

3. Mr K. Stadel Nielsen of Dansk Sojakage fabrik, Denmark invited Mr Pearson to predict what might be the distribution between membrane cells and other types in 10 years.

   This was very difficult to predict. He felt that one paper (see Chapter 22) had been too optimistic but he did feel that there would not be another grass-roots diaphragm cell plant built anywhere in the world.

CHAPTER 12

Considerations in Conversion of Existing Chlor-alkali Plants to Membrane Cell Operation

T. F. O'BRIEN, Catalytic Incorporated, Philadelphia, PA, USA

12.1 INTRODUCTION

The membrane cell revolution in the chlor-alkali industry has begun. With very few exceptions, membrane cells will be considered in all new projects. In addition, as the world market strengthens and energy prices continue to increase, more and more producers will find it economic to convert existing capacity to membrane cell operation.

This paper will present some of the technical and engineering considerations in such conversions. 'Conversion' will be taken to mean a change in operation of an existing plant to include new membrane cells. This definition does not require removal of all the existing cells from production. The discussion will necessarily be general and will be based on 'typical' plant arrangements. We will assume that no increase in capacity occurs, although it will be appreciated that, because of the membrane cells' lower power consumption, more cell room capacity could be obtained with the same electrical gear.

To highlight some of the problems which might be met during a plant conversion is in some ways a negative approach. It must be emphasized at the start that the problems are not fundamental and that sound engineering and planning will make a plant conversion simple and successful.

Both mercury and diaphragm cell plants will be included in this study. While we expect both types to be candidates for conversion, we feel that in the near future, there will be more opportunities for conversion of mercury cell plants. First, their operators will continue to meet severe environmental pressures and, in much of the world, a long-term threat of enforced shutdown. Second, some of the major advantages enjoyed by diaphragm cells over mercury cells will also apply in a comparison with membrane cells. Economically, the most important of these may be the fact that diaphragm-cell plants can operate with brine as the basic raw material. Membrane plants may require a solid salt feed or brine evaporation to keep a water balance.

In addition, a diaphragm-cell plant will not already have the brine dechlorination system required by the membrane cells. Finally, the advantage held by membrane cells in total energy consumption might not translate into an equivalent cost benefit. Many diaphragm plants operate on waste or exhaust steam which is assigned a low value.

We will therefore aim the discussion primarily at mercury-cell plants but will bring in other considerations which apply especially to diaphragm plants.

The cost data to be presented will assume the use of ICI's FM21 membrane electrolyzer. As preferred contractors for this new technology, we at Catalytic are convinced that it is outstandingly cost-effective.

12.2 OVERVIEW

First, consider the major components of a chlorine plant (Fig. 12.1). Besides the cell room itself, there are systems to handle the brine and the three products of electrolysis.

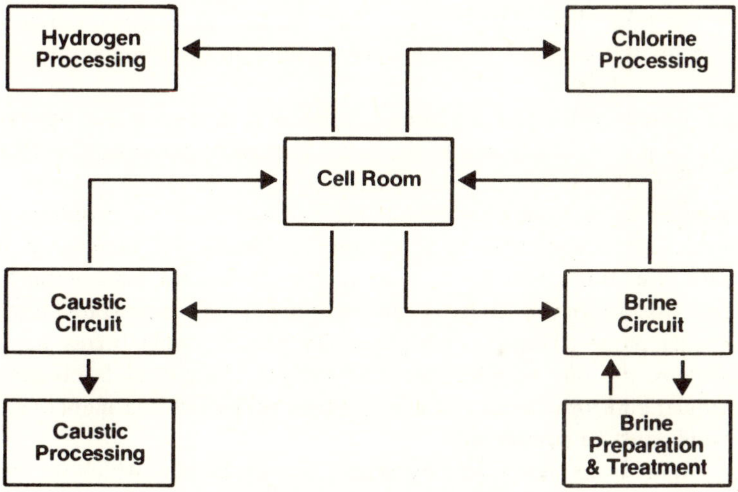

Fig. 12.1 – Chlorine plant components.

We will not discuss hydrogen processing here. There are many variations in existing systems but probably very few complications to consider in a conversion.

Outside the cell room, the most interesting and most involved system is that associated with the brine.

12.3 BRINE PROCESSING

Figure 12.2 shows the components of a membrane-cell plant's brine processing system. Depleted brine from the cells, which is approximately saturated with chlorine, must be collected and then pumped on to the processing system, which is described below.

12.3.1 Brine treatment and resaturation

Much of the brine recirculation process in a membrane-cell plant will be similar to that existing in a mercury-cell plant. Depleted brine must be dechlorinated and then resaturated. Treatment of resaturated brine for calcium and magnesium is common practice. The brine next can be acidized and fed to the cells.

Because a much greater depletion of brine is taken in a pass through a membrane cell, the brine circuit will be larger than necessary after conversion.

Much of the equipment will operate all the better for the extra capacity. Some parallel units might then be shut down or removed. In other cases, the turndown capability will have to the reviewed. For example, process pumps will be well oversized and will at best waste power. Bypasses may have to be installed or pumps replaced.

A limit of 5 or 10 g/l of sodium sulphate is desirable with most cells. This may be much less than that in the brine circulating through existing mercury cells. Further chemical treatment or an increased purge on the brine system may be necessary.

Silica content of the existing brine circulation system must also be reviewed. A limit of 60 p.p.m. is allowed at the cells. The use of diatomaceous filter aids should be avoided and even the use of sand bed filters should be reviewed.

The lower brine flow after conversion can aggravate the problems caused by impurities. Residence time in all existing equipment will increase markedly. In the saturator, this might allow more of the calcium sulphate present in the raw salt to dissolve. In the settler, it can result in a more concentrated underflow and a smaller purge stream (this is physically possible; what finally happens is determined by the rate of withdrawal of sludge). If impurity levels cannot be brought under control with an acceptable purge, additional treatment or a change in processing must be considered.

The sulphate can be removed by treatment with barium, which is expensive. Alternatives would include systems which suppress dissolution of calcium sulphate. Control of operating conditions, the use of additives, or a short contact time can make it possible to dissolve nearly all the NaCl while rejecting most of the $CaSO_4$.

Some chlorate will be destroyed during acidification, but preliminary estimates should assume that it is removed only by purging. If chlorate problems are critical, we can consider the use of other anode coatings. At the same overall anode current efficiency, these can produce more oxygen and less chorate. More

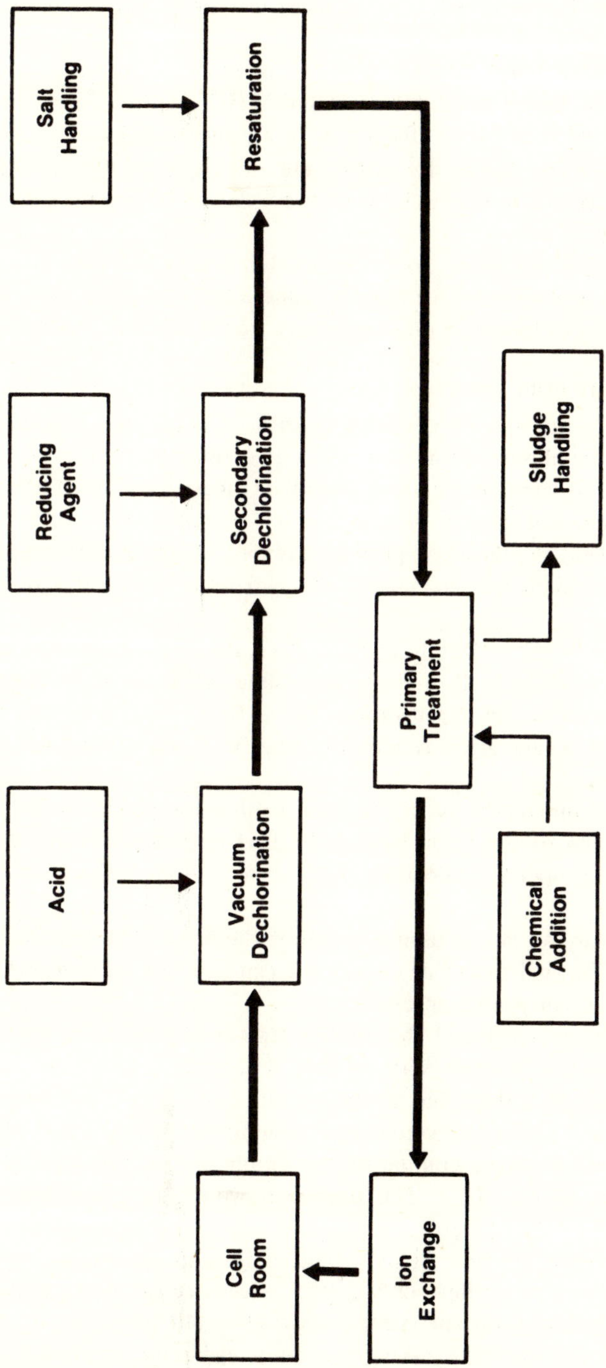

Fig. 12.2 – Brine processing.

acid addition at the cells will also increase efficiency and suppress chlorate formation.

There has been much concern over the presence of mercury in the brine after a conversion. Heavy metal cations, including Hg^{++}, seriously affect membrane performance at levels of 50 or 100 p.p.b. In a brine loop, however, mercury is present as a complex anion. Its effects are therefore greatly reduced. Most evidence today is that concentrations of the order of 10 p.p.m. total mercury can be tolerated.

In a diaphragm cell plant, no dechlorination system will exist and the brine treatment plant probably will be sized to handle only the fresh feed. The latter might still be used to remove impurities from a side stream.

12.3.2 Secondary brine treatment

Where other cells can operate consistently with brine having a total hardness of 5 or 10 p.p.m., membrane cells require much lower numbers for best performance. A total hardness of less than 50 p.p.b. should be the goal, and this is best obtained through use of an ion-exchange process.

The equipment should be supplied by a vendor designing to a resin producer's standards. The system we recommend has two resin beds and a programmed controller.

Although each bed is sized to operate alone, the two will nomally operate in series. Either sequence, A–B or B–A, is possible. Approximately daily, the first bed is taken off line for regeneration. For a time, then, only one bed is in operation. When regeneration is complete, the fresh bed is put back on line as the second in series.

We also recommend the use of automatic hardness analysers. One will monitor the total hardness of the brine leaving the first exchanger. It will detect any incipient breakthrough and allow the operator to act before the second bed also becomes loaded.

Another analyser can be used to monitor the brine from the primary treatment system. This will detect any upset and can also be used to divert the brine for retreatment if considered necessary.

This system, when it and the primary treatment process are operated and maintained properly, will be very reliable. For each project, we would also consider addition of a third bed as a guard unit. This would not be automated with the others in full 'merry-go-round' fashion but would stand alone. Full integration would seriously complicate the simple process valving shown for the two-bed system in Fig. 12.3. The third bed would be regenerated on some regular schedule by the operator.

Chelating resins are now being produced for this application, because the conventional cation resins have only a weak selectivity for most mutivalent ions over sodium. Several different types are available, but they have in common a relatively high cost and the need for a fairly elaborate regeneration process.

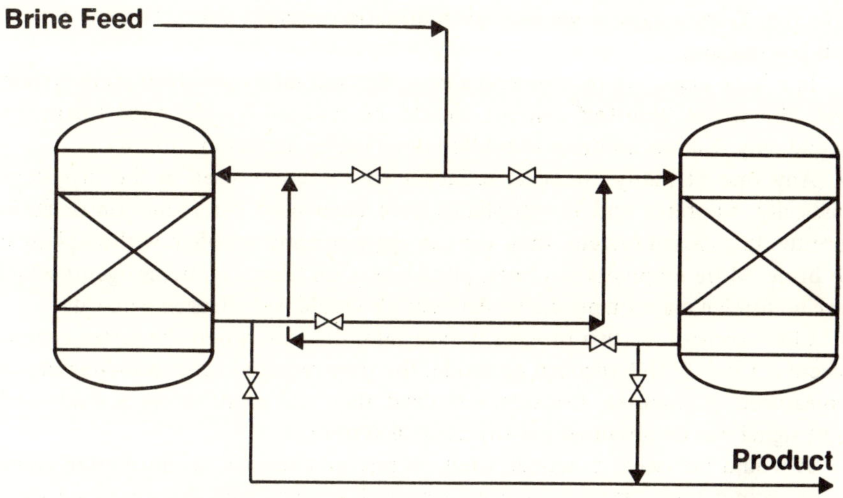

Fig. 12.3 – Ion-exchange system two-column operation.

Typical sequences include:

(1) displacement of brine
(2) backwash
(3) regeneration with HCl
(4) displacement of acid
(5) conversion to active form with NaOH
(6) displacement of caustic
(7) sweetening-on with brine.

There are large changes in the bulk density of the resin during this process and a substantial expansion of the bed during backwashing. The vessels must be sized and the internals designed to accommodate these volume changes. At the same time, efficient operation requires that the various steps be carried out with a minimum volume of chemicals. Efficient displacement from the bed and removal of dead volumes of fluid from the vessel are very important. In several steps of the regeneration sequence, differences in density can be used to promote plug-flow displacement. In others, process considerations are more important. Proper nozzle arrangement can reduce the volumes to be handled. Careful attention must be paid to the mechanical design and to arrangement of distributors, collectors, piping, etc.

12.3.3 Secondary dechlorination

The resins used for brine treatment are quite sensitive to oxidation. The available chlorine content of the brine fed to the resin beds should be less than 1 p.p.m., while that from a typical vacuum dechlorination process contains some tens of parts per million.

In a large plant, we recommend a two-stage secondary dechlorination system. First, the active chlorine content should be reduced by chemical treatment. Second, any residual chlorine should be removed by activated carbon.

Any one of many chemical reducing agents can be used in the first stage. Classically, sulphites and thiosulphates have been used for reduction of active chlorine. In a closed system, their use can aggravate any problem with sulphate in the brine. More emphasis has been placed here on 'clean' reducing agents which leave no foreign reaction products. An example would be hydrogen peroxide.

Close control of this process is especially difficult when the latter type of reagent is used. With hydrogen peroxide, the time required for complete reaction is measured in minutes. Considerable dead time will exist before a stable end-point signal can be obtained for any control system.

It would not seem practical, then, to rely on chemical dechorination alone. In the second stage, another classical approach is taken with the use of activated carbon. Probably its most familiar application in dechlorination is in the treatment of potable water.

With brine, there is not yet available a wide range of commercial experience. The following points should be kept in mind when considering a system:

(1) The carbon will be attacked by the chlorine and will need periodic replacement.
(2) Fines will be produced by this attack and must be removed by regular backwashing.
(3) The carbon system can be installed anywhere between the chemical addition and the ion exchange system.

Because of the need for backwashing, the advantages of a normally low space velocity, and the problems of construction of large vessels, we would recommend the installation of multiple beds in parallel in a large plant. Each bed should be able to take more than its normal flow while another is backwashing. The exact deisgn will depend on the configuration of the plant and the brine storage philosophy.

One would usually prefer to run the carbon system on acidic brine and so would always place it before the primary treatment system. This is based on the observation that carbon works best in HOCl and most poorly on OCl^-. The best operation would then be expected under mildly acidic conditions (pH 5 to 7). The second choice would be more strongly acidic, and the last choice alkaline, conditions.

We would not recommend that a system whose function is to protect the resin, which in turn protects the integrity of the cells, be designed to operate properly only when controlled at near-neutral pH. This is a notoriously difficult operating problem. Our approach, therefore, would be to provide enough carbon to handle high- or low-pH operation in any case. Having this, we can then avoid the problems of intermediate pH adjustment.

Figure 12.2 shows the secondary dechlorination system immediately following vacuum dechlorination. All downstream equipment will then benefit from handling fully dechlorinated brine. The best location, however, will depend on plant layout and the method of handling of backwash liquid and any fines carried through during normal operation. The two stages of dechlorination might be separated in order to provide some time for reaction.

A problem which might arise in treatment of acidic brine is gassing in the beds because of decomposition of chlorate. This should be reviewed case by case, and it may well be better to treat on the alkaline side.

One last word on the use of a chemical reducing agent: its purpose is to reduce the load on the carbon bed and so to reduce the expense and trouble connected with that operation. In most large plants, this would seem to be worthwhile. In a very small plant, there is less justification. It is also possible that the reduced brine circulating rate might allow an existing vacuum dechlorinator to operate so effectively that chemical reduction can be eliminated.

12.4 CHLORINE PROCESSING

The next major area to consider is chlorine processing. No differences exist among the three types of cells which would call for any basic changes in the process. Cooling, drying, compression, and liquefaction are common to all (Fig. 12.4). In some plants, there may be a shortage of cooling capacity. The other areas stand to gain from membrane cell operation.

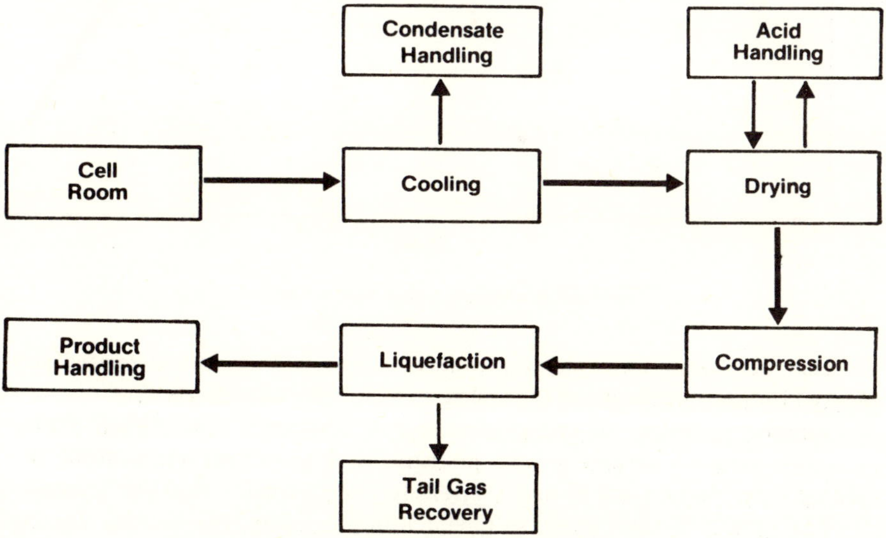

Fig. 12.4 – Chlorine processing.

12.4.1 Chlorine cooling

Unless an existing mercury plant has some excess cooling capacity, this area might be a problem. Since membrane cells operate at higher temperatures and lower anolyte concentrations than many mercury cells, the chlorine gas will contain more water.

The total heat duty may therefore be considerably higher after conversion. The amount of water left uncondensed will be less so.

The performance of a typical chlorine cooler is shown in Fig. 12.5. As this process moves along the condensing curve and the water content is reduced, the heat transfer coefficient and the condensing temperature both fall off. When a wetter gas is fed to a cooler, then, both the coefficient and the temperature driving force are enhanced at each point along a tube. The result is a considerable increase in heat flux. When a secondary cooler is used, with chilled water as the cooling medium, the increased temperature driving force may be even more significant because of widening of the usually close temperature approach.

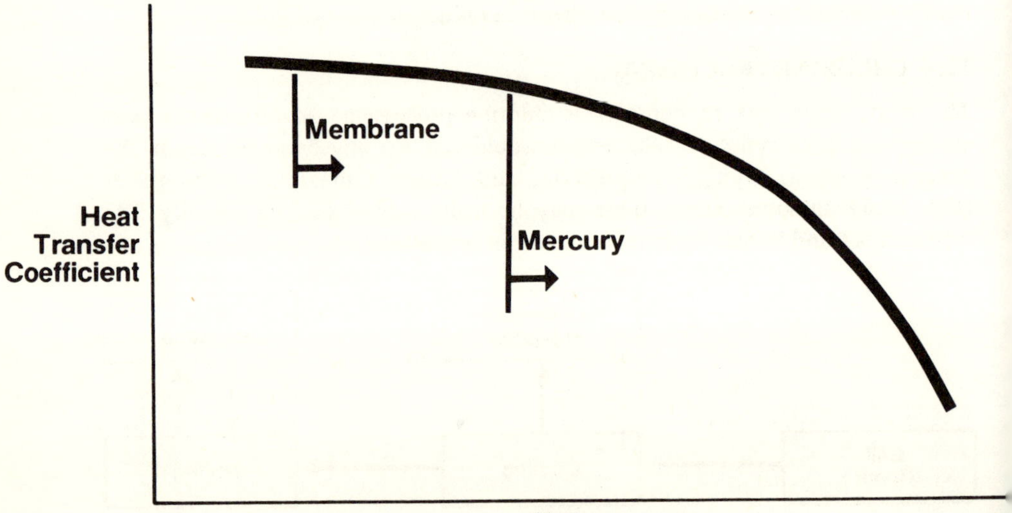

Fig. 12.5 – Chlorine cooler performance.

The exchangers, especially the primary cooler, will have to be rerated and checked for pressure drop and the ability to handle the condensate flow.

Another potential offsetting advantage of membrane cells is their ability to operate under a modest postive pressure. At a given final temperature, the residual water content will be nearly inversely proprotional to the total pressure.

The upshot is that, given sufficient utility capacity, the existing cooling system will remove a greater volume of water from membrane cell gas than from mercury cell gas. The remaining difference might well be taken up in the drying system. The only alternative to this is to add more cooling capacity.

Some plants use direct contact cooling in place of the closed system assumed above. The same effects will apply. The liquid-flow capacity of the equipment, however, probably will not permit much of an increase in load.

The greater volume of highly chlorinated condensate must next be taken care of. The preferred method depends on the overall plant water balance. It may be possible to send this water to the brine circulating system for dechlorination, or it may be necessary to strip the chlorine before disposal or reuse. Whatever system already exists should be checked for its ability to handle the load.

Note that, in a diaphragm cell plant, the problem of cooling capacity may not exist and the use of the brine system may present a new option for handling of the condensate.

12.4.2 Chlorine drying

If a problem exists in chlorine cooling (see discussion above), it may be possible to increase the water load on the drying towers.

Unless the system is already marginal, this should not present a problem in mass transfer. It will require an increase in acid flow or a decrease in spent acid concentration. The latter may not be acceptable but is one way of reducing the total fresh acid consumption. The former will present no problem to the conventional packed towers, because the net acid flow is only a small fraction of the total circulation.

Heat evolution will also increase with the added water load. Much of this will be taken up by the higher throughput of acid, but some higher temperatures will result. Many old systems use brick-lined towers, and again, the higher temperature practically is no problem. Where cooled systems are used, the existing coolers almost certainly can cope.

The gas-handling capacity of the drying system, like that of the coolers, will increase along with the operating pressure.

12.4.3 Acid handling

No change in concept will be required in the existing system. If the acid consumption increases because of a cooling problem, certain equipment will need checking.

First, if metering pumps are used for acid feed, their capacity will have to be checked. Many plants use centrifugal pumps with a bypass or a head tank arrangement. These should have no capacity problem.

Most plants will have a system for dechlorination of spent acid. This might be semi-batch or might use a continuous stripper. In either case, the design should be checked against the new requirements.

Finally, adequacy of flow-through instruments and gravity lines should be verified.

12.4.4 Compression

Given dry chlorine, the existing compression system should be more than adequate. The volume of the cell gas will be slightly less, because of the higher purity obtained with the membrane cells and the absence of dilution air. If the cells operate under pressure, this will also reduce the compression load.

There will probably be a recycle valve responding to a suction pressure or compressor surge controller; with most compressors, this will have to handle a greater flow.

12.4.5 Liquefaction

Like compression, liquefaction will also be simpler. The higher cell gas purity and absence of dilution air will reduce the gas flow. The absence of inerts will increase the partial pressure of chlorine gas and allow higher liquefaction efficiency at the same conditions. If desired, the operating pressure also can be increased, because of the smaller load on the compressor(s).

If secondary liquefaction is used, some thought should be given to turn-down capability.

12.4.6 Tail gas recovery

The nature of the uncondensed gas from liquefaction will change greatly. As pointed out above, a higher degree of liquefaction can be obtained at the plant's standard operating conditions, so that less chlorine will remain. The composition of the inerts will be quite different. Hydrogen will be nearly absent, and oxygen will be one of the major constituents. A different set of safety problems results.

Tail gas recovery systems can be divided into those which are designed for recovery of chlorine as such and those in which chlorine is transformed to another product (e.g. HCl or NaOCl). The latter can be subdivided into those which are profitable operations and those which are necessary evils. In the profitable cases, improved results in chlorine liquefaction may be of little real value. In the other cases, the load will be reduced. Less byproduct will result, but at lower operating efficiency. In an HCl unit, for example, the higher oxygen content of the tail gas will increase the consumption of hydrogen and the generation of heat for each unit for production. Some changes in operation may be required to maintain good control simultaneously of acid strength and scrubber operation.

One advantage of the use of HCl process is that all the acid produced from normal membrane cell tail gas can be used internally.

Among systems designed for recovery of chlorine, absorption in a solvent with subsequent stripping is the most familiar. Carbon tetrachloride is the standard solvent. This is a potential air pollutant which must be removed from the off-gas before venting. Carbon adsorption is frequently used to do this. This process would have to be approached very carefully with membrane-cell tail gas.

Many plants which require nearly complete recovery of chlorine as liquid use secondary or tertiary liquefaction with pressure staging. Here, the off-gas from one stage of liquefaction is compressed and returned to the unit at the same temperature level. This can be done in a separate exchanger or in an independent tube bundle in a larger shell. Unless space problems were critical, such a system could easily be added during a conversion. The available refrigeration capacity would be used more effectively. An existing liquefier could be modified to handle two tube passes at different pressures, but the installation of a separate exchanger will be simpler and will require much less downtime.

At very high degrees of liquefaction, even 0.1% hydrogen in the feed gas will be concentrated to dangerous levels. Such systems require careful design from a sound basis developed along with the membrane and cell suppliers.

12.4.7 Other systems

Other systems already in place should include those for storage and loading of liquid chlorine, evacuation of equipment and piping, and scrubbing of chlorine vents and emergency releases.

No changes should be required in storage and loading systems. The existing purge/evacuation system should also be adequate, although some repiping may be necessary.

Normal loads on a scrubbing system will not increase. If there is an emergency dump of wet gas, vapor flow to the system in a mercury-cell plant will be higher. as will the heat load on the circulating liquor. A close review of the materials of construction should be made.

12.5 CAUSTIC PROCESSING

This is one major area which differs fundamentally from both mercury- and diaphragm-cell plants. The caustic processing system in a mercury plant remains the simplest because of the cells' unique ability to produce 50% NaOH of high purity. The membrane cells produce a caustic of similar purity but at lower concentration. In addition, control of temperature and composition in the cathode chambers requires a heavy recycle of caustic to the cells. With diaphragm cells, the need for an elaborate evaporator and salt recovery system makes this the most complex and most expensive caustic process of the three.

Figure 12.6 shows the basic components of a membrane-cell plant's caustic processing system.

12.5.1 Caustic recycle

Caustic from the cells must first be collected in a tank. Such a collection system probably exists in a mercury-cell plant. The piping still should be checked for the larger flow which will have to be handled. In a diaphragm-cell plant, the cell liquor handling system performs a similar function, but must be checked for both size and material of construction.

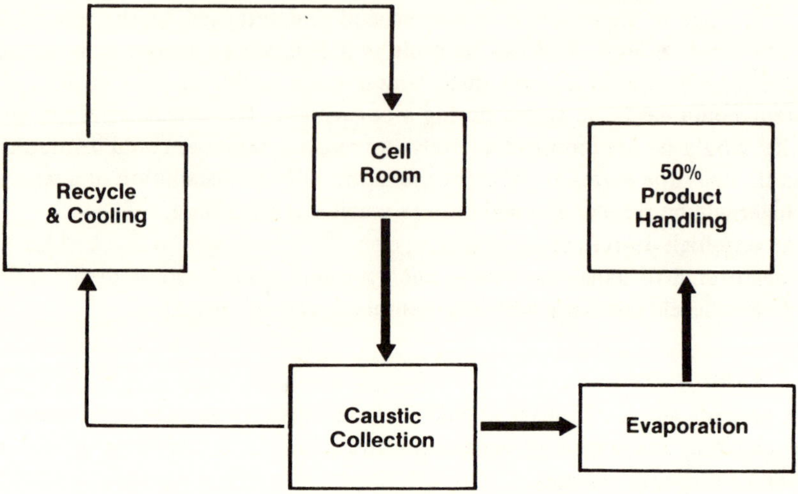

Fig. 12.6 – Caustic soda processing.

The caustic recycle system will not exist in the plant being converted. For each volume of net production sent on to evaporation, several volumes are recycled to the cell room. This stream is cooled and diluted with water to control the process. For the best economy, some of the cooling may be obtained by interchange with the feed brine.

As developments continue to reduce membrane cell energy consumption, it becomes possible to consider the elimination of caustic recycle. The waste heat would be taken up by the brine and water fed to the cells. It would still be important to maintain uniform cell temperatures and catholyte concentrations.

12.5.2 Caustic evaporation

This discussion assumes that the final product is 50% NaOH. An evaporator must therefore be added when mercury cells are replaced. This is a straightforward requirement and not really a 'conversion'. Other approaches are possible in a partial conversion, which will be treated later.

In a diaphragm plant, the existing evaporator system will be much more elaborate than necessary. The duty will far exceed that of the membrane plant, and the solid and slurry handling equipment will not be needed at all. With the much lower duty when handling membrane cell caustic, evaporator turndown will be a serious problem. Part of the evaporator system may have to be bypassed. A single effect might be used, for example, from a two- or three-effect system. Steam economy will then not be as good as in a custom-designed system.

With proper redesign, the excess capacity in the evaporators can be used to concentrate the brine if solid salt is not available.

12.5.3 Product storage and handling

If all product is to be 50%, no significant changes should be required here.

Some intermediate storage should be provided for caustic from the cells. In a mercury plant, this will probably mean new facilities. In a diaphragm plant, the cell liquor storage and boilout tanks can be reviewed for use.

12.6 CELL ROOM

The membrane electrolyzers must suit the available current supply and fit within the space made available.

Because of the usual stacked-cell arrangement, designs are very flexible. Most monopolar systems can be adapted to all but the very high current circuits (i.e. several hundred kiloamperes). At these very high loads, and in many other cases with bipolar arrangements, the available current might have to be divided and controlled, or the arrangement of multiple rectifiers changed.

In a straight replacement project, space will not be a problem. With FM21 cells, a typical (not overly compact) layout gives a yearly capacity of 140 tonnes of NaOH for each square meter of floor space, at a moderate current density of 3 kA/m^2. This is approximately double the capacity obtained in most existing plants. Unit-for-unit replacement is easily carried out with the sacrifice of this potential advantage and the redesign of intercell connectors.

If we are to plan to save space for present or future expansion, the existing arrangement of aisles, trenches, and pipe supports will constrain the design. The existing headers will have to be modified or replaced to allow expansion. Figure 12.7 indicates the piping services required with each type of cell; the membrane conversion will also require the addition of new lines. New headers should be prefabricated, and their installation can be timed for a planned shutdown prior to changeout of the cells themselves.

	mercury	membrane	diaphragm
chlorine	X	X	X
hydrogen	X	X	X
brine feed	X	X	X
depleted brine	X	X	
recycle caustic		X	
product caustic (or cell liquor)	X	X	X

Fig. 12.7 – Required piping services.

It is possible to replace the exiting cells in stages by shorting out part of the circuit. Membrane cell assembly and installation might be handled by the existing plant crew. This procedure will require the installation of temporary bus work which can be moved as work proceeds. Careful planning and installation are essential both for efficiency and for continued operating safety.

Work on the cell room floor should keep pace with the installation of the cells. Other building work can be planned for a convenient time. The adequacy of a natural ventilation system should be checked in the early stages.

In most cases, it will be important to control the downtime caused by the conversion. We would strongly recommend that the cell room work be thoroughly engineered beforehand.

In extreme cases, with buildings in dilapidated condition or when under heavy pressure to maintain production, we might recommend that a new cell room be built.

12.7 PARTIAL CONVERSIONS

A partial conversion might take the form of a hybrid plant which combines some of the advantages of two types of cell. This is a familiar concept in the industry, several plants at present combining mercury and diaphragm cells.

The most obvious disadvantage of membrane cells relative to mercury cells is the low concentration of the caustic product. Assuming that 50% solution is required for sales, a full switch to membrane cells will mean the added investment and operating expense of an evaporator. In some cases, a boiler expansion will also be required.

In a combined plant, the caustic from the membrane cells could be used as part of the feed to the existing decomposers (Fig. 12.8) and no evaporator would be required. Up to 50% of the total production can be made in the membrane cells.

Turning to diaphragm cells, their ability to operate on a brine feed was pointed out in the Introduction as their main economic advantage. In a combined plant, with either mercury or membrane cells, one can have the favourable cost of a brine feed for the entire production and the advantage of a pure caustic product from part of the plant. At least in the case of the membrane cell combination, there is also the advantage of lower energy consumption for that part of the production.

In the more familiar diaphragm–mercury combination (Fig. 12.9) the usual arrangement involves a closed brine loop in which the depletion in the cells is offset by dissolving salt recovered from the evaporators. The closed loop is necessary to avoid contamination of the diaphragm cell system by mercury. Impurities in the recovered salt can be removed from the mercury system only by purging.

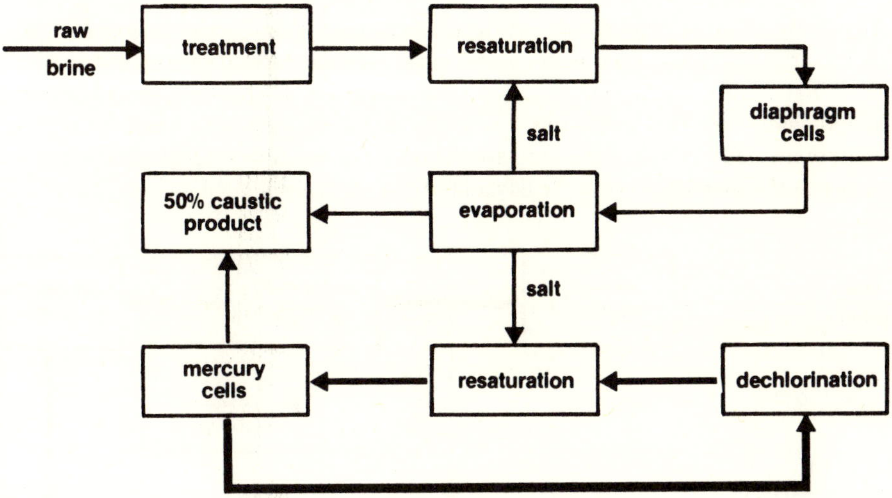

Fig. 12.8 – Combined operation of membrane and mercury cells – 50% caustic production.

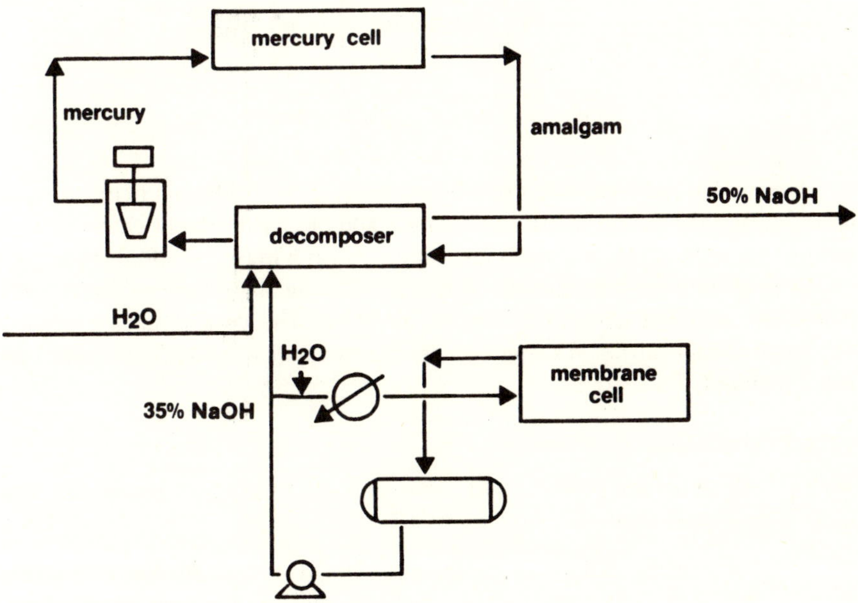

Fig. 12.9 – Mercury/diaphragm combined plant.

Membrane and diaphragm cells can be used with the same process configuration. Alternatively, a once-through arrangement is also prossible (Fig. 12.10). Brine from primary treatment would flow through secondary treatment, the membrane cells, dechlorination, and resaturation. It would then be fed to the diaphragm cells. This eliminates the accumulation of recycled impurities in the membrane cells. It also can eliminate the need for secondary dechlorination and will give the diaphragm cells the advantage of a purer brine feed.

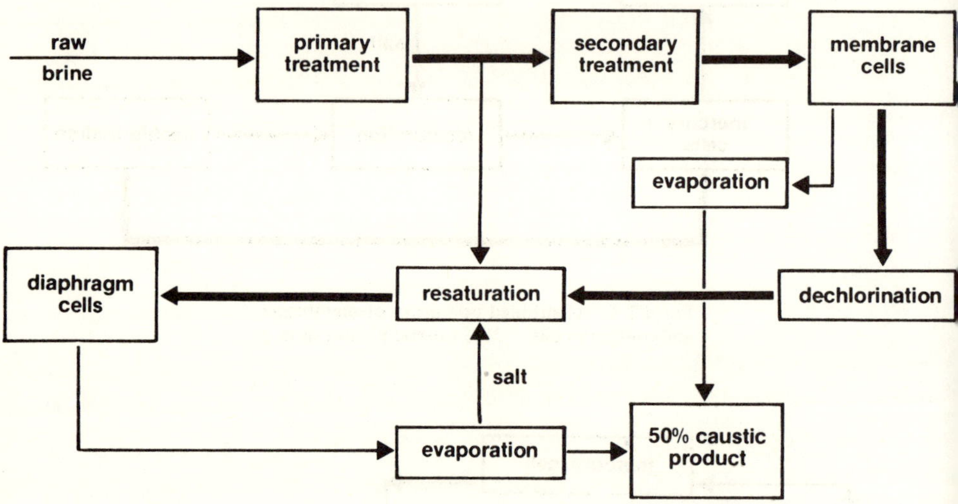

Fig. 12.10 – Membrane/diaphragm combined plant.

In a partial conversion, the number of existing cells to be left in operation is another variable for optimization. If space and electrical supply characteristics permit, the cells can be left intact and simply run at a lower current density.

A final note on cell renewal: This is a relatively simple process with many membrane cells and remarkably so with an FM21. The work space in a mercury cell room should suffice. A diaphragm cell renewal area will have many times the space required.

12.8 ECONOMICS

We have calculated the economics of conversion as a function of power cost. Our basic assumptions were:

(1) An existing cell room producing 100 000 tonnes of NaOH annually would be fully converted to membrane cells. As was pointed out above, partial conversion may be more attractive in some cases or may be considered as a first step.

(2) No expansion in capacity would be included. This is the most reasonable assumption for a general study, but we have tried to point out the ease with which cell room capacity can be increased. If a market exists for the products, this can have a powerful effect on the economics.

(3) Inflation is ignored. More accurately, uniform inflation is assumed and results are presented in today's money. We have also done some calculations on the effects of differential inflation in energy costs.

(4) There are no projects planned which are essential to continued operation of the existing plant. This may be quite the reverse of the truth. If such projects can be eliminated, the real incremental investment for conversion can be reduced.

Specific technical assumptions used in the evaluation are given in Fig. 12.11. Diaphragm- and mercury-cell energy requirements are representative of those plants which we have studied as candidates for conversion. Note that steam cost has been taken as proportional to electrical energy cost.

- FM 21 cells used
- current density 3 KA/M^2
- membrane life 2½–3 yr
- electrode coating life 5–6 yr
- salt feed to mercury plant
- brine feed to diaphragm plant
- power requirements, DCKWH/tonne NaOH

membrane	2200
mercury	3400
diaphragm	2700

- steam requirements, tn/tn NaOH

membrane	1.2
mercury	0.4
diaphragm	4.2

- 1 tonne steam – 200 KWH

Fig. 12.11 – Technical assumptions.

Figure 12.12 shows the financial assumptions used. These are typical of current US practice. From these sets of assumptions, we have calculated the operating cost savings offered by membrane cells at different unit energy costs. Figures 12.13 and 12.14 illustrate these calculations at $50/MWh.

The costs of conversion were estimated to be $20 million for a mercury-cell plant and $21 million for a diaphragm-cell plant. These allow for supply and installation of cells in a US Gulf Coast plant, modification and replacement of cell room piping, the addition of brine treatment and caustic circulation systems, and the addition or modification of evaporation equipment.

The savings and the estimated investment costs were then used to calculate common measures of profitability. Both the simple return on investment and the internal rate of return are shown in Figs. 12.15 and 12.16. In the supporting cash flow calculations, major expenditures for membrane replacement and electrode recoating were delayed until the third and fifth years, respectively. Compared to a constant level of expenditure, this increases the present worth of the savings by about 5%.

If energy costs increase at more than the overall inflation rate, conversion will be more attractive. For each 1% in differential inflation rate from a starting price of $50/MWh, the ROI and IRR each increase by roughly 1%.

- no salvage value
- 10-year book depreciation
- 5-year tax depreciation (S.Y.D.)
- 46% income tax
- 13% interest
- 10-year project life
- 10% tax credit

Fig. 12.12 – Financial assumptions.

	membrane	diaphragm	difference
Chemicals	6.00	6.00	–
IE Resin	1.50	–	(1.50)
Electrodes	5.00	10.00	5.00
Membranes	10.00	–	(10.00)
Diaphragms	–	1.00	1.00
Steam-Evap.	10.00	30.00	20.00
–Aux.	6.00	13.00	7.00
Cooling Water	1.00	1.70	0.70
Elec.-Cells	115.00	140.50	25.50
–Aux.	9.15	14.15	5.00
Royalty	2.00	–	(2.00)
			50.70

Fig. 12.13 – Variable cost comparison membrane versus diaphragm cells.

	membrane	mercury	difference
Chemicals	6.00	8.00	2.00
Hg Control	–	6.00	6.00
IE Resin	1.50	–	(1.50)
Electrodes	5.00	10.00	5.00
Membranes	10.00	–	(10.00)
Steam-Evap.	6.00	–	(6.00)
–Aux.	6.00	4.00	(2.00)
Cooling Water	0.70	1.00	0.30
Elec.-Cells	115.00	180.00	65.00
–Aux.	9.15	12.50	3.35
Royalty	2.00	–	(2.00)
			60.15

Fig. 12.14 – Variable cost comparison membrane versus mercury cells.

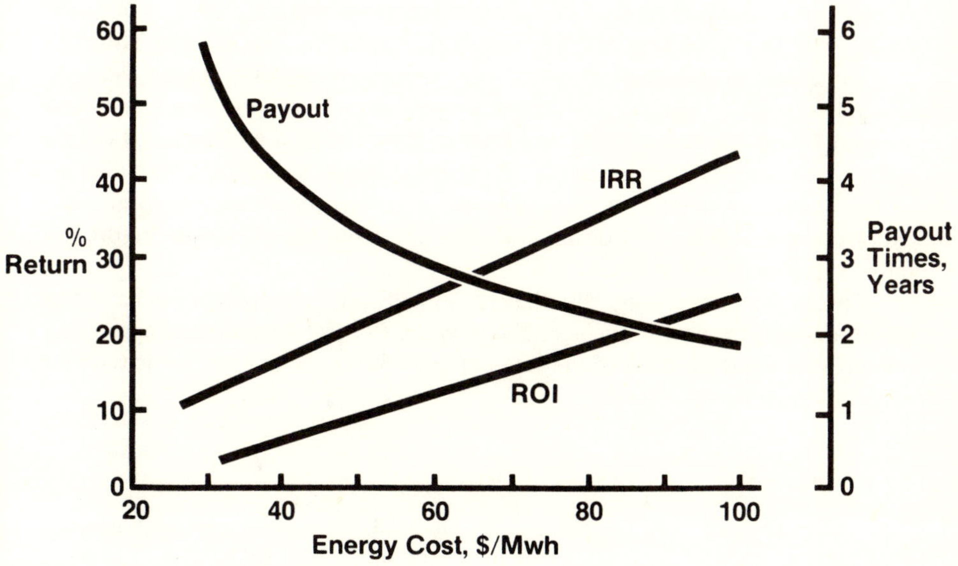

Fig. 12.15 – Diaphragm cell conversion economics.

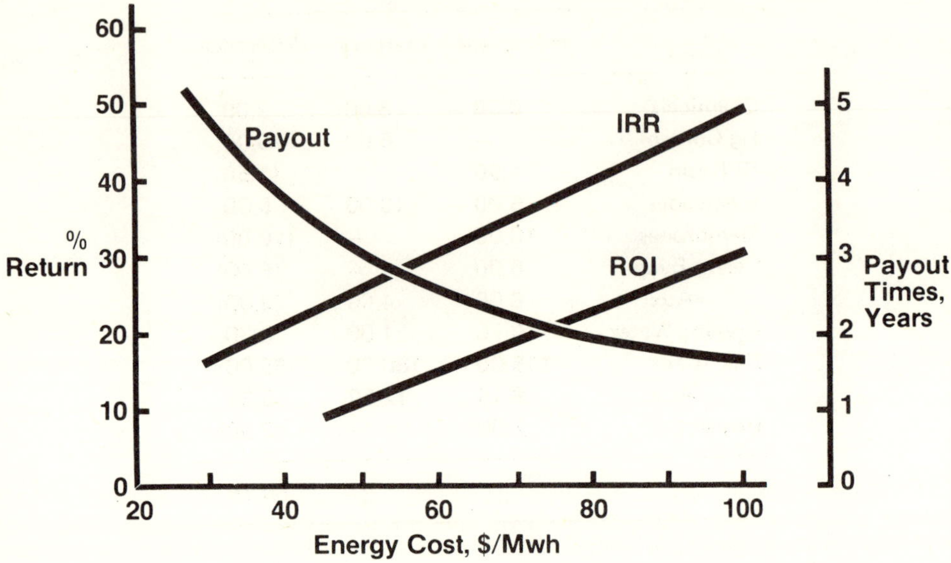

Fig. 12.16 – Mercury cell conversion economics.

At the higher power costs, a project begins to look quite attractive. Several factors not considered here can improve the outlook for conversion even more. First, relief from problems with environmental regulations is a definite, although hard to quantify, advantage. Second, improvement in product quality can help a diaphragm cell operator's competitive position. Third, the economic data presented here are based on US conditions. They assume a 46% income tax rate and a 5-year depreciation schedule. In most countries, policies are more liberal, with a lower tax rate and a more generous depreciation allowance. As an indication of the effects of taxation, we have therefore shown simple before-tax returns on investment in Fig. 12.17.

Finally, we would point out that FM21 cells can be supplied to most of the world under government finance. The Export Control Guarantee Department will provide loans for most of the value of any UK equipment and services under very attractive terms.

12.9 SUMMARY

Conversion projects will be highly individualized. Each plant will have its own problems, its own constraints, and its own economics. With today's low operating factors and uncertain markets, many operators cannot seriously consider conversion. In a strong industry operating near capacity, the case for conversion is strong and will become more compelling as energy costs increase.

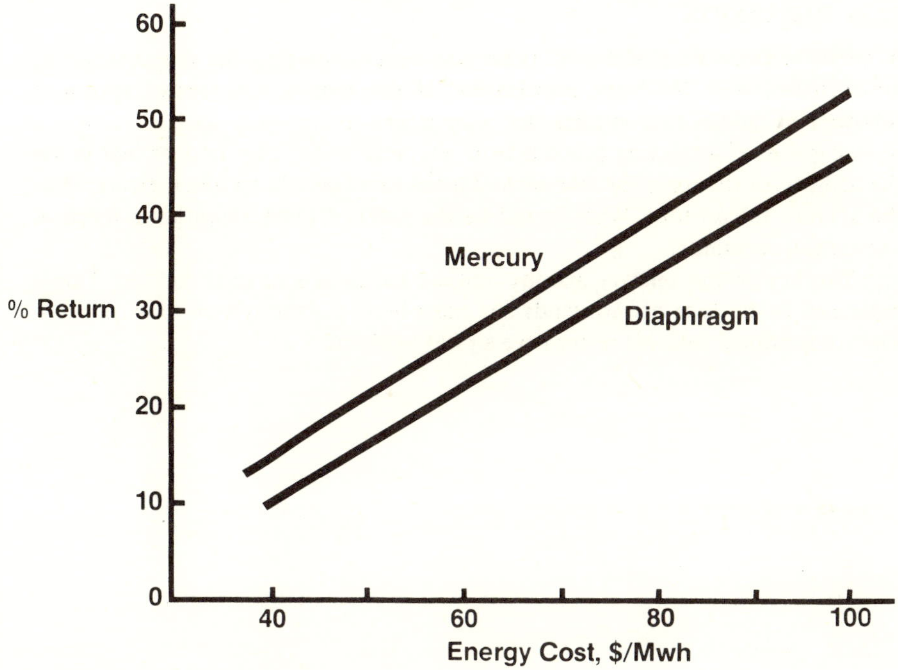

Fig. 12.17 – Before-tax return on investment.

Without minimizing the problems in engineering and construction, we find that conversion projects are not difficult in concept and that most major problems are easily identified. Detailed engineering and planning will reveal the more subtle problems.

We expect the 1980s to see the beginning of a thorough shakeout in the chlor-alkali industry. Only the efficient producers will survive, and stopgap improvements will not see the others through these difficult times.

The membrane cell offers a fundamentally new approach to production of chlorine and co-products with lower energy consumption. Developments now in progress will make operation with solution-mined brine more attractive. We think that the status of the membrane cell is remarkably similar to that of the permanent anode shortly after its commercial introduction.

12.10 ACKNOWLEDGEMENT

Arthur Boyce and Graham Dibble of Catalytic International in London reviewed parts of the text, provided background information, and contributed many helpful comments.

12.11 DISCUSSION

A question from Prof. Kahn of India was posed regarding the regeneration of ion-exchange resin for brine purification. If the resin is first treated with acid followed by alkali, this implies the resin is in a completely ionized form – a form capable of attracting protons from the acid brine. This implies that in use the affinity of the resin for calcium and magnesium should be much greater than the affinity for protons. What would be the ratio of these affinities in terms of association constants?

The key is the relative affinity for the hardness ions over sodium. This is measured in the tens of thousands and must be so for the system to work at all. The conventional cationic resins have a ratio of about 5:1.

SECTION D

Electrodote Coatings

CHAPTER 13

Developments in Coatings for Metal Anodes

D. A. DENTON, J. T. DESOUZA, J. H. ENTWISLE, D. LEE, H. G. WILSON,
ICI Mond Division, Cheshire, England

13.1 INTRODUCTION

This paper describes the development of titanium anode coatings in ICI and their performance in mercury, diaphragm and membrane chlorine cells.

13.2 DEVELOPMENT OF COATINGS FOR MERCURY CELLS

ICI has been actively involved in the development of anode coatings since the pioneering work of H. B. Beer and J. B. Cotton whose first patents covering coatings of precious metals appeared in 1957. At this time the main area of interest for application was in mercury cells and the coatings were prepared either by the electrodeposition of platinum or by the thermal decomposition of paint formulations which contained platinum compounds. Full scale trials of both these types of coating were made during 1959–62 but the technical performance was not sufficiently good to justify the replacement of graphite anodes, particularly since the quality of graphite available at this time was also improving. The main problem encountered was a decrease in the electrocatalytic activity of the platinum surface as a result of gradual passivation, leading to a high chlorine overpotential and hence to high cell-energy consumption; the electroplated coating was the worse in this respect. Additionally, the resistance of the coatings to the cell environment was poor. The basic corrosion rate of the coatings under the cell conditions was high, while occasional anode to cathode contacts caused amalgamation which added substantially to coating loss. These effects were worse with the paint platinized coatings. Amalgamation of the platinum during shorting led to a low contact resistance with the mercury cathode and heavy currents flowed causing substantial anode and coating damage. Coating loss was also promoted by the reverse flow of current which occurs whenever a cell is switched out of circuit. As a consequence, the overall cost of anode consumption was unacceptably high, even at the relatively large interelectrode spacings employed.

The problem of passivation is exemplified in Fig. 13.1 which shows chlorine overpotential as a function of current density for platinum, graphite and a modern precious metal oxide coating. A transition from a state of low to high overpotential is apparent for the platinum coating there being a hysteresis effect as current is raised and lowered. This transition is affected by other electrolysis parameters and the increase occurs more readily the greater the temperature and pH of the anolyte.

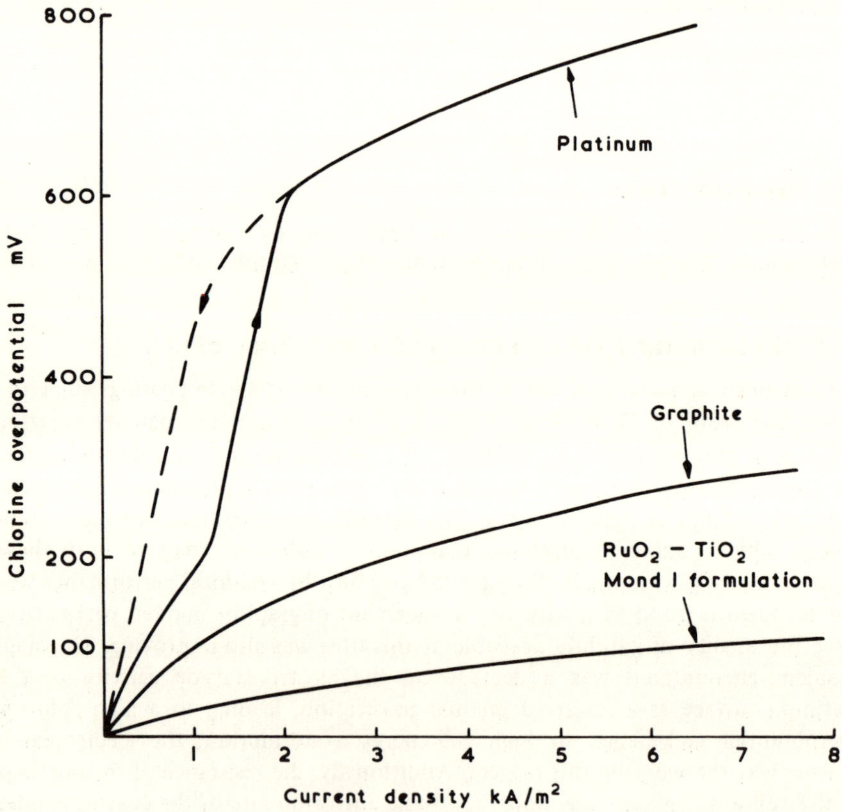

Fig. 13.1 – Chlorine overpotential on graphite and on titanium anode coatings.

Research to overcome these problems continued throughout the 1960s. At first, effort was concentrated on remedying some of the faults of platinum coatings and substantial progress was made in reducing the passivation effects and the basic platinum corrosion rate. Thus platinum coatings containing iridium or prepared by electrodeposition in the presence of organic compounds such as gelatin proved to have low chlorine overpotentials and corrosion rates. However, they still suffered from amalgamation and extensive damage upon shorting, even with an improved and more robust design of anode structure. Emphasis then switched to the use of oxides of precious metals as coatings. It became apparent that the best oxide was ruthenium dioxide which was found to be obtained instead of the metal when using the traditional paint thermal decomposition procedures. This material exhibits metallic conduction and has a low chlorine overpotential but when used alone has poor adhesion characteristics and consequently a high consumption rate. This deficiency was first improved by the technique of covering with a glaze and later by mixing into the paint formulation compounds of metals giving rise to inert oxides such as titanium dioxide. Ruthenium dioxide and the rutile form of titanium dioxide are isomorphous and can form solid solutions which show strong adhesion to the titanium substrate. A coating based on this mixture, termed the Mond 1 coating, was developed which had the unique blend of properties required – a low chlorine overpotential and corrosion rate and resistance to current reversal. It also does not amalgamate and consequently shows greater resistance to damage by shorting than the metallic coatings. (This type of isomorphous mixed oxide coating was first patented by H. B. Beer in 1967). A number of other coatings e.g. SnO_2–RuO_2 were developed by ICI around this time as shown in Table 3.1 but these were not sufficiently durable for use in mercury cells and effort was concentrated on the large scale application of the Mond 1 coating. Successful development of this coating depended upon studies of the effects of the various preparative parameters on performance. These included titanium surface pretreatment, paint formulation, coating loading, firing tempartures, etc., the laboratory work being followed by full-scale trials. A large production plant was built at Northwich, Cheshire, which commenced operation in 1972.

A new anode design shown in Fig. 13.2 had also been developed, and this came into use throughout the ICI Group. Additional performance data for the Mond 1 coating on these anodes are shown in Table 13.2. Large scale experience revealed that significant operational problems still arose from anode to cathode shorting. Shorting results mainly from the build up of deposits of 'thick mercury' on the cell baseplate which cause the cathode thickness to increase with time. It leads to current efficiency loss, coating loss and ultimately to damage to the anode structure. A new coating formulation termed the Mond 5b was developed to reduce these effects which are illustrated in Figs. 13.3 and 13.4. Figure 13.3 shows how current efficiency declines as the interelectrode gap falls below 1 mm for the two types of coating. The loss is due first to the increased rate of reduction of dissolved chlorine at the cathode and also to the development of some electronic conduction. However, electrolysis still occurs in the immersed region as a result of surface tension effects which allow brine to creep down the interface

Table 13.1
ICI Mond Division anode coatings

Formulation reference	Main constituents	Patent reference	Comments
Mond Pt, Pt–Ir metal	Electroplated Pt metal Pt or Pt–Ir from paint platinizing formulation	– –	Problems with passivation etc.
Improved Pt metal	Pt electroplated in the presence of organic compounds	GB 1 237 077	Non-passivating under mercury cell conditions
Mond 1	RuO_2, TiO_2	–	Coating composition depends on type of cell. Preferred coating for diaphragm and membrane cells.
Mond 2	RuO_2, with overglaze	GB 1 206 863 GB 1 294 373 GB 1 352 872	
Mond 3	RuO_2, SnO_2	GB 1 244 650 GB 1 277 033 GB 1 354 897	
Mond 5a Mond 5b	RuO_2, TiO_2, with added refractories	GB 1 402 414 GB 1 484 015	Preferred coating in mercury cells. Candidates for membrane cells.

Fig. 13.2 – The ICI mercury cell anode.

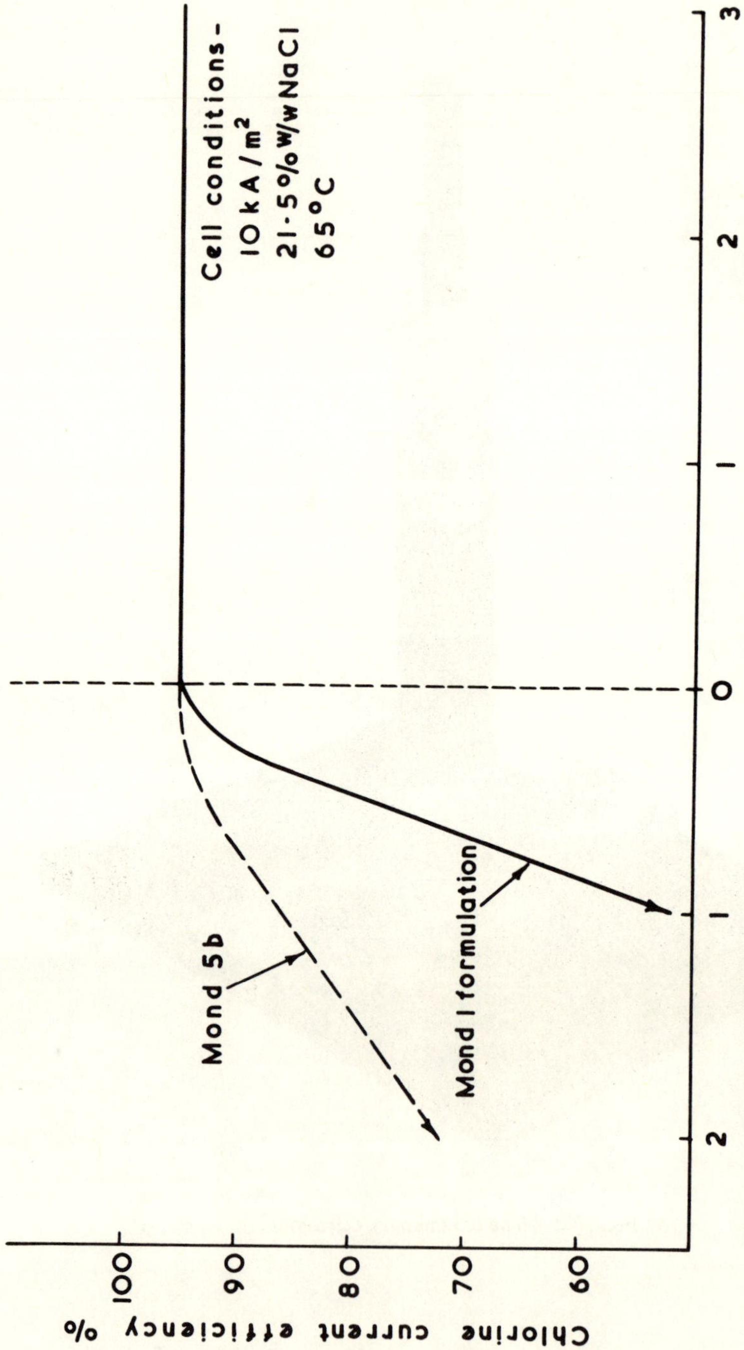

Fig. 13.3 – Relationship between current efficiency and interelectrode gap – depth of shorting contact for ICI mercury cell anode.

Table 13.2
Life of Mond 1 and Mond 5b coatings, and gas composition, in mercury cells

Life at 10 kA/m²† years	Gas composition, % v/v		
	Cl_2	H_2	O_2
2–3	>99	0.1	0.2

† Depends on preparative conditions, and on conditions of cell operation.

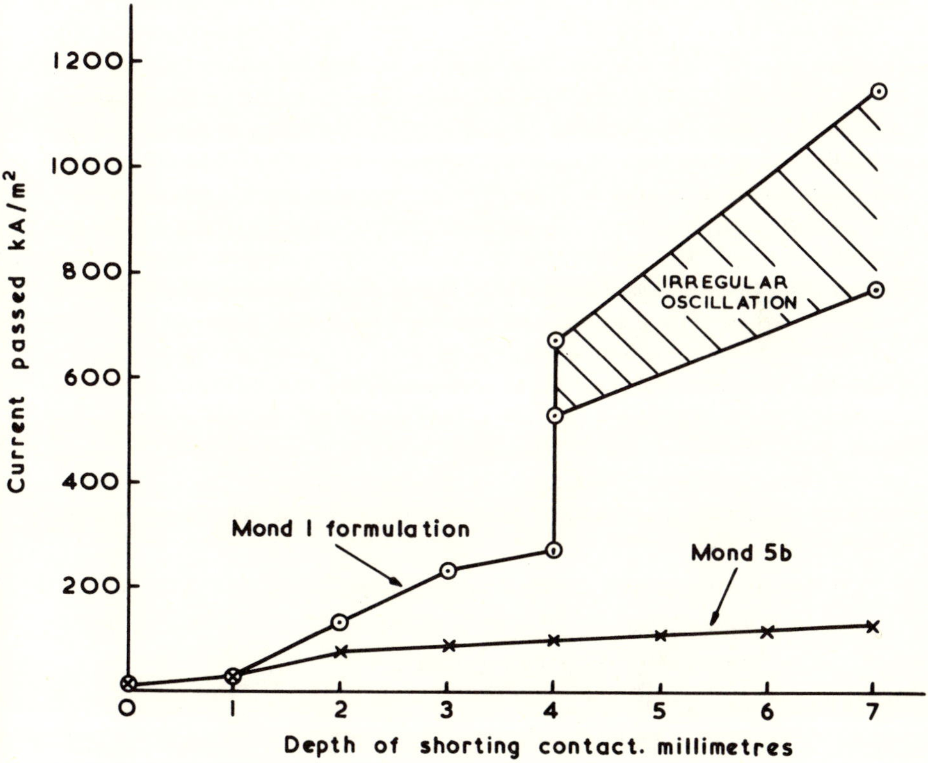

Fig. 13.4 – Current passed related to depth of shorting contact of ICI mercury cell anode into flowing cathode.

between anode coating and amalgam. The chlorine bubbles formed also help to separate the aqueous and amalgam phases. At a certain critical depth which varies with the mercury surface velocity the contact pressure becomes sufficient to collapse the chlorine bubble layer and to prevent electrolysis almost entirely. When this occurs, a steep rise in current results as ohmic contact between coating and amalgam and arcing develop. The effect is illustrated for the Mond 1 coating in Fig. 13.4. It is prevented in the case of the Mond 5b coating by virtue of its structure which effectively imparts a high contact resistance with the amalgam. This facilitates operation at lower interelectrode gap and hence improves cell energy consumption. Under normal conditions of operation the life of the Mond 5b coating is similar to that of the Mond 1 coating.

13.3 ANODE COATINGS FOR DIAPHRAGM CELLS

The advantages of metal anodes in reducing cell energy consumption and anode consumption costs quickly led to their replacing graphite anodes in both mercury and diaphragm cells. However, the worldwide concerns about mercury in the environment which began in the 1960s caused a fundamental change in the direction of cell development with diaphragm cells becoming the preferred choice of technology for new chlorine plants. The first modern diaphragm-cell plant in ICI was commissioned at Lostock in 1977 and the most advanced design of ICI diaphragm cell, the DMT cell, is illustrated in Fig. 13.5. The anode coatings and anode design principles developed for mercury cells found application in the new designs of diaphragm cell. The emphasis on performance characteristics was, however, changed. The problem of anode–cathode shorting no longer applied, but the need to avoid passivation effects and to ensure an acceptable life for the coating remained. The platinum coatings were used with some success in the early prototype cells but overall were found to be less attractive than the coatings based on ruthenium dioxide. As a consequence of the different electrolysis conditions in the diaphragm cell, the extent of chlorine current efficiency loss owing to the formation of oxygen, chlorate and hypochlorite and the balance between these reactions also became of greater importance. The main anolyte reactions contributing to chlorine current efficiency loss may be summarized as follows:

(a) Chlorine hydrolysis and reaction with alkali.

$$Cl_2 + H_2O \longrightarrow H^+ + Cl^- + HOCl$$
$$HOCl + OH^- \longrightarrow OCl^- + H_2O$$

(b) Chlorate formation.

$$2HOCl + OCl^- \longrightarrow ClO_3^- + 2H^+ + 2Cl^-$$

(c) Oxygen formation.

$$2H_2O \longrightarrow O_2 + 4H^+ + 2e$$
$$2HOCl \longrightarrow O_2 + 2H^+ + 2Cl^-$$

Fig. 13.5 – The ICI DMT43 diaphragm cell.

These reactions are strongly pH-dependent and generally increase in extent as more alkali enters the anolyte from the back migration of hydroxyl ions or in the feed brine. The kinetics of the oxygen evolution reactions also depend on the nature of the anode coating and are influenced by various anode design and electrolyte parameters. The effect of the nature of the coating is illustrated in Fig. 13.6 which shows the change in oxygen formation as the proportion of ruthenium dioxide in the Mond 1 coating is varied between 35 and 60% w/w. As the RuO_2 content increases the rate of oxygen evolution also increases while at the same time the increased acidity generated by this reaction reduces the rate of chlorate formation. The kinetic parameters of the chlorine evolution reaction show little dependence on the RuO_2 content of the coating over this range and the main effect results from changes in the electrolytic oxygen evolution reaction. This is illustrated in Fig. 13.7 where the oxygen overpotential is seen to reduce as the concentration of RuO_2 increases.

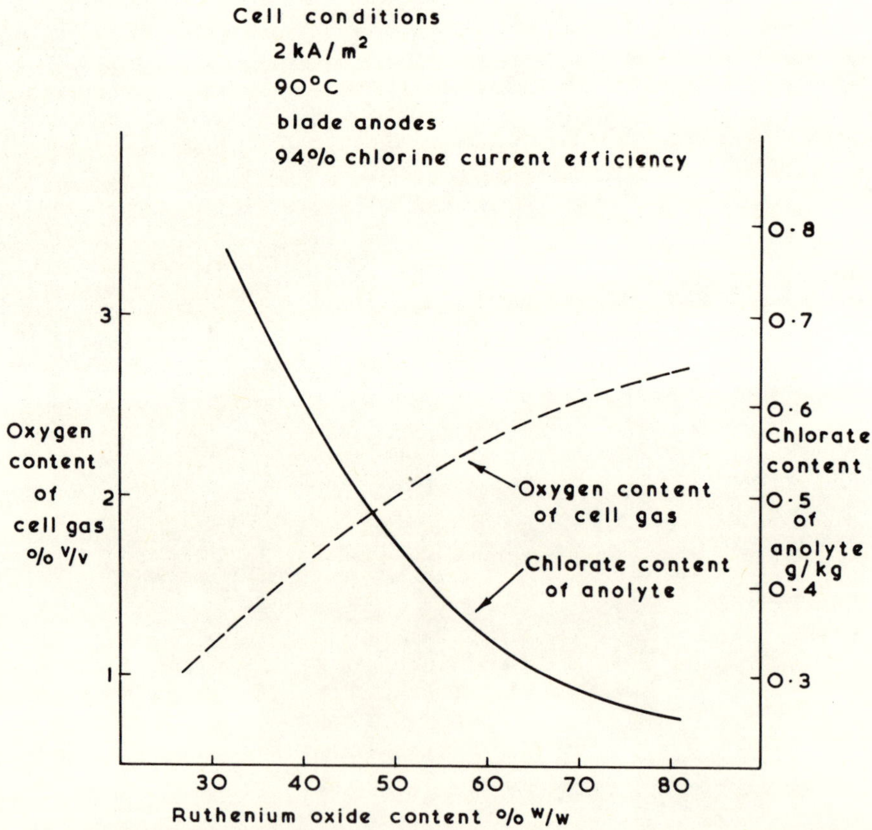

Fig. 13.6 – Diaphragm cell – effect of ruthenium dioxide content of Mond 1 coating on oxygen and chlorate levels.

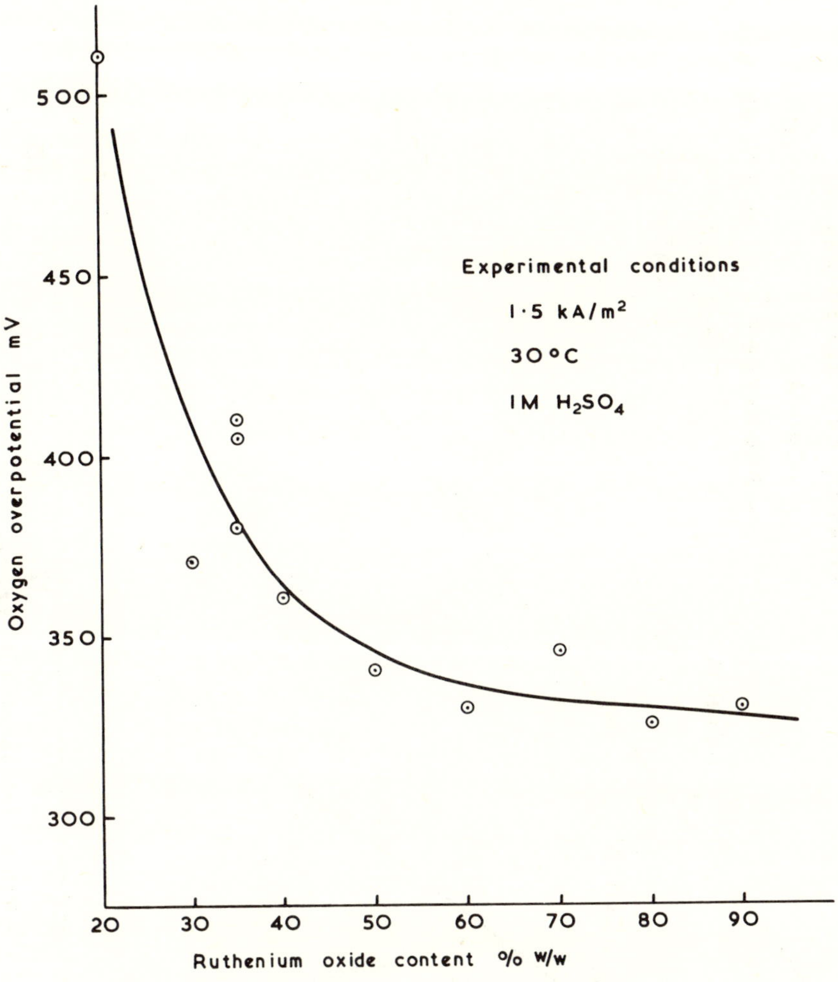

Fig. 13.7 – Effect of ruthenium dioxide content of Mond 1 anode coating on oxygen overpotential.

An increase in the proportion of RuO_2 in the coating has the undesirable effect of increasing its consumption rate and hence its useful life. This effect can be countered by suitable choice of the total RuO_2 loading and the coating preparative conditions. The consumption pattern for the Mond 1 coating formulation containing 35% RuO_2 is shown in Fig. 13.8. A life in excess of 10 years is indicated by extrapolation.

The shape of the anode structure also influences the proportion of oxygen in chlorine, as shown in Fig. 13.9. Under corresponding conditions of surface area

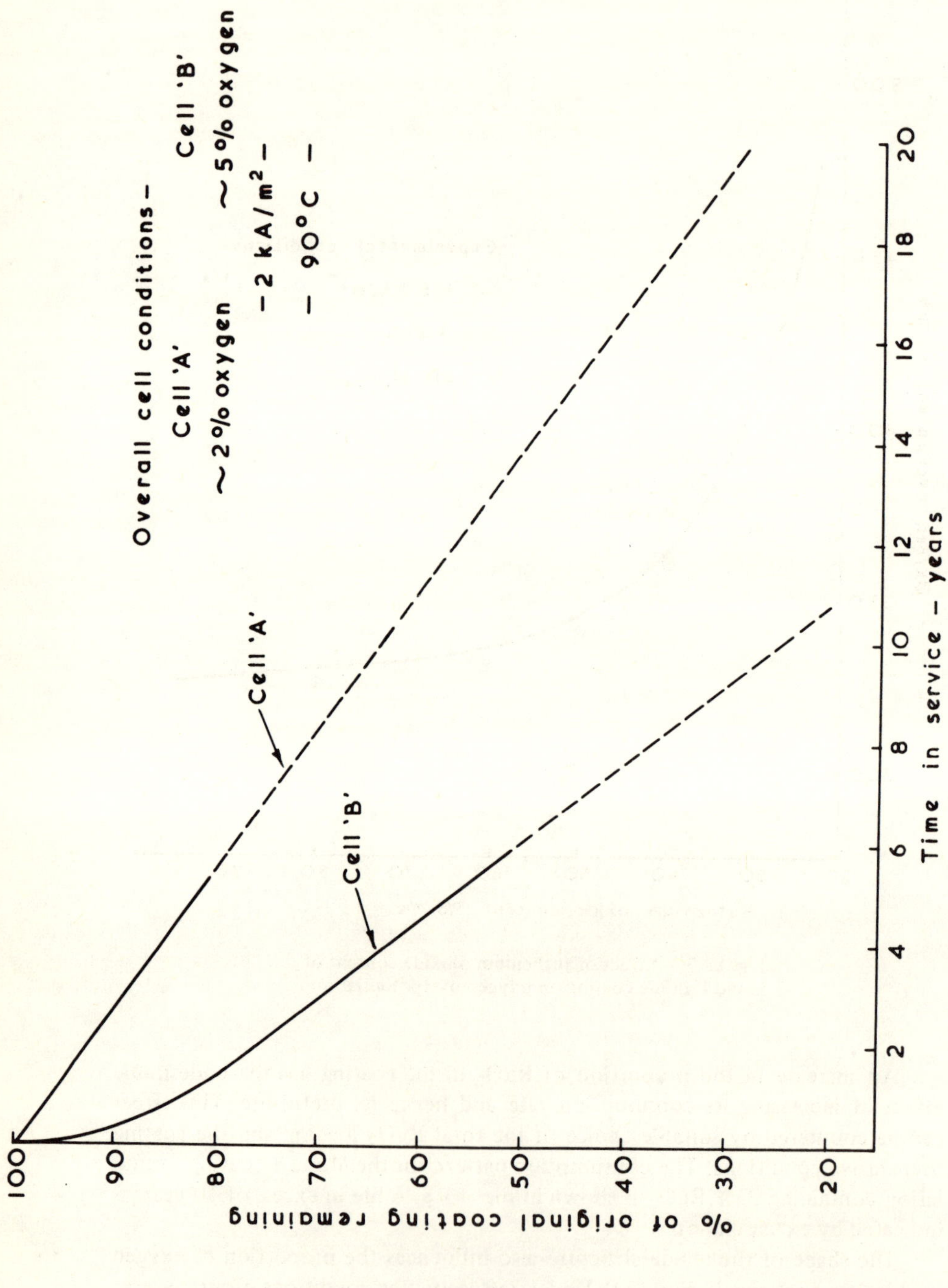

Fig. 13.8 – Pattern of coating loss for Mond 1 coating (35% w/w RuO_2)

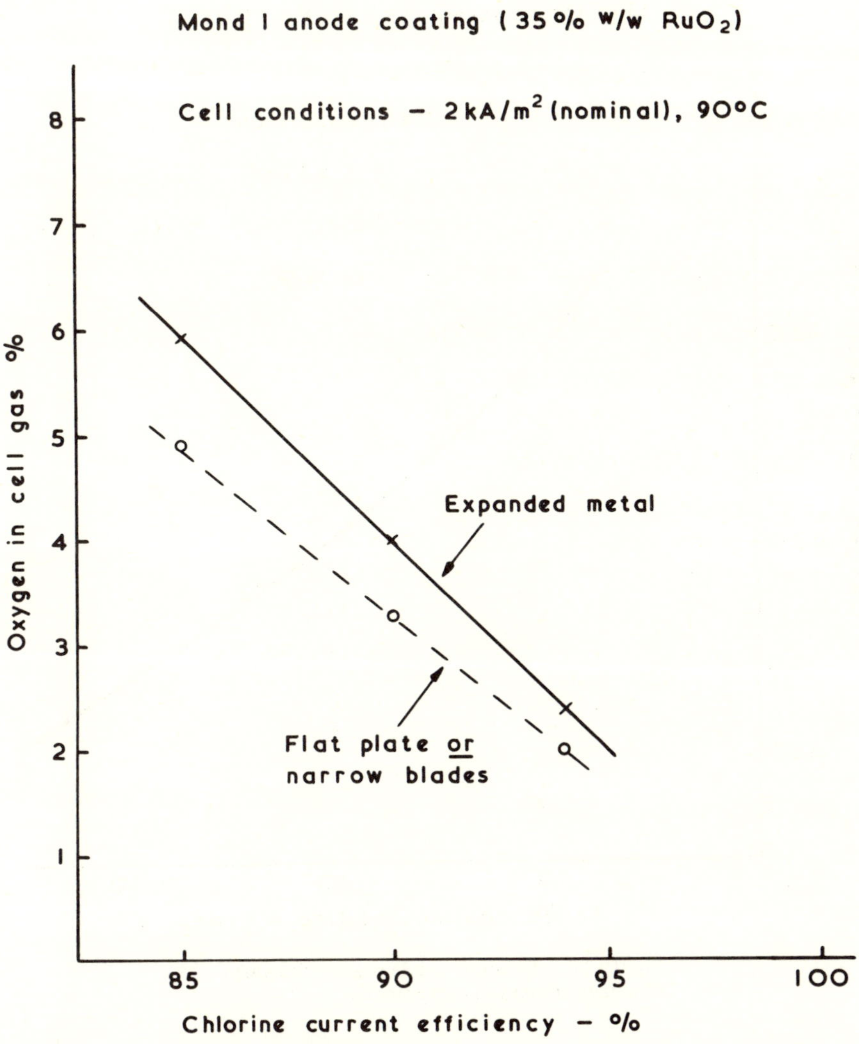

Fig. 13.9 – Effect of nature of anode substrate on the level of oxygen in diaphragm cell chlorine gas, and relationship to chlorine current efficiency.

and other parameters, an expanded metal structure produces more oxygen than a flat plate or narrow blade structure. This effect may be related to the evenness of coating achievable with these different structural forms since it has been found that this parameter influences the kinetics of the oxygen evolution reaction. This is illustrated by the oxygen overpotential measurements shown in Fig. 13.10.

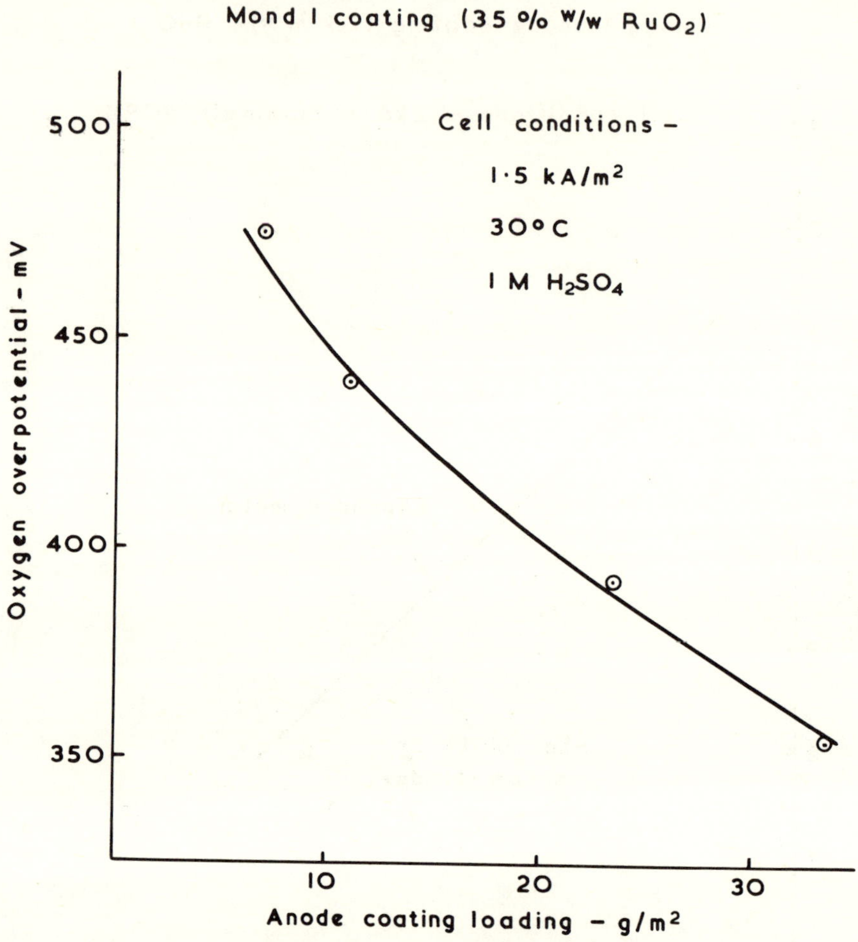

Fig. 13.10 – Effect of anode coating thickness (loading) on oxygen overpotential.

In addition to these effects, other electrolysis parameters may influence the balance of products from side reactions and the effects of pH, current density, anolyte temperature and salt concentration are indicated in Table 13.3. Performance data for the Mond 1 coating in diaphragm cells are summarized in Table 13.4.

13.4 ANODE COATINGS FOR MEMBRANE CELLS

Developments in the last 5 years have mainly been concerned with membrane cells and the latest ICI design is illustrated in Fig. 13.11. The standard coating for membrane cells is at present the Mond 1 coating which has a predicted life of

Table 13.3

Diaphragm and membrane cells – effect of changes in cell operating conditions on balance of anolyte reactions

Nature of change	Effect of change in levels of:		
	Oxygen in chlorine	Chlorate	Dissolved (available) chlorine Cl_2, HOCl, OCl^-
Increase OH^- level entering anolyte (thin diaphragm, more alkaline brine, etc.)	Increase	Increase	Increase
Reduce cell temperature	Reduction	Reduction	Increase
Reduce anode current density	Increase	Increase (small)	Reduction
Reduce anolyte sodium chloride concentration	Increase	Reduction	Increase (small)

Table 13.4

Summary of performance of Mond 1 coating in diaphragm cells (ICI DMT cell, 90°C, 2 kA/m^2 (nominal), anolyte 21% w/w NaCl)

Formulation % w/w RuO_2	Projected life at 2 kA/m^2 years	Levels of side products at 94% chlorine current efficiency	
		Oxygen in chlorine % v/v	Chlorate in anolyte g/kg
35	>10	1.4	0.70
60	>5	2.3	0.35

Life and levels of side products depend upon coating preparative conditions, and on cell design and operating factors.

Levels of side products are given for the same cell system.

Fig. 13.11 – The ICI-FM21 membrane cell.

6 years in the FM21 cell. This is less than in the diaphragm cell owing to the more demanding electrolysis conditions. Current trends in cell design are towards minimum electrode separation for which duty other coating formulations such as the Mond 5a type are seen to have potential advantages. As indicated in Table 13.5 electrolysis conditions and cell geometric factors can significantly affect coating life and other performance criteria and it is usually necessary to establish these for each particular cell design under consideration and to match the coating formulation accordingly.

The influence of the electrolysis conditions on the life of certain RuO_2 coatings has recently been investigated in the laboratory using a radiochemical technique in which the rate of loss of ruthenium to anolyte and cell gas were

Table 13.5

Some factors affecting anode coating performance in membrane cells

1. Anolyte alkalinity
2. Localised high pH – damaged membrane
3. Gradient in NaCl concentration along anode between inlet and exit
4. Ratio $\frac{\text{electrode area}}{\text{liquor volume}}$
5. Cell geometry (e.g. anode–membrane contact)
6. Anode design
7. Gas disengagement pattern

studied under both diaphragm and membrane cell conditions. The results obtained are summarized qualitatively in Table 13.6. An interesting finding was a direct relationship between ruthenium loss rate and the oxygen content of chlorine, which is shown in Fig. 13.12.

Table 13.6

Diaphragm and membrane cells – effects of changes in cell operating conditions on anode coating wear rate

	Effect of change in level of ruthenium found in:		Overall effect on anode coating wear rate
	Cell gas	Anolyte	
Increase alkali level entering anolyte	Increase	Increase	Increase
Reduce cell temperature	Reduction	Little change	Reduction
Reduce anode current density	Reduction	Increase	Increase
Reduce anolyte sodium chloride content	Increase (small)	Little change	Increase

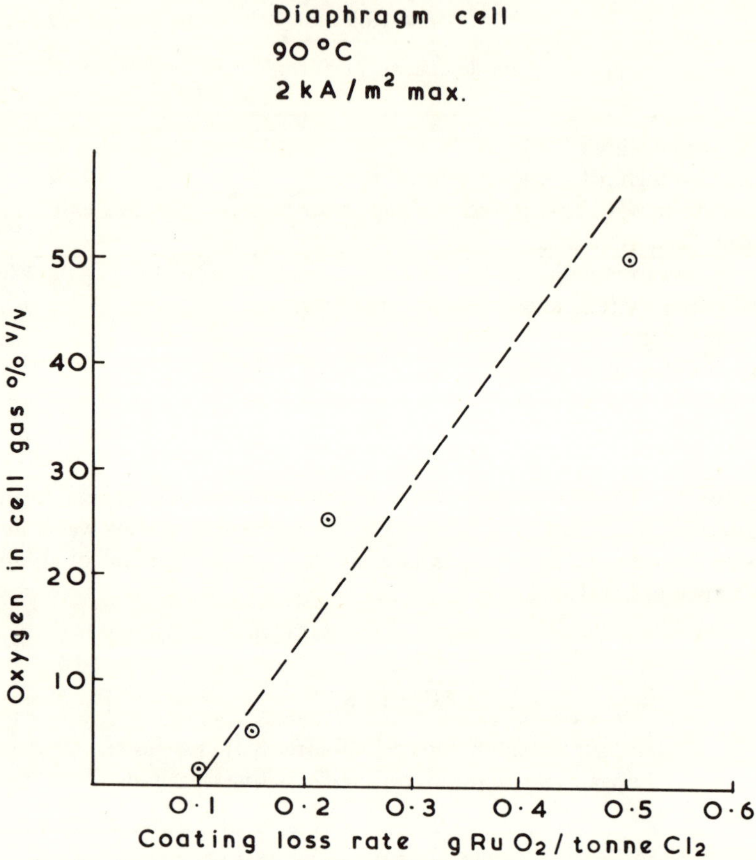

Fig. 13.12 – Relationship between anode coating loss rate (by radiochemical measurement) and level of oxygen in cell gas. Mond 1 coating (35% w/w RuO_2).

13.5 CONCLUSIONS

The ICI anode coating technology originally developed for mercury cells has proved to be directly transferable to diaphragm and membrane cells and all three types of anode are now coated and recoated at the Northwich plant. This plant now supplies anodes on a worldwide basis and can adapt the coating preparative conditions to meet the performance criteria required for most designs of cell.

Many of the factors affecting anode coating life and gas consumption in chlorine cells have been identified but considerable work remains to be done to quantify their influence fully.

CHAPTER 14

The Development of Low Overvoltage Cathodes

D. E. BROWN, P. O. FOGARTY, M. N. MAHMOOD, A. K. TURNER,
BP Research Centre, Sunbury-On-Thames, Middlesex, England

14.1 INTRODUCTION

The development by BP of a low cost electrocatalyst for hydrogen evolution has been reported previously. This earlier work showed that coated electrodes had a very low hydrogen overvoltage, and a long operating life in alkaline water electrolysis [1, 2]. The high intrinsic activity of the catalyst is attributed [3] to the presence of a face centred cubic nickel–molybdenum alloy in the coating. More recently the coating process has been scaled up and the electrocatalyst has been successfully applied to substrates of 1 m^2 area.

In this paper we report the preliminary results of a series of tests to investigate the suitability of the BP electrocatalyst in chor-alkali cells. Lowering the overvoltage for hydrogen evolution at the cathode would significantly reduce the power consumption of both diaphragm and membrane-type cell. The results described below do not refer to tests in operating diaphragm or membrane cells, but were determined by measurements of cathode half-cell potentials. A range of conditions have been examined. Firstly, coated mild steel electrodes were operated in 15% NaOH–17% NaCl electrolyte using an oxygen evolving (Pt) anode to study possible attack of the electrode by chloride ions. The second test looked at the effect on electrode performance of an environment containing hypochlorite ions, such as might be experienced during interruptions in diaphragm cell operation. In the third test an asbestos sheet was soaked in a NaOH–NaCl solution and baked on to the coated electrode to simulate the conditions used in the application of polymer reinforced diaphragms. A further series of experiments was carried out to study the electrode performance in NaOH electrolyte to determine the suitability of the BP electrocatalyst for use in membrane cells.

14.2 EXPERIMENTAL TECHNIQUES

The electrocatalyst coating was applied to the substrate using a spray pyrolysis technique. Test pieces of perforated mild steel plate were successively coated with an aqueous solution of nickel and molybdenum salts and the coatings was then thermally decomposed to an oxide mixture. This procedure was repeated until the desired loading has been achieved. The oxide mixture was then heat-treated in a hydrogen atmosphere to produce the active electrocatalyst. Details of the electrocatalyst preparation are published elsewhere [4].

Long term tests of coated cathodes have been carried out in a series of electrolytes. In all these experiments the anodic reaction was oxygen evolution; this enabled a simple facility to be established for long term operation. Plastic cells containing the electrolyte were maintained at 70°C and had continuous distilled water top-up to replace water lost by evaporation and electrolysis. The anode was contained in a separate compartment in the cell to prevent mixing of the evolved gases. Nickel mesh anodes were used in the tests using NaOH electrolyte; platinum foil anodes were employed for experiments in the NaOH–NaCl electrolyte. Throughout these studies the saturated calomel electrode was used as the reference electrode; it was calibrated against the reversible hydrogen electrode (Pt/H_2) for each electrolyte. The electrode potentials were IR corrected using the interruptor technique and therefore represent true overvoltages for hydrogen evolution under the specified conditions. The polarization curves were determined by measuring the steady-state electrode potential over a range of current densities.

14.3 PERFORMANCE OF THE BP ELECTROCATALYST

14.3.1 Hydrogen evolution in NaOH–NaCl electrolyte

Figure 14.1 shows the performance of a coated electrode operated continuously at 250 mA cm^{-2} for 2200 h in 17% NaOH–15% NaCl (70°C). The overvoltage for hydrogen evolution was approximately 65 mV and its activity did not change significantly over the test period. Polarization curves measured after 264 h and

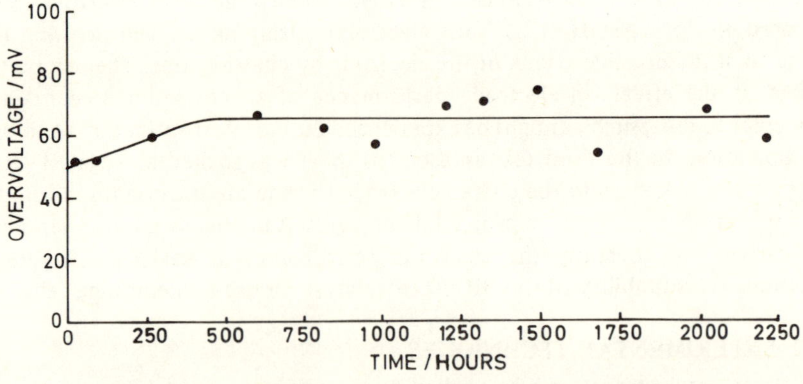

Fig. 14.1 – Long term test in 17% NaOH–15% NaCl (70°C) at 250 mA cm^{-2}.

1680 h electrolysis also indicate that there was no significant change in the electrode performance over this period (Fig. 14.2).

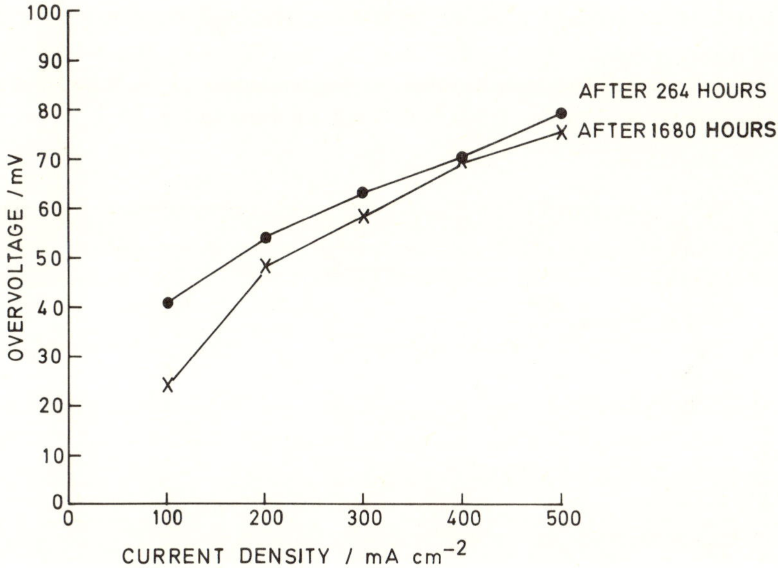

Fig. 14.2 – Polarization curves in 17% NaOH–15% NaCl (70°C).

In these experiments using NaOH–NaCl electroyte, a platinum anode was employed (nickel anodes were used in all other tests). The possibility of contamination of the cathode by anodically dissolved platinum was investigated by X-ray photoelectron spectroscopy (XPS or ESCA). After electrolysis for 240 h, no trace of platinum was detected at the electrode surface. This implies that the behaviour of the operating cathode was not influenced by the use of platinum anodes in the cells.

14.3.2 The effect of hypochlorite on coated electrodes

The possibility of corrosive attack of an unpolarized coated electrode by hypochlorite ions was investigated using the following procedure. The electrode, coated with the electrocatalyst, was operated at 250 mA cm^{-2} in 15% NaOH (70°C) for 24 h; it exhibited a very low hydrogen overvoltage of 70 mV. It was then removed from the cell and immersed for 2 h is a freshly prepared solution of sodium hypochlorite (2 gl^{-1}) at pH 9. The electrode was then returned to the cell and run for a further 24 h. After this time the overvoltage had risen to 95 mV. This procedure of electrolysis for 24 h followed by immersion in hypochlorite solution was repeated three times. The overvoltage dropped to 72 mV after the second dip treatment, then measured 81 mV and 74 mV after subsequent dips;

this is very close to the initial value. Thus, there appears to be no significant change in the activity of the electrode after this series of treatments, compared to its initial performance. After further continuous operation at 250 mA cm^{-2} a steady overvoltage of about 70 mV was observed for a further 450 h. The electrode potential then rose slightly to about 99 mV, after a total of 1100 h, and remained at this plateau value for the remainder of the test (up to 1600 h). The behaviour of the electrode is shown graphically in Fig. 14.3.

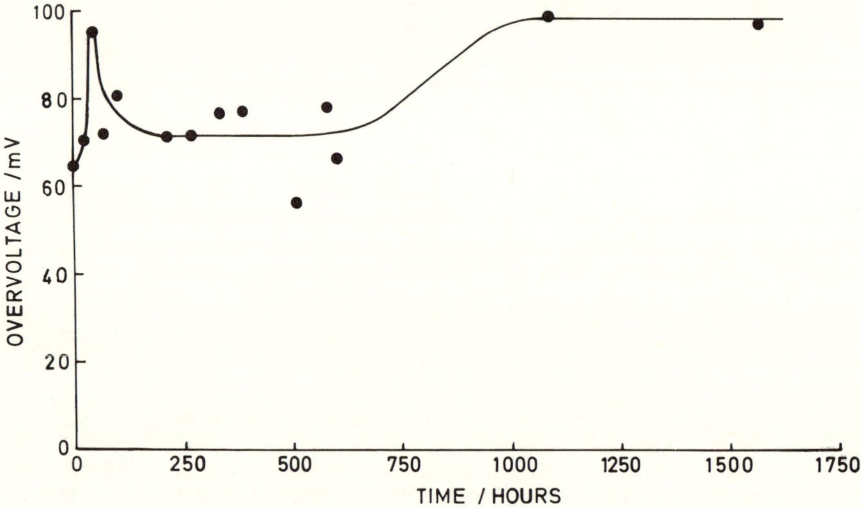

Fig. 14.3 – Immersion in pH 9 hypochlorite solution after 24, 48, 72 and 96 h. Long term test in 15% NaOH (70°C) at 250 mA cm^{-2}.

Subjecting the coated electrode to this procedure of dipping in pH 9 sodium hypochlorite solution appears to have produced only a slight deterioration in the electrode performance after about 1600 h operation. Extrapolation of these test results to give an indication of the expected behaviour of coated electrodes in a diaphragm cell under normal operating conditions, with power interruptions, is difficult. However, the laboratory results are encouraging.

14.3.3 Simulated diaphragm application

In cases where polymer reinforced asbestos diaphrams are used, it is important that any electrocatalyst coating applied to the cathode can withstand application of the diaphragm without significant loss of catalyst performance. The high temperatures used during the baking or curing step could be potentially damaging, especially since the electrocatalyst is non-noble metal based. In this work the conditions experienced by the electrocatalyst coating during the application of a polymer reinforced diaphragm have been simulated. The procedure used [5] was first to soak a 1-mm thick asbestos sheet in 15% NaOH–17% NaCl solution for 2 months to allow the asbestos to swell. It was then pressed onto the catalyst coated electrode, dried at 105°C for 24 h, then baked in air at 270°C for 24 h. In separate experiments, the asbestos layer was applied to just one side, and to both sides of the catalyst coated electrodes. It should be emphasized that the aim here was not to apply a functioning diaphragm to the electrode, but was merely to simulate the conditions to which the electrocatalyst would be subjected during diaphragm application.

In the first test, the asbestos sheet which had been baked onto just one side of the coated electrode became detached after only a few hours electrolysis. The cell was operated with the treated side of the electrode facing the Luggin capillary, using 30% NaOH electrolyte at 70°C. Initially the overvoltage at 250 mA cm^{-2} was 32 mV, rising over the first 200 h on test to 50 mV and remaining at this level for a further 1000 h (see Fig. 14.4). The polarization curve measured after 168 h is shown in Fig. 14.5. The electrode performance after applying the asbestos sheet is comparable to that shown in Figs. 14.1 and 14.2; the activity and long term stability of the electrocatalyst has not been affected.

Application of an asbestos sheet to both sides of the elctrode produced, after baking, a strongly adherent asbestos envelope around the test piece. The electrode was tested in 15% NaOH–17% NaCl electrolyte (70°C) at 250 mA cm^{-2}. For the first few hundred hours of the test the asbestos remained attached to the cathode and the measured overvoltage was 150 to 180 mV (see Fig. 14.6). When the asbestos became detached the overvoltage dropped to its expected value of about 70 to 80 mV and remained at this level. In this case, it also appears that the simulated diaphragming procedure has not affected the activity or long term stability of the electrocatalyst coating.

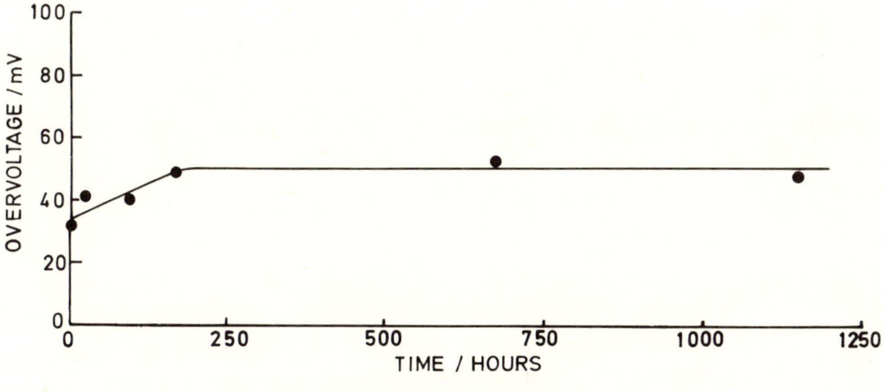

Fig. 14.4 – Asbestos sheet applied to one side of coated electrode. Long term test in 30% NaOH (70°C) at 250 mA cm^{-2}.

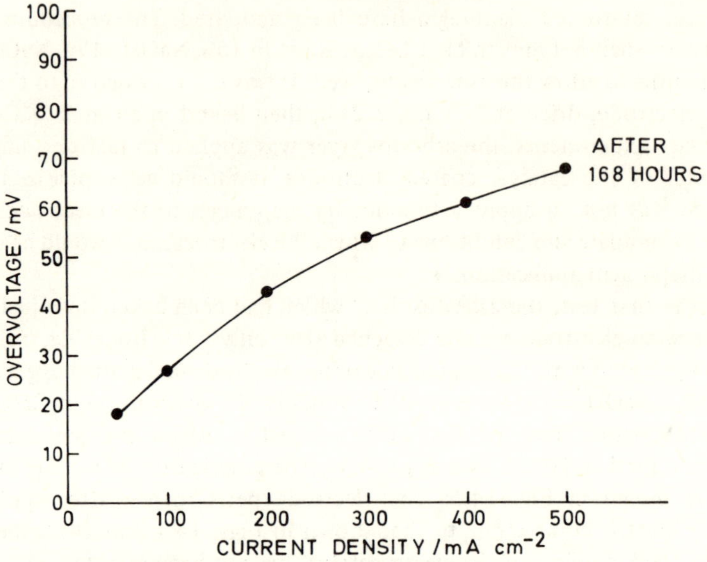

Fig. 14.5 – Asbestos sheet applied to one side of coated electrode. Polarization curve in 30% NaOH (70°C).

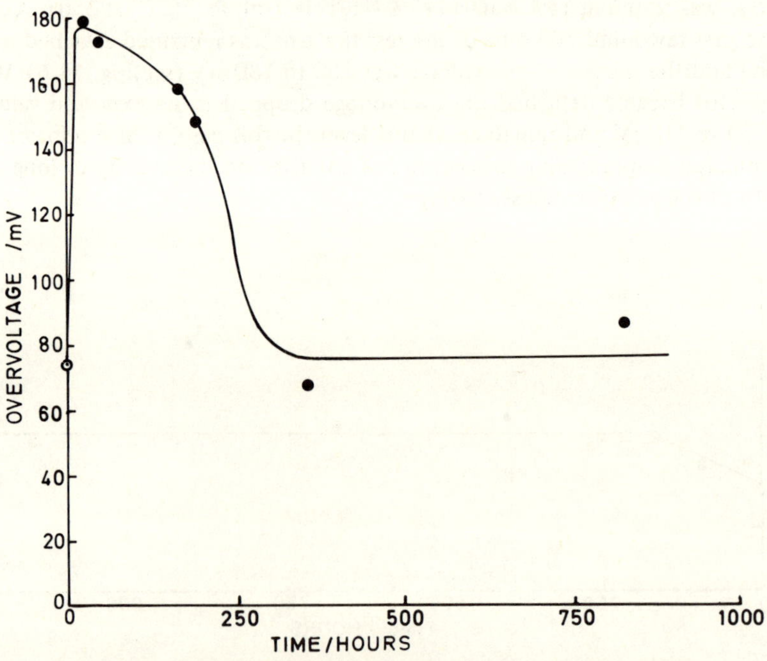

Fig. 14.6 – Asbestos sheet applied to both sides of coated electrode. Long term test in 15% NaOH–17% NaCl (70°C) at 250 mA cm^{-2}.

The susceptibility of the electrocatalyst to oxidation was studied by thermogravimetric analysis. A sample of the catalyst was scraped from a coated electrode and its weight change was monitored when heated in air at 20°C per minute. The results are shown in Fig. 14.7. The initial weight loss up to 225°C agreed quantitatively with the total water vapour content of the out-going air stream (determined by Karl Fischer titration). The sample started to gain weight at about 280°C and then steadily increased in weight with temperature. The results imply that the electrocatalyst is not oxidized at temperatures below about 280°C. This is consistent with the electrochemical evidence presented above indicating no loss of catalyst performance after the simulated diaphragming procedure which incorporated a baking step at 270°C in air.

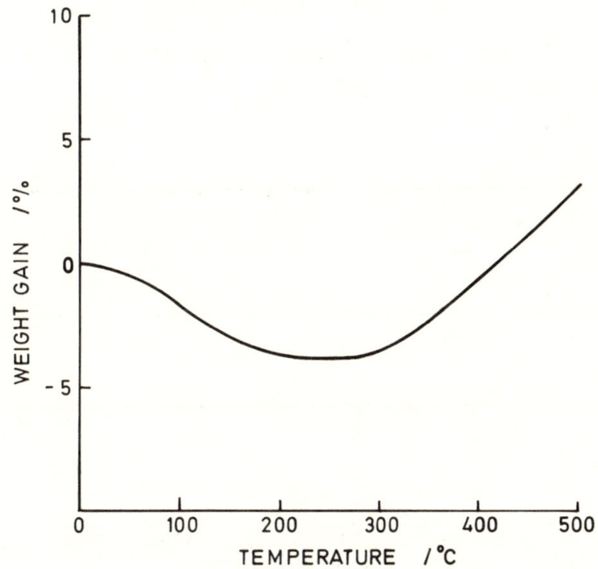

Fig. 14.7 – Thermogravimetric analysis of electrocatalyst in air.

14.3.4 Hydrogen evolution in NaOH electrolyte

Coated electrodes were operated in 15% and 30% NaOH electrolytes at current densities of 250 mA cm^{-2} and 500 mA cm^{-2}. In each case the hydrogen overvoltages were very low and their performance was maintained over the 2000 h test period. In 15% NaOH solution the overvoltage was between 60 mV and 80 mV at both current densities. In 30% NaOH the overvoltage was about 60 mV at 250 mA cm^{-2}, and about 80 mV at 500 mA cm^{-2}. The results are shown graphically in Figs. 14.8 and 14.9.

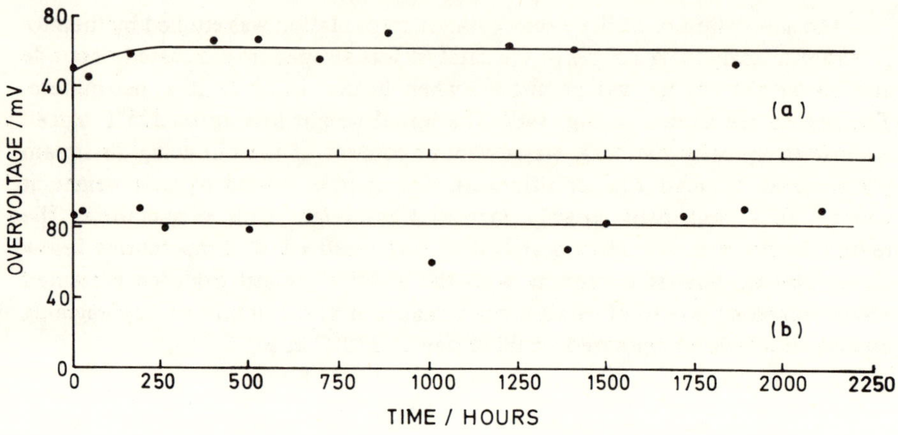

Fig. 14.8 – Long term tests in 15% NaOH (70°C):
(a) 250 mA cm^{-2}; (b) 500 mA cm^{-2}.

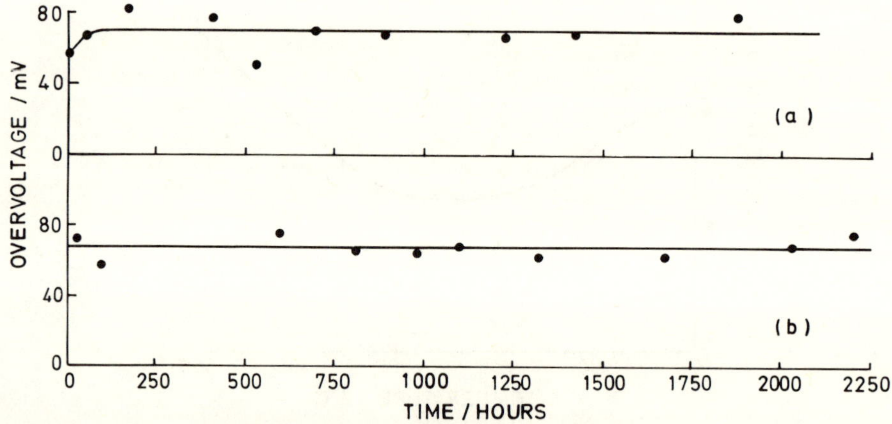

Fig. 14.9 – Long term tests in 30% NaOH (70°C):
(a) 250 mA cm^{-2}; (b) 500 mA cm^{-2}.

Polarization curves measured after 48 h and 888 h electrolysis in 15% NaOH at 250 mA cm^{-2} are shown in Fig. 14.10. The results are closely similar, confirming that there was no significant change in the activity of the electrocatalyst with time. In Fig. 14.11 the polarization curve is plotted on a logarithmic scale. In the region of typical operating current densities, the measured Tafel slope is 62 mV per decade. The low Tafel slope accounts for the small difference in the overvoltages recorded at 250 and 500 mA cm^{-2}.

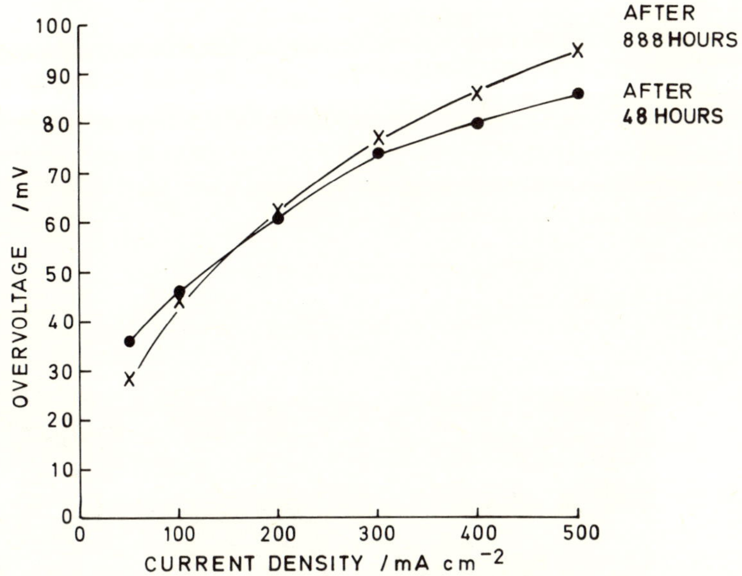

Fig. 14.10 – Polarization curves in 15% NaOH (70°C).

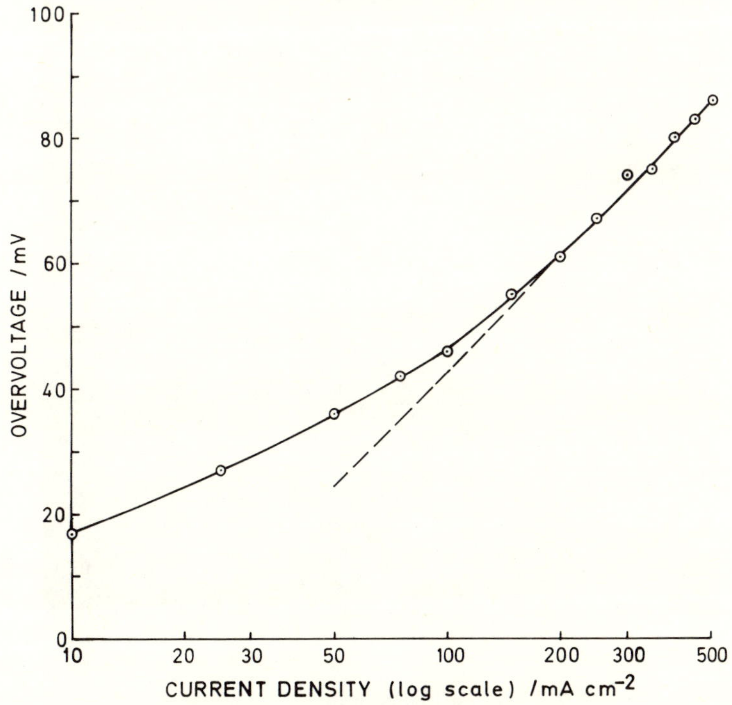

Fig. 14.11 – Tafel plot, 15% NaOH (70°C).

14.3.5 Electrocatalyst morphology

Scanning electron microscopy of the electrocatalyst coating reveals a fairly uniformly distributed layer with a 'mudcracked' appearance (Fig. 14.12). This morphology is typical of coatings produced via a thermal decomposition route. After 3000 h operation in 30% NaOH (70°C) at 500 mA cm^{-2}, scanning electron microscopy indicated that there were no significant changes in the morphology of the electrocatalyst coating (Fig. 14.13).

Fig. 14.12 – Scanning electron micrograph of the electrocatalyst, as prepared.

Fig. 14.13 – Scanning electron micrograph of the electrocatalyst, after 3000 h in 30% NaOH (70°C) at 500 mA cm^{-2}.

14.4 CONCLUSIONS

The results indicate that the BP electrocatalyst exhibits a very low overvoltage for hydrogen evolution in both NaOH and NaOH–NaCl electrolytes. Furthermore, there is no significant change in the overvoltage during operation for at least 2000 h (3 months), the duration of the tests so far. The electrocatalyst appears to be resistant to the conditions experienced in the application of polymer reinforced asbestos diaphragms, and during interruptions in diaphragm cell operation. In these cases, the results are based on a simulation of the worst conditions to which a coated electrode might be subjected.

The demonstrated performance of the BP electrocatalyst indicates the potential for considerable savings in the power consumption of both diaphragm and membrane chlor-alkali cells.

14.5 ACKNOWLEDGEMENTS

Permision to publish this paper has been granted by The British Petroleum Company plc.

The authors would like to thank Mr R. J. Howes and Mr D. Thornley for the TGA measurements, Dr D. Perry for the ESCA studies, Mr K. Meade for the scanning electron microscopy and Mr M. Williams for the manufacture of equipment.

14.6 REFERENCES

[1] Brown, D. E., Mahmood, M. N., Turner, A. K., Hall, S. M., and Fogarty, P. O., *Hydrogen Energy Progress*, **1**, 151 (1980); *Int. J. Hydrogen Energy*, **7**, 405 (1982).

[2] Murray, J. N., *Alkaline Solution Water Electrolysis: '81*, presented at 16th IECEC Mtg Atlanta, Georgia, 17th–21st June 1981.

[3] Brown, D. E., Hall, S. M., Mahmood, M. N., Man, M. C. M., Turner, A. K., Wood, D., and Anderson, S., *Proceedings of Symposium on Electrocatalysis Proc.*, **82–2**, 145 (1982); W. E. O'Grady, P. N. Ross, Jr, and F. G. Will, eds. Electrochemical Society.

[4] Brown, D. E., and Mahmood, M. N., *Electrodes and Method of Preparation Thereof for Use in Electrochemical Cells*, European Patent Application No. 79 301963.9 (1979).

[5] For example, UK Patents 1 410 313 and 1 498 733.

14.7 DISCUSSION

1. A question as to what is the fate of the molybdenum which is part of the alloy, particularly in open circuit conditions was asked by Dr R. D. Giles of Inco.

   Dr Turner replied that the molybdenum that is part of the alloy remains in the alloy in the coating for as long as the test proceeds. The alloy contains between 13 and 15% molybdenum but the initial starting solution contains a little more than this. Consequently, there is an excess of molybdenum in the coating. That excess is removed over the first two days of operation after which no further loss occurs. Indeed, numerous experiments have been undertaken where the electrolyte has been changed in the cell and there has been no loss of activity. The catalytic compound of the coating is the nickel/molybdenum alloy and that remains stable in the coating.

2. Prof. Fleischman asked whether there was a continuous series of solid solutions from nickel up to 13–15% nickel alloy with no phase changes thereafter.
   At the higher molybdenum contents there are a number of intermetallic phases which can form. The molybdenum is going into the alloy as a solid solution and is randomly substituted into the nickel positions in the lattice,

thereby expanding the lattice parameter of the material from about 3.52 to 3.58, was Dr Turner's reply.

3. An anonymous questioner asked whether any problems had been seen due to corrosion of the mild steel substrate if the coating was not uniformly applied.

   Dr Turner replied that no corrosion has been detected and had not interfered with any of the tests.

CHAPTER 15

Electrocatalysis and the Chlorine Evolution Reaction

J. A. HARRISON, Department of Physical Chemistry,
University of Newcastle upon Tyne

15.1 INTRODUCTION

Electrocatalysis has always been a difficult subject for electrode kinetic investigation, because of the dominant effect of the physical and chemical properties of the surface on the kinetics. This introduces an extra dimension into the problem, in addition to the usual ones of transport of reactants and products and the electroactive species to and from the surface. The investigation of gas forming reactions, in particular, presents some considerable difficulties. For a particular experimental configuration it is necessary to be able to calculate the ideal result, say a stationary current–potential or impedance–potential curve. Only by this means is it possible to compare with any certainty different electrocatalytic electrodes. It is also necessary to be able to measure the ohmic loss and to be able to control the bubble flow from the electrode, using a rotating disc electrode or a flowing solution system. The measurement of the electrode impedance is also essential to reinforce the interpretation of the steady *i–E* curve and to measure the double layer capacity of the electrode. Constant potential current–time transients may also help to decipher the time effects, due, for example to bubble changes with time. In this paper an attempt is made to describe a comprehensive approach to the problem of electrocatalysis. Experiments of this kind have been made possible by the development of extensive computer controlled instrumentation and data analysis.

The previous literature on the mechanism of the Cl_2 and O_2 evolution reaction is not extensive. It has been reviewed in previous publications from this group [1–4,6], and in two recent articles [7,8].

15.2 THE COMPUTER METHOD

Although there have been a number of attempts to computerize electrochemical methods, these all seem to have run into difficulties. I believe that the present system, described here, is the first successful and large-scale system for making and interpreting electrode kinetic measurements. The aim has been to develop a powerful tool for fundamental investigations of rate processes in electrode kinetics, and also to develop a system which could be used in industry.

Many methods have been developed for the study of the electrode kinetics of electrode reactions. Figure 15.1(a) shows the methods which have been chosen in the present work [5]. These have been chosen for their intrinsic usefulness. Under computer control all can be implemented equally easily. The electrical methods shown in Fig. 15.1(a) involve measuring and generating potential signals and time. This combination is eminently suitable for computer exploitation. However, experience shows that it is preferable to incorporate dedicated, well designed instruments where this is possible. Figure 15.1(b) shows a functional diagram of the organization of a system which will implement the methods of Fig. 15.1(a). In addition it is necessary to have a system of controlling the steady state diffusional state of the system, by a flow system or a rotating or other similar type of system. This can be reasonably primitive as the primary investigation tool in this system is the electrical response.

15.3 DATA HANDLING

A large amount of work has been carried out [9] on the calculation of various reaction schemes. Almost parallel developments have taken place in polarography [10], linear potential sweep methods [11, 12], double potential pulse methods [13], and in a.c. polography [15] and in the a.c. method [14, 16–20]. These are recent or representative examples. This field is now so extensive that it would need a longer article to cover it adequately. However the details of the calculations in electrode will not be described here, as they have been comprehensively discussed in a recent paper [5]. Figure 15.1(c) details the reaction types which are included in the present data handling system. These have been chosen to represent common reaction types and many other reaction schemes which lead to sets of linear equations can be calculated by a trivial extension of the present method. It is necessary to calculate both the steady state and the relaxation response. This is an aspect which has been neglected in some previous treatments. Also shown diagrammatically in Fig. 15.1(c) is the important step of efficiently comparing experiment and theory. In the present computer system this is carried out using well designed subroutines which minimise the difference between experiment and theory by varying selected parameters. In this method of approach it is the art of the researcher to select, from the total parameters list, the parameters which can sensibly be varied [5]. This method encompasses all the special methods of electrode kinetics [5]. In this work it has not been necessary to employ the full capabilities of the system as shown in Figs. 15.1(a), (b) and (c). The most powerful methods have turned out to be the steady state

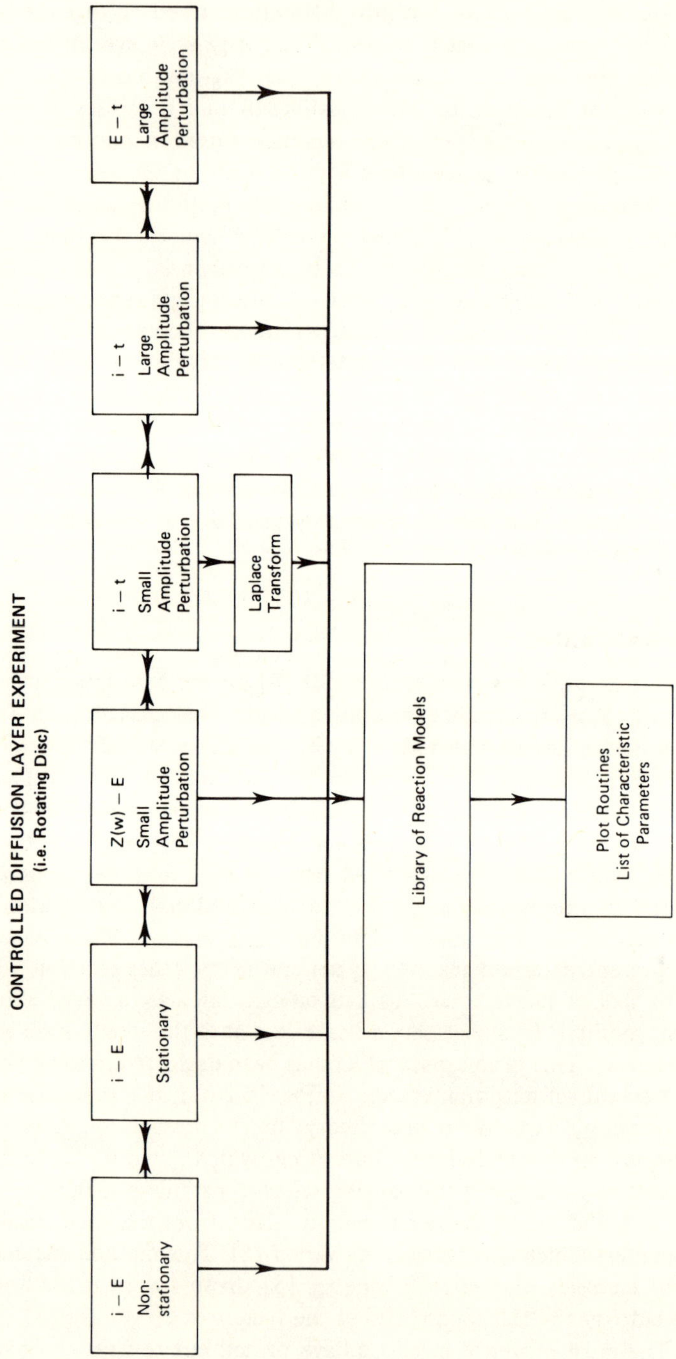

Fig. 15.1(a) – The electrochemical methods which are available in the computer controlled investigation system.

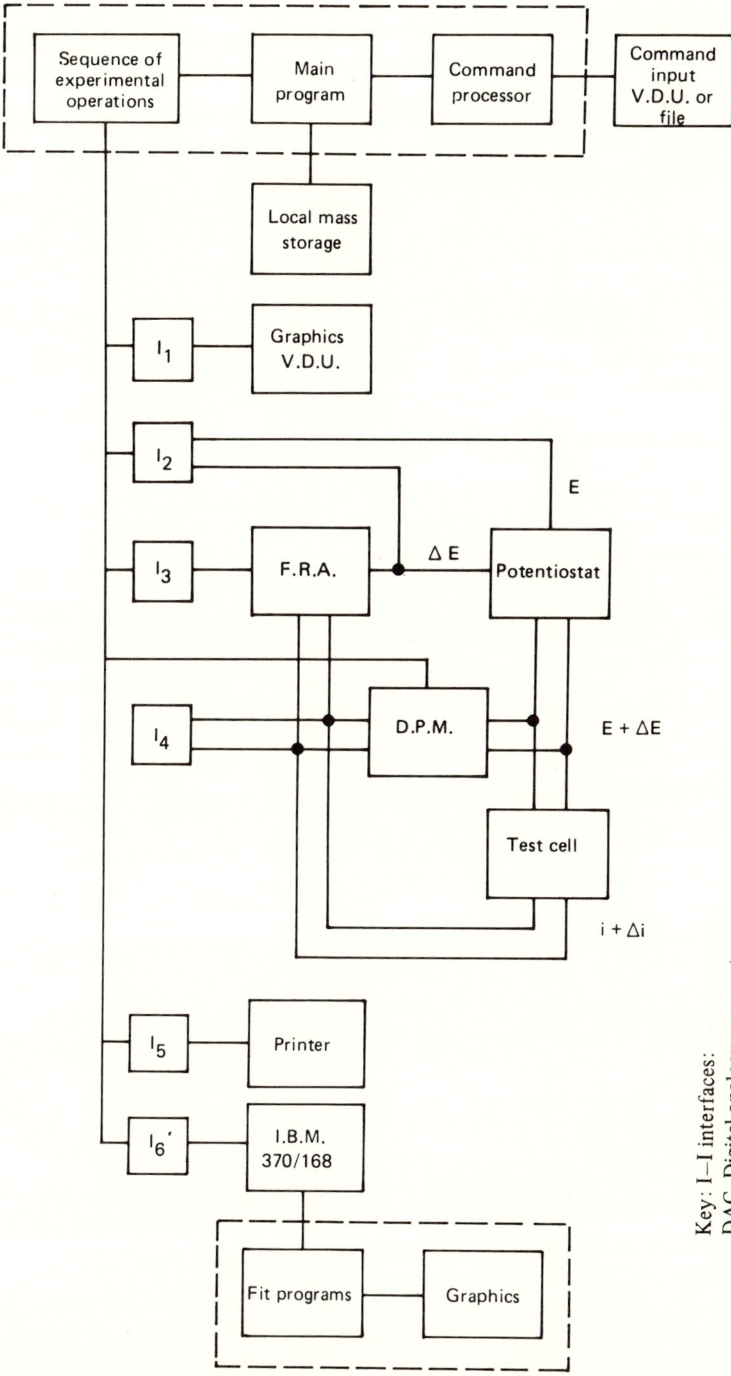

Key: I–I interfaces:
DAC Digital-analogue convertors
ADC Analogue-digital convertors
DPM Digital voltmeter
FRA Frequency response analyser, S.E Labs SM 2001 or more recently SE 2450.

Fig. 15.1(b) – The layout of the system used to implement the methods of Fig. 15.1(a). The dashed boxes enclose software functions.

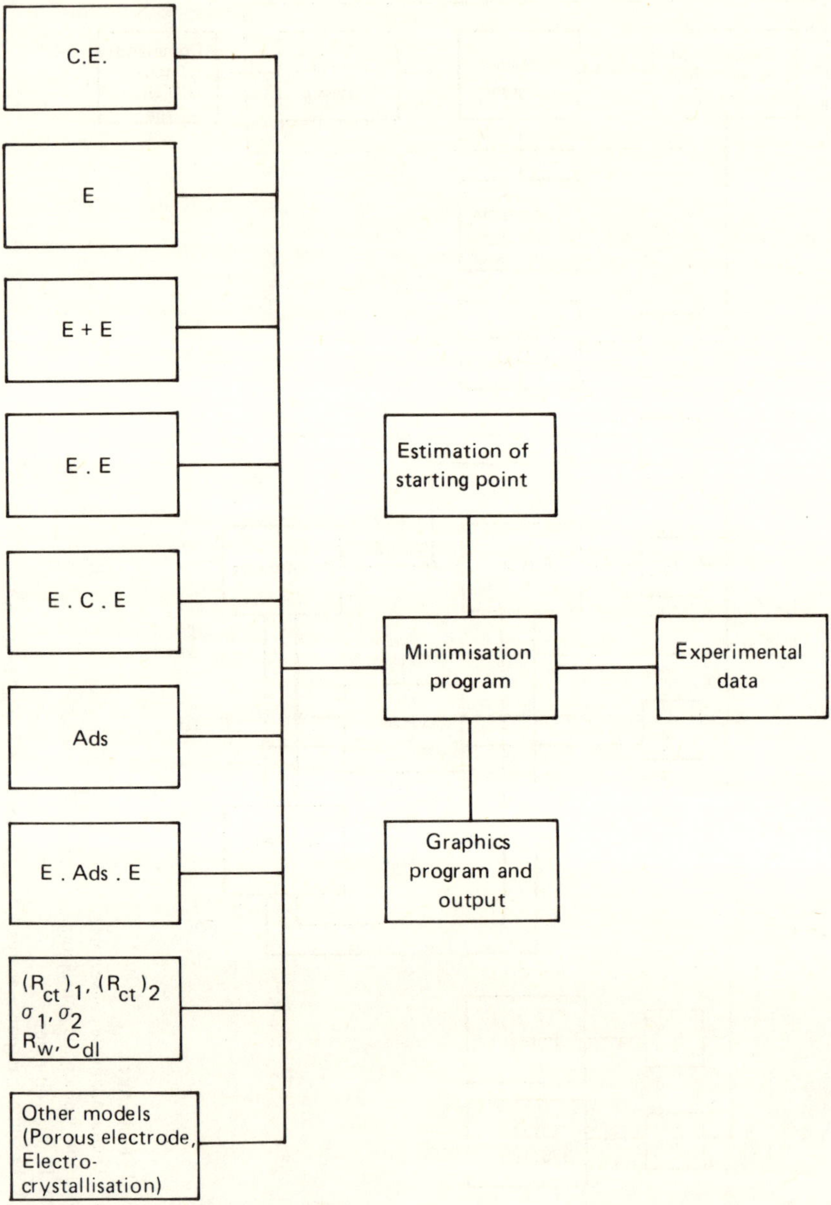

Fig. 15.1(c) – Some examples of electrochemical reaction schemes [5] which can be immediately analysed and handled by the available software. The analysis is based on the steady state (i–E) and impedance ($Z(E)$–E) calculations although it could be extended to include other experimental methods.

i–E curve, the *Z*(*w*)–*E* curve and the shape of the individual *Z*(*w*) diagrams, and the separation of the impedance into the components, as the box in Fig. 15.1(c) and as dicussed in section 15.8 Appendix.

The full capabilities of the methods have not yet been applied to the chlorine reaction, for which a comparative analysis based on the E, C.E, E. Ads. E, reactions would be most appropriate. In this work the main method of data analysis has been based on E reaction type (see the discussion of *i–E* curves in section 15.8 Appendix) and on the equivalent circuit analysis marked (R_{ct}), etc. in Fig. 15.1(c).

15.4 CHLORINE FORMATION

15.4.1 Manual results

The previous investigations [1–4,6] of the electrode kinetics of the Cl^- oxidation reaction have been carried out at the rotating disc and at a horizontal plate electrode. Most well defined experiments were at the rotating disc electrode, and at currents less than 43 mAcm^{-2}. This value was determined by the solubility of the Cl_2 and the diffusion layer thickness $\delta = 2.3 \times 10^{-3}$cm. With this configuration it is easily possible to calculate the optimum *i–E* curves for the oxidation. Figures 15.2(a), (b) and (c) show sample curves. These constitute the mark against which

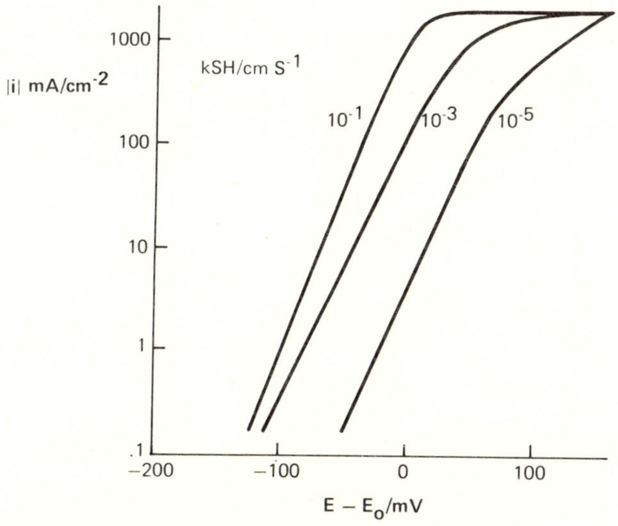

Fig. 15.2(a) – Theoretical *i–E* curves, calculated assuming that the Cl reaction behaves like a redox reaction: $A + 2e = B$ for the listed parameters:

$C_A^b = 0$; $C_B^b = 10^{-3}$/mol cm^{-3}

$D_A = D_B = 10^{-5}$/cm^{2}s^{-1}; $\delta = 10^{-3}$/cm

$a = 120$; $b = 40$/mV

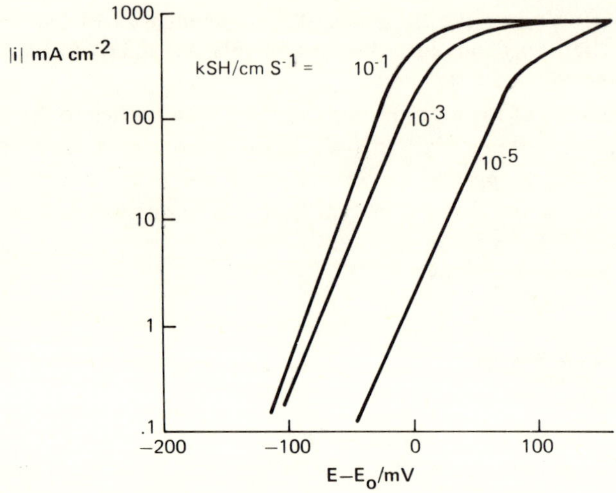

Fig. 15.2(b) – Theoretical i–E curves, calculated assuming that the Cl reaction behaves like a redox reaction: $A + 2e = B$ for the listed parameters:

$C_A^b = 0$; $C_B^b = 10^{-3}$/mol cm^{-3}

$D_A = D_B = 10^{-5}$/cm^2 s^{-1}; $\delta = 2.3 \times 10^{-3}$/cm

$a = 120$; $b = 40$/mV

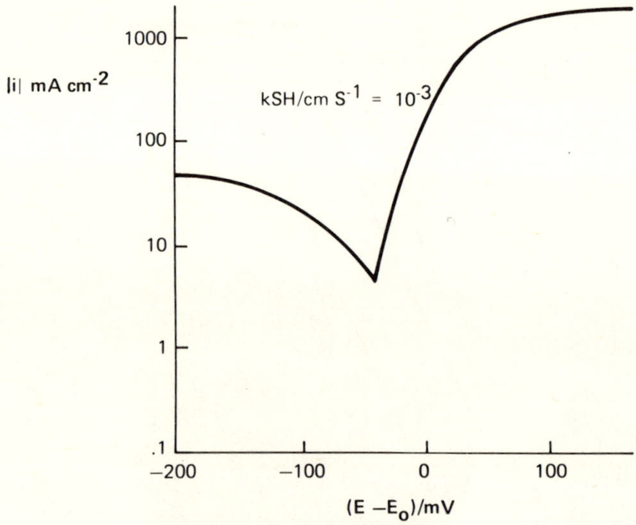

Fig. 15.2(c) – Theoretical i–E curves, calculated assuming that the Cl reaction behaves like a redox reaction: $A + 2e = B$ for the listed parameters:

$C_A^b = 3 \times 10^{-3}$; $C_B^b = 10^{-3}$/mol cm^{-3}

$D_A = D_B = 10^{-5}$/cm^2 s^{-1}; $\delta = 10^{-3}$/cm

$a = 120$; $b = 40$/mV

the success of an electrocatalyst can be judged. The curves at high rate are the 'reversible' curve, and curves for any conditions are rapidly calculated by computer for any condition.

Experimental *i–E* data are shown in Figs. 15.3(a), (b) and (c) for chlorine formation on smooth Pt, porous Pt and for RuO_2/TiO_2 rotating disc electrodes. Also shown in Figs. 15.3(a), (b) and (c) is the measured concentration dependence of the steady state *i–E* curves. As usual it is essential to measure the solution ohmic resistance from a high frequency impedance analysis (see Figs. 15.6(a)–(e)) and to make a correction to the applied potential to obtain the true potential.

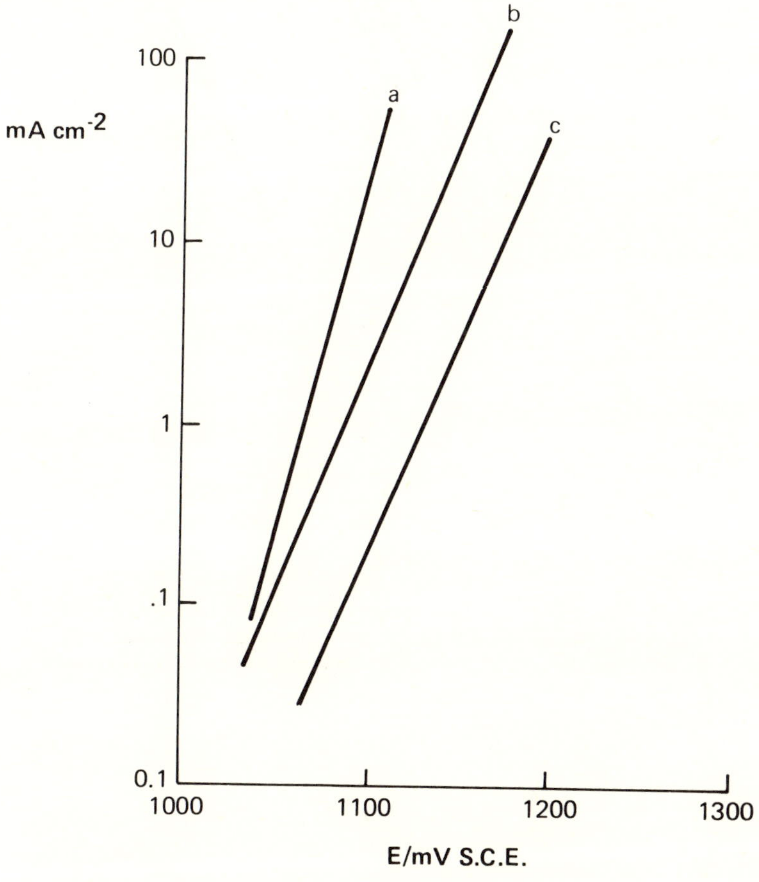

Fig. 15.3(a) – Experimental *i–E* curve on: 'a' porous Pt (7.5 Hz rotation speed); 'b' smooth Pt (7.5 Hz rotation speed); and 'c' RuO_2/TiO_2. The solution is 1 M HCl.

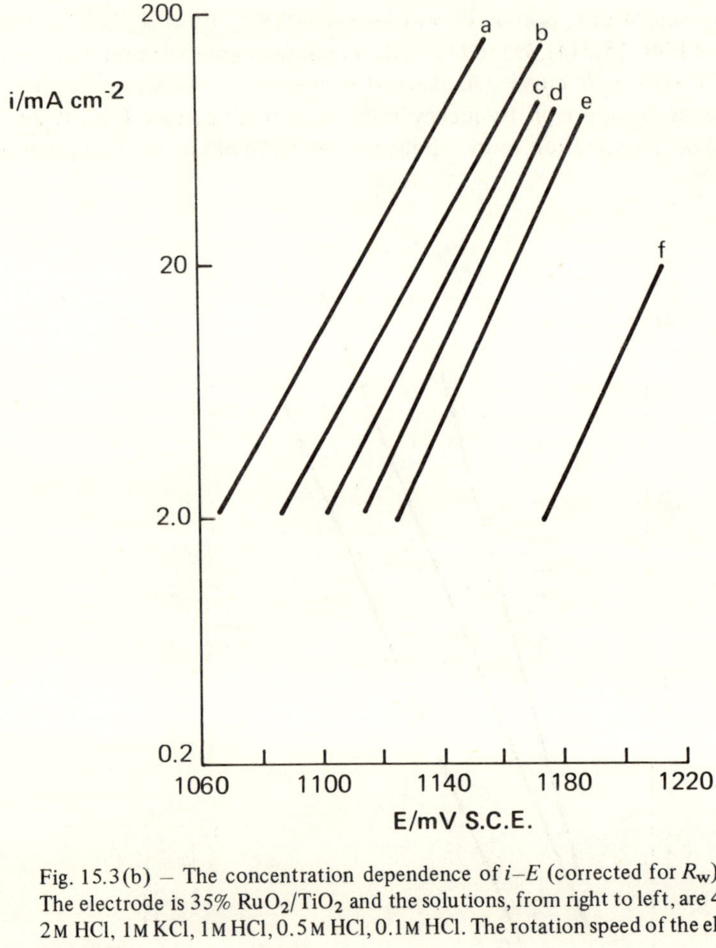

Fig. 15.3(b) – The concentration dependence of i–E (corrected for R_w) curves. The electrode is 35% RuO_2/TiO_2 and the solutions, from right to left, are 4 M HCl, 2 M HCl, 1 M KCl, 1 M HCl, 0.5 M HCl, 0.1 M HCl. The rotation speed of the electrode was 15 Hz.

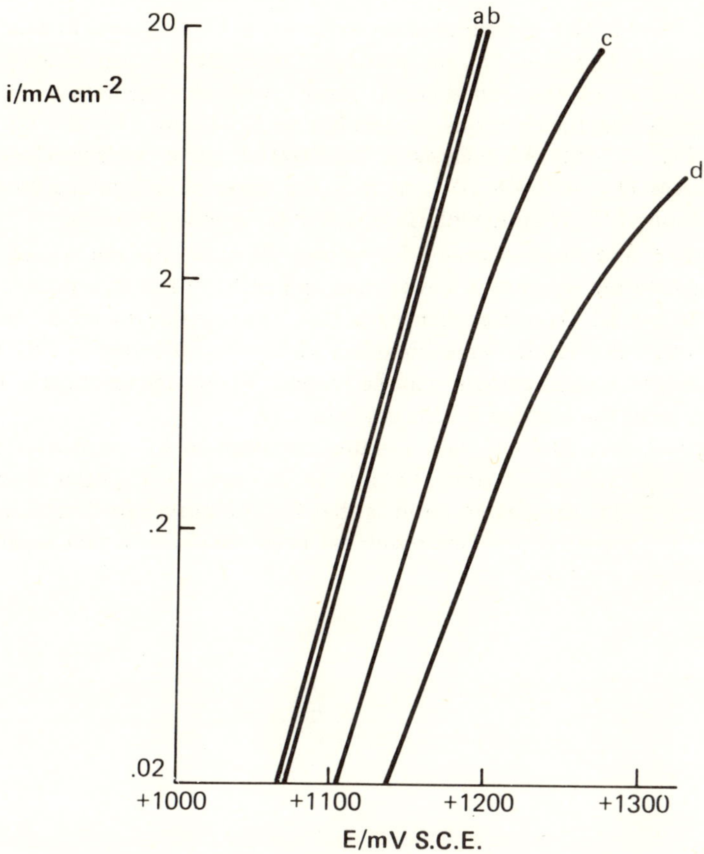

Fig. 15.3(c)— The composition dependence of the *i–E* curves: 'a' 80%; 'b' 30%; 'c' 10%; 'd' 5%. The rotation speed of the electrode was 7.5 Hz.

Figures 15.4(a), (b) and (c) show examples of the $Z(w)$–E data for similar conditions to those of Figs. 15.3(a), (b) and (c). The interpretation of the data at this point is as follows:

(1) The i–E and $Z(w)$–E results for the oxidation of Cl^- on porous Pt indicate 'reversible' behaviour for the particular experimental configuration. A comparison of the experimental results with the theoretical curves indicates that the electrode process has an E_0 of some 1130 mV S.C.E. (see Fig. 15.3(a)). An analysis of the $Z(w)$–E curve (see Fig. 15.4(a)) indicates $D_{Cl_2} = 7.4 \times 10^{-6} cm^2 S^{-1}$. The curve is determined by the diffusion of Cl_2 from the interface, as the Cl^- is in large excess.

(2) The i–E and $Z(w)$–E results for smooth Pt show that the rate of the Cl_2 formation reaction is slower than on porous Pt and is controlled by the interfacial electron exchange reaction. The C_{dl} value is some $\mu F\ cm^{-2}$. Although the results were obtained where the influence of PtO was minimized a small amount may be present. When large amounts of PtO are present the reaction is certainly inhibited.

(3) The i–E and $Z(w)$–E data for the oxidation of Cl^- on RuO_2/TiO_2 electrodes indicates a slow reaction. In this case the C_{dl} value is some mF cm^{-2}. An unexpected result is that the i–E curves are first order in the Cl^- concentration. This result has been obtained on two separate occasions.

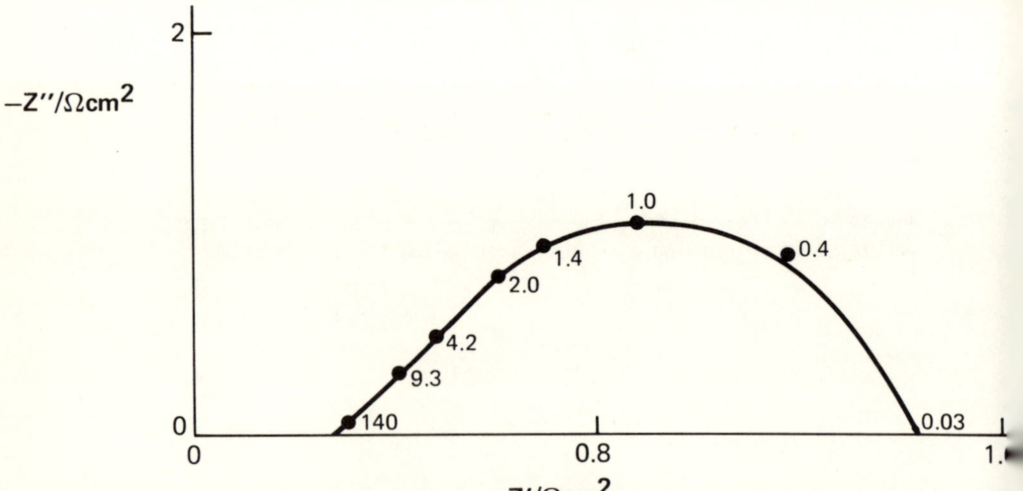

Fig. 15.4(a) – Example of $Z(w)$–E diagram for porous Pt in 1M HCl. The rotation speed of the eclectrode was 7.5 Hz.

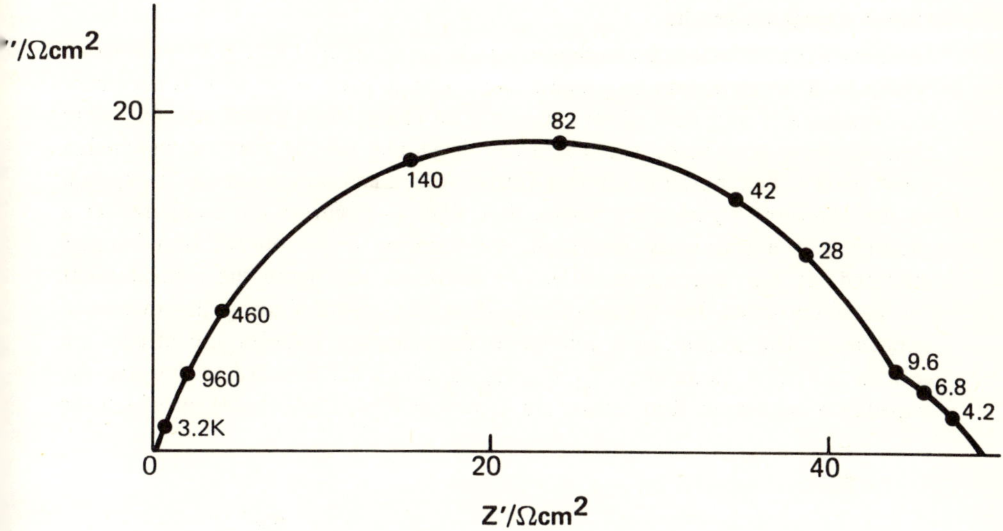

Fig. 15.4(b) – Example of $Z(w)$–E diagram for smooth Pt in 1M HCl. The rotation speed of the electrode was 7.5 Hz.

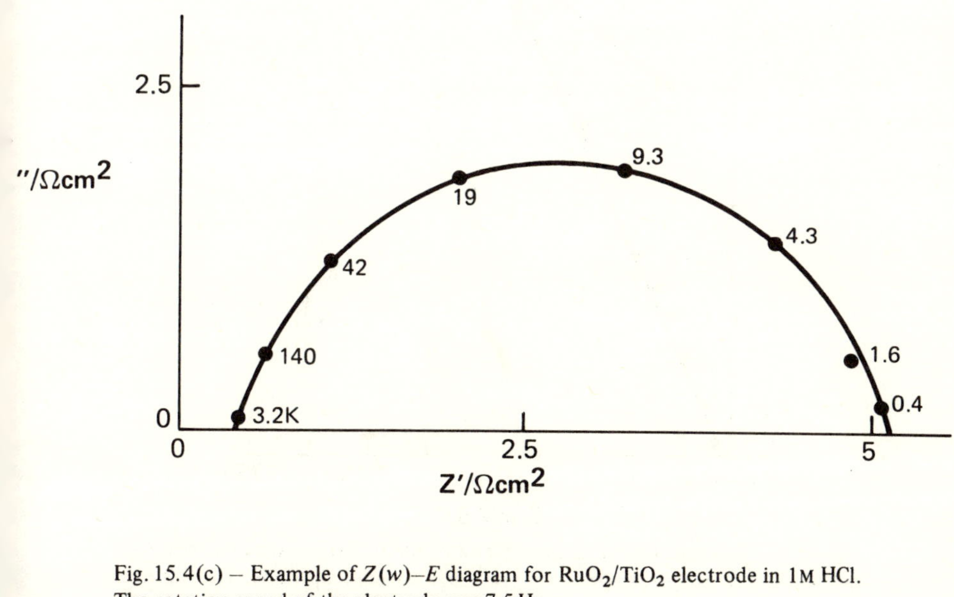

Fig. 15.4(c) – Example of $Z(w)$–E diagram for RuO_2/TiO_2 electrode in 1M HCl. The rotation speed of the electrode was 7.5 Hz.

15.4.2 Computer results

These have concentrated on characterising a larger number of RuO_2/TiO_2 electrode types. As it is convenient to prepare electrodes in plate form so it is necessary to investigate if any new effects appear if an electrode is stationary. The plate electrode have been investigated in a cell developed for the purpose, in which a 4 cm² plate is held horizontally in a Teflon holder, and only a portion is exposed to the solution. Figure 15.5 shows that when *i–E* curves are compared at a rotating and a stationary electrode, the chlorine evolution reaction beomes inhibited, at high current densities. As shown in the figure the measurement has been corrected for the ohmic drop in the solution using the measured impedance. The inhibition is related to the effect of bubbles and effects are seen in the other parameters C_{dl}, R_{ct}, R_w, σ, which can be used to describe the impedance behaviour. Run results are shown in Figs. 15.6(a)–(e), in which the

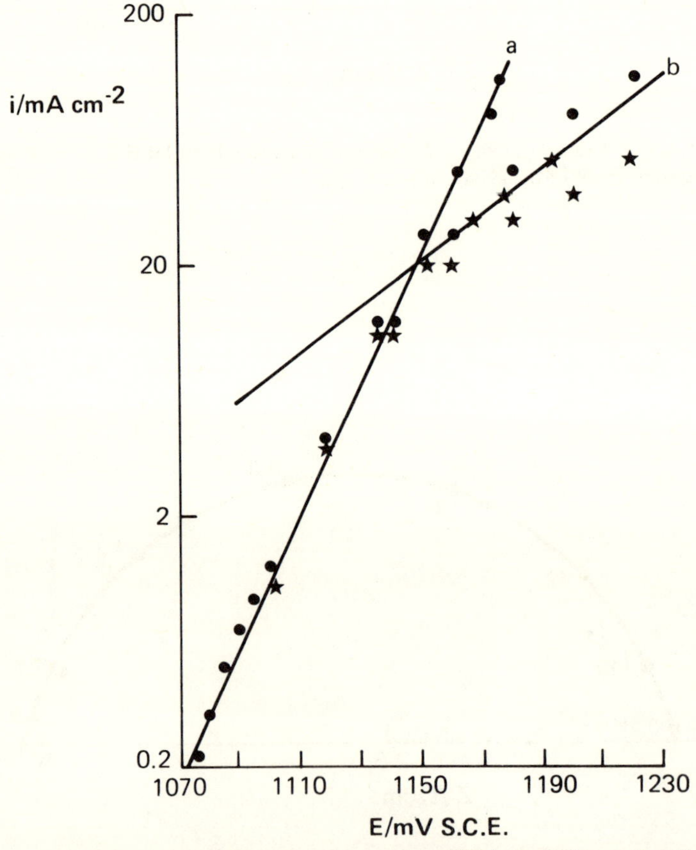

Fig. 15.5 – Example of an *i* *E* curve for a 35% RuO_2/TiO_2 rotating disc electrode in 1 M HCl. 'a' Rotation speed 12.8 Hz; 'b' stationary electrode. The full lines are corrected for the measured ohmic resistance.

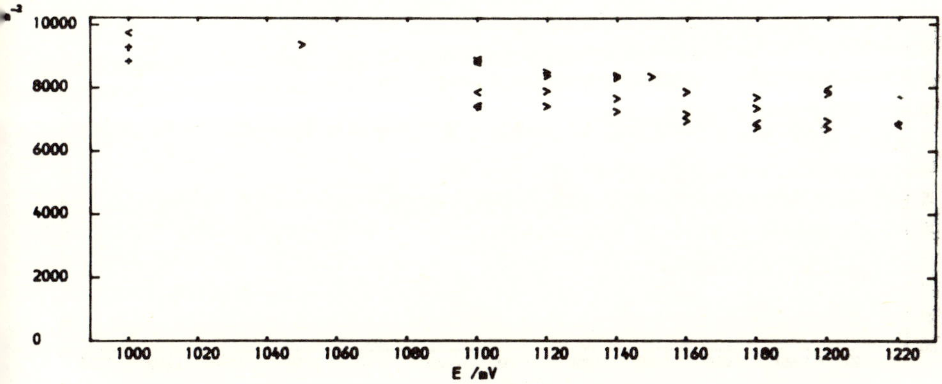

Fig. 15.6(a) – The C_{dl}–E curve obtained during an automatic run. The electrode was a rotating disc (40 Hz) 55% RuO_2/TiO_2 (BRIKA) electrode in 1 M HCl.

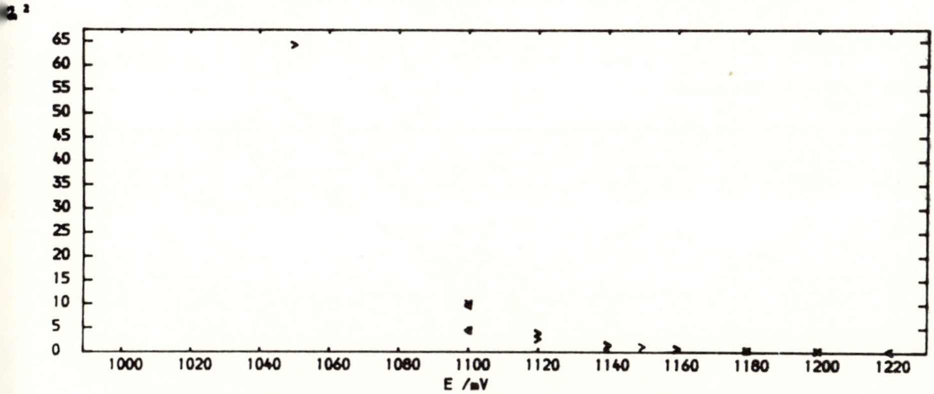

Fig. 15.6(b) – The R_α–E curve obtained during an automatic run. The electrode was a rotating disc (40 Hz) 55% RuO_2/TiO_2 (BRIKA) electrode in 1 M HCl.

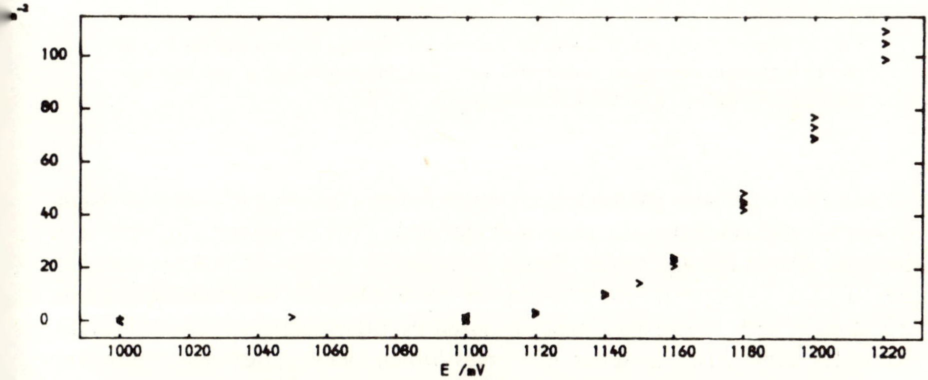

Fig. 15.6(c) – The i–E curve obtained during an automatic run. The electrode was a rotating disc (40 Hz) 55% RuO_2/TiO_2 (BRIKA) electrode in 1 M HCl.

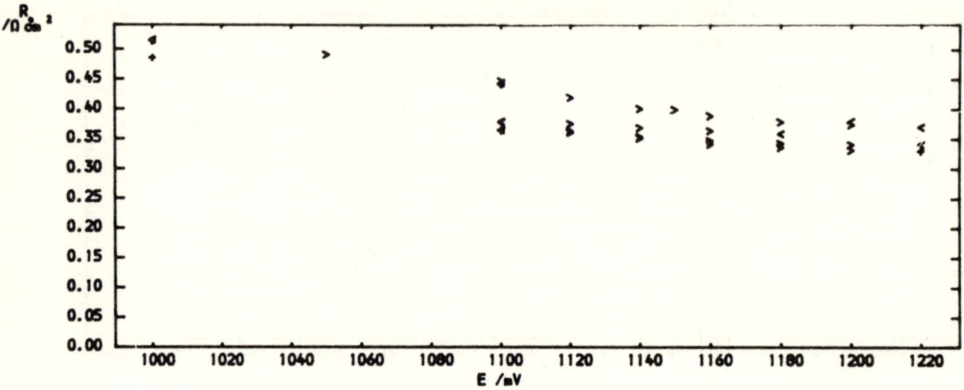

Fig. 15.6(d) – The R_w–E curve obtained during an automatic run. The electrode was a rotating disc (40 Hz) 55% RuO_2/TiO_2 (BRIKA) electrode in 1M HCl.

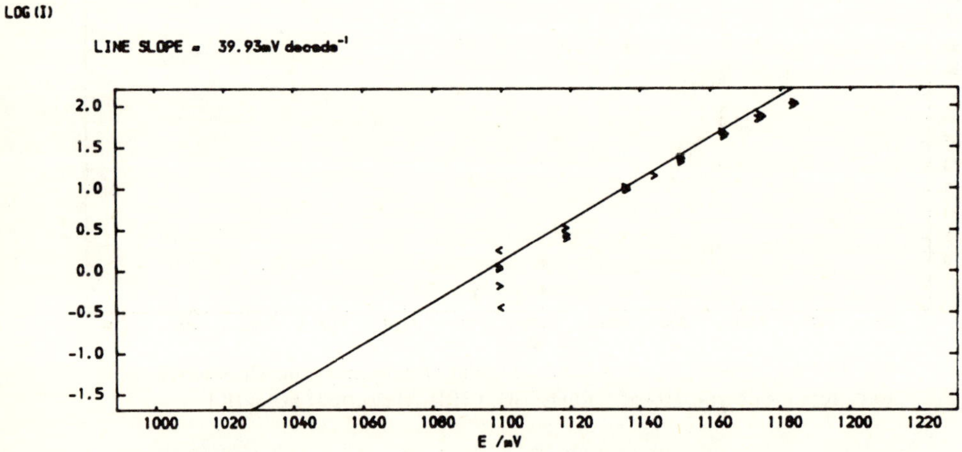

Fig. 15.6(e) – The log i–E curve (corrected for ohmic effects using the R_w–E curve) obtained during an automatic run. The electrode was a rotating disc (40 Hz) 55% RuO_2/TiO_2 (BRIKA) electrode in 1M HCl.

i–E and the impedance parameters are displayed as a function of potential. Runs of this kind characterize the particular electrode. The parameter C_{dl}, when the electrode obeys the equivalent circuit discussed in section 15.8 Appendix, has been used to characterize the real area of the RuO_2/TiO_2 electrode as it seems to correlate with the surface area measured by other methods [22–25]. Some electrodes show $Z(w)$–E series of curves which obey a different equivalent circuit, of the type expected for a porous electrode [26]. The use of impedance to identify electrode by structural type has practical significance.

15.5 CHLORINE REDUCTION

It is in principle not necessary to separate off this topic as the theory applies equally to oxidation and reduction. Figures 15.7(a) and (b) show i–E curves for

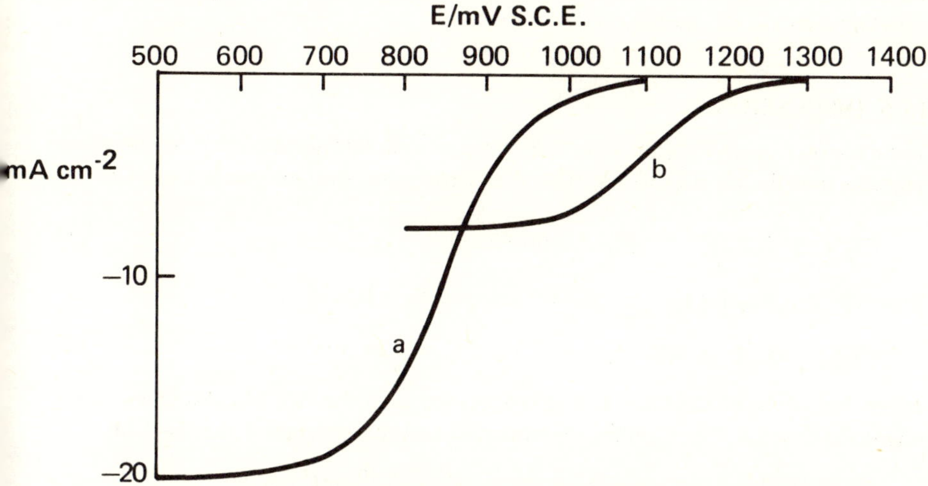

Fig. 15.7(a) – Cl_2 reduction in: 'a' 1M HCl on a 35% rotating RuO_2/TiO_2 electrode (14 Hz); and 'b' 1M HClO on smooth Pt rotating disc electrode (7.5 Hz). The standard rates are similar on both electrodes.

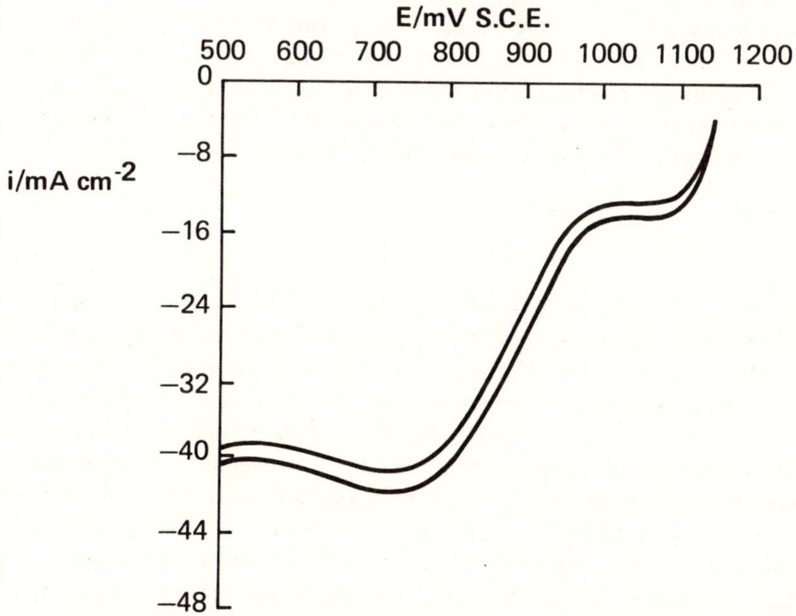

Fig. 15.7(b) – Cl_2 reduction in 0.5M HCl on a 35% RuO_2/TiO_2 electrode (14 Hz) showing the two waves.

the reduction of Cl_2 on Pt and on RuO_2/TiO_2 electrodes. In strongly acid solution only a single wave is observed. The reaction seems to have a similar standard rate on both electrodes. In less acid solution two waves are observed, Fig. 15.7(b). This result has still to be confirmed (see [27]), however, it seems to be consistent with the data for Cl^- oxidation.

15.6 DISCUSSION

The data on porous Pt where the i–E curve is completely controlled by diffusion suggests that the E_0 is 1130 mV. This is not too close to the expected value of

$$Cl_2 + 2e \rightleftarrows 2Cl^- \quad E_0 = 960\,\text{mV} \tag{1}$$

However, it is much closer to that for the overall reaction

$$HOCl + H^+ + 2e \rightleftarrows H_2O + Cl^- \quad E_0 \simeq 1241\,\text{mV} \tag{2}$$

which has an E_0 of 1241 mV. If this is accepted then the two Cl_2 reduction waves might make sense. The equilibrium constant and the kinetics of the reaction

$$Cl_2 + H_2O \rightleftarrows HOCl + H^+ + Cl^- \quad K \cong 10^{-10}\,\text{mol}^2\,\text{cm}^{-6} \tag{3}$$

have been investigated [21]. The kinetic data suggests that it will be possible to reduce Cl_2 and HClO separtely. The waves in Fig. 15.7(a) and (b) could then be assigned to these species, the more negative wave being that of Cl_2. The kinetic observations above could be consistent with this overall reaction especially the fact that the oxidation reaction is observed to be first order in Cl^-. This result has been confirmed by working where the influence of Cl_2 bubbles is a minimum and is probably a genuine result. A possible reaction scheme would then be

$$HOCl + H^+ + e \longrightarrow H_2O + Cl^{\cdot} \tag{4}$$

$$Cl^{\cdot} + e \rightleftarrows Cl^- \,. \tag{5}$$

It is interesting to compare these results with those for the O_2 reaction alone on RuO_2/TiO_2 because of the possibility of a common intermediate. The O_2 evolution reaction in HClO, say, has a 60–70 mV slope at a rotating electrode. This reaction is more difficult to investigate satisfactorily because of the low solubility of O_2. The implication from this data is surely that the rate determining step is located at the first or second electron exchange of the 4-electron overall reaction. Because of the large difference in rates (at $E = 1200$ mV) a common intermediate between the two reactions is probably unlikely. A further point is that the impedance response does not provide any evidence for the participation in the reaction of $Cl^{\cdot}$. It might be expected that the value of C_{dl} would increase at potentials in the Cl_2 evolution region. However, this is not the case. C_{dl} is essentially the same in the double layer region as in the reaction region. It appears to be determined primarily by the structure of the electrode surface, the solution composition and by the influence of the Cl_2 bubbles.

15.7 ACKNOWLEDGEMENT

I am indebted to Drs I. R. Burrows and R. I. Knowles for many discussions on this topic.

15.8 APPENDIX

Chlorine evolution and reduction has similar characteristics to other reactions in which a gas is involved. Consider the reactions

$$Cl_2 + 2e \rightleftarrows 2Cl^- \qquad E_0' = 1360\,\text{mV} \tag{6}$$

$$2H^+ + 2e \rightleftarrows H_2 \qquad E_0' = 0\,\text{mV} \tag{7}$$

$$O_2 + 4H^+ + 4e \rightleftarrows 2H_2O \qquad E_0' = 1229\,\text{mV} \tag{8}$$

where the value of E_0' is the usual one taken from Tables. A more useful value of E_0 (1 mol cm^{-3} reactant and product), for kinetic purposes, can be derived from the Nernst equation, for example for the reaction H^+/H_2

$$-E_0 = 243 + \frac{RT}{F} \ln 10^{-3} - \frac{RT}{2F} \ln 8 \times 10^{-7} \tag{9}$$

where the first term on the right-hand side is the E_0' value from Tables, corrected to S.C.E., and the solubility of the gas is the number in the second ln term. The resulting values of E_0 are 960mV for chlorine, –246mV for hydrogen and 1257 mV for oxygen. Kinetic measurements can then be interpreted using these values.

15.8.1 Consider as a one step reaction

The subject of complex reactions is covered by a limited number of papers on stoichiometric coefficients. It is useful here to consider some concrete examples which are then used to calculate the i–E characteristic shown in Fig. 15.8.

***i–E* curve**

Consider the gas forming reaction

$$Cl_2 + 2e \underset{k_b}{\overset{k_f}{\rightleftarrows}} 2Cl^- \tag{10}$$

where the kinetic rates for the forward and backward direction are given by

$$k_f = k\text{SH}\exp(-\beta(E - E_0)) \tag{11}$$

$$k_b = k\text{SH}\exp(\alpha(E - E_0)) \tag{12}$$

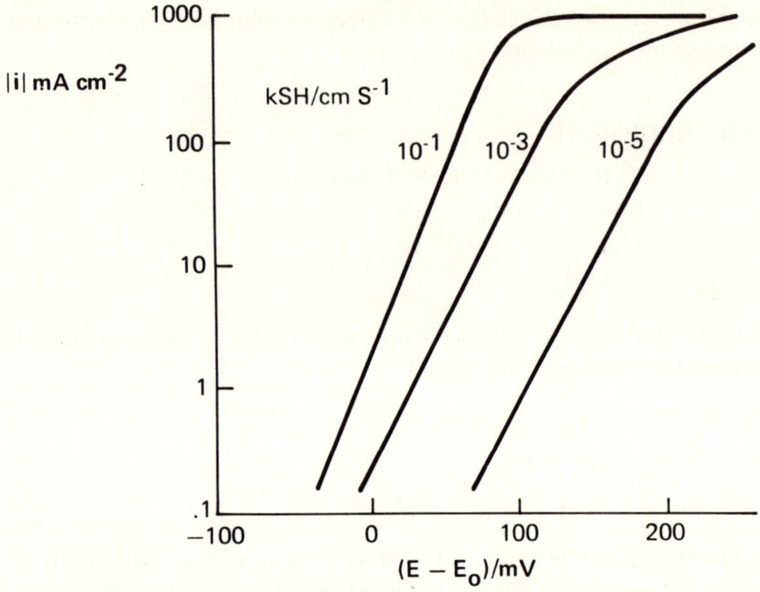

Fig. 15.8 – Theoretical curves for the reaction: $A_2 + 2e = 2B$ for the listed parameters:
$C^b_{A_2} = 0$; $C^b_B = 10^{-3}$/mol cm^{-3}
$D_{A_2} = D_B = 10^{-5}$/cm^2 s^{-1}; $\delta = 10^{-3}$/cm
$a = 120$; $b = 40$/mV

with

$$\alpha = 2.303/a \; ; \quad \beta = 2.303/b \tag{13}$$

and the current would be given by

$$i = 2\mathrm{F}((C^S_{Cl^-})^2 k_b - C^S_{Cl_2} k_f) \tag{14}$$

as is usual the 2e overall reactions demands that the Tafel slopes are related by

$$1/a + 1/b = 1/30 \; . \tag{15}$$

The usual Ficks law considerations lead to

$$i = 2\mathrm{F}D_{Cl_2}(C^s_{Cl_2} - C^b_{Cl_2})/\delta \tag{16}$$

$$i = \mathrm{F}D_{Cl^-}(C^b_{Cl^-} - C^s_{Cl^-})/\delta \tag{17}$$

and the surface concentrations are given by

$$C^s_{Cl_2} = C^b_{Cl_2} + i\delta/2\mathrm{F}D_{Cl_2} \tag{18}$$

$$C^s_{Cl^-} = C^b_{Cl^-} - i\delta/\mathrm{F}D_{Cl^-} \; . \tag{19}$$

In the reversible limit the surface concentrations obey equation and

$$C^{s}_{Cl_2} = (C^{s}_{Cl^-})^2 \exp((E-E_0)\,2.303/30) \tag{20}$$

It is a simple matter to substitute for the surface concentrations and to solve for *i*. In general the expression for *i* is

$$i = 2FkSH((C^{s}_{Cl^-})^2 \exp((E-E_0)\,2.303/40) - \\ - C^{s}_{Cl_2} \exp(-(E-E_0)\,2.303/120)) \tag{21}$$

The usual problem of the units of the quantities is recognized.

Z(w)–E curve

As discussed earlier it is usual to interpret the impedance by equivalent circuit and equation

$$\frac{1}{Z(jw)-R_w} = \frac{1}{R_{ct}+(1-j)6w^{-1/2}} + jwC_{dl}\ . \tag{22}$$

This is the basis of calculation for the box marked (R_{ct}), etc. in Fig. 15.1(c). It would then be further possible to interpret the parameters (R_{ct}), and by a reaction mechanism and the relations, say

$$\frac{1}{R_{ct}} = \frac{\partial i}{\partial E}\ ;\ \sigma_A = \frac{\partial i}{\partial C^s_A}\left(F\,\frac{\partial i}{\partial E}\,\sqrt{2D_A}\right). \tag{23}$$

It would also be convenient to put in particular reaction mechanism by writing the impedance as

$$\frac{1}{Z(jw)-R_w} = \frac{\partial i}{\partial E} + \frac{\partial i}{\partial C^s_A}\,\frac{\overline{dC^s_A}}{\overline{dE}} + \frac{\partial i}{\partial C^s_B}\,\frac{\overline{dC^s_B}}{\overline{dE}} \tag{24}$$

and inserting the coefficients [5].

15.8.2 Consider as a two step reaction

i–E curve

Consider the overall equation (00) to be split into steps

$$Cl_2 + e \underset{k_{b1}}{\overset{k_{f1}}{\rightleftarrows}} Cl^{\cdot} + Cl^- \qquad (E_0)_1 \tag{25}$$

$$Cl^{\cdot} + e \underset{k_{b2}}{\overset{k_{f2}}{\rightleftarrows}} Cl^- \qquad (E_0)_2 \tag{26}$$

then assuming that each step has a characteristic 120 mV slope

$$k_{f1} = (kSH)_1 \exp(-(E-(E_0)_1)1\ 2.303/120) \quad (27)$$
$$k_{b1} = (kSH)_1 \exp((E-(E_0)_1)\ 2.303/120) \quad (28)$$
$$k_{f2} = (kSH)_2 \exp(-(E-(E_0)_2)\ 2.303/120) \quad (29)$$
$$k_{b2} = (kSH)_2 \exp((E-(E_0)_2)\ 2.303/120) \quad (30)$$

with the requirement

$$2E_0 = (E_0)_1 + (E_0)_2 \quad (31)$$

the partial current are then given by

$$i_1 = F(k_{b1}C^s_{Cl^-}[Cl^\cdot] - k_{f1}C^s_{Cl_2}) \quad (32)$$
$$i_2 = F(k_{b2}C^s_{Cl^-} - k_{f2}[Cl^\cdot]) \quad (33)$$

If it is assumed, by way of demonstration, that the first step is the slow step, $[Cl^\cdot]$ is low, and the surface concentrations are given by equations (18) and (19) then

$$i = 2F\,kSH((C^s_{Cl^-})^2 \exp((E-E_0)\ 2.303/40) - - C^s_{Cl_2} \exp(-(E-E_0)\ 2.303/120)) \quad (34)$$

with

$$kSH = (kSH)_1 \exp(((E_0)_1 - E_0)\ 2.303/120) \quad (35)$$

Similar arguments can be used when the second step is rate determining. It is also possible to incorporate adsoption isotherms to describe the $Cl^\cdot$ coverage, if there are indications that this is necessary. If both reactants are present in the bulk then E_e can be measured and E_0 calculated, or the i–E curve can be written in terms of i_0 and E_e.

$$i = i_0\left(\frac{(C^s_{Cl^-})^2}{(C^b_{Cl^-})^2} \exp((E-E_e)\ 2.303/40) - \frac{C^s_{Cl_2}}{C^b_{Cl_2}} \exp((E-E_e)\ 2.303/120)\right) \quad (36)$$

Figure 15.8 shows a plot of equation (21) for some trial parameters.

15.9 REFERENCES

[1] Denton, D. A., Harrison, J. A., and Knowles, R. I., *Electrochim. Acta*, **25**, (1980), 1147.

[2] Denton, D. A., Harrison, J. A., and Knowles, R. I., *Electrochim. Acta*, **24**, (1979), 521.

[3] Burrows, I. R., Denton, D. A., and Harrison, J. A., *Electrochim. Acta*, **23**, (1978), 493.
[4] Burrows, I. R., Entwisle, J. H., and Harrison, J. A., *J. Electroanal. Chem.*, 77, (1977), 21.
[5] Harrison, J. A., Automation in Electrode Kinetics, Part 7, *Electrochim. Acta*, 27, (1982), 1113.
[6] Denton, D. A., Harrison, J. A., and Knowles, R. I., *Electrochim. Acta*, **26**, (1981), 1197–1207.
[7] Trasatti, S., and O'Grady, W. E., *Advances in Electrochemistry and Electrochemical Engineering*, **12**, (ed. H. Gerischer and C. W. Tobias), Wiley–Interscience, 1981.
[8] Caldwell, D. E., *Comprehensive Treatise of Electrochemistry*, **2**, (ed. J. O. M. Bockris, B. E. Conway, E. Yeager, and R. E. White), Plenum, 1981.
[9] See for example: Bard, A. J., and Faulkner, L. J., *Electrochemical Methods*, Wiley, New York, 1980.
[10] For example: Galvez, J., and Molina, A., *J. Electroanal. Chem.*, **110**, (1980), 49.
[11] Mastragostino, M., Nadjo, L., and Saveant, J. M., *Electrochim. Acta*, **13**, (1968), 72.
[12] Nicholson, R. S., and Shain, I., *Anal. Chem.*, **36**, (1964), 706.
[13] Hanafey, M. F., Scott, R. L., Ridgeway, T. H., and Reilley, C. N., *Anal. Chem.*, **50**, (1978), 116.
[14] Delahay, P., *New Instrumental Methods in Electrochemistry*, Interscience, New York, 1954.
[15] Smith, D. E., *Electroanalytical Chem.*, (ed. A. J. Bard), **1**, Marcel Dekker, New York, 1966.
[16] de Levie, R., and Propisil, L., *J. Electroanal. Chem.*, **22**, (1969), 277.
[17] Moreira, H., and de Levie, R., *J. Electroanal. Chem.*, **29**, (1971), 353.
[18] Sluyters-Rehbach, M., and Sluyters, J. H., *Electroanal. Chem.*, (ed. A. J. Bard), **4**, Marcel Dekker, New York, 1970.
[19] Rangarajan, S. K., *J. Electroanal. Chem.*, **55**, (1974), 297, 329, 337, 363.
[20] Armstrong, R. D., Bell, M. F., and Metcalfe, A. A., *Electrochemistry*, **6**, The Chemical Society, London, 1978.
[21] Eigen, M., and Austin, K., *J. Amer. Chem. Soc.*, **84**, (1962), 1355.
[22] Kuhn, A. T., and Mortimer, C. J., *J. Electrochem. Soc.*, **120**, (1973), 231.
[23] O'Grady, W., Iwakura, C., Huang, J., and Yeager, E., *Electrocatalysis*, (ed. M. W. Breiter), Electrochem. Soc. Inc., 1974.
[24] Burke, L. D., and Murphy, O. J., *J. Electroanal. Chem.*, **96**, (1979), 19.
[25] Burke, L. D., Murphy, O. J., and O'Neill, J. F., *J. Electroanal. Chem.*, **81**, (1977), 391.
[26] de Levie, R., in *Advances in Electrochemistry and Electrochemical Engineering*, **6**, (ed. P. Delahay), Interscience, 1967.
[27] Kadija, I. V., *J. Electrochem. Soc.*, **127**, (1980), 599.

15.10 OPEN DISCUSSION

1. In opening the discussion from the Chair, Prof. Fleischmann asked for clarification as to whether the polarization graphs presented were predicted or actual measured values.

   Dr Harrison replied that all the data was automatically measured – the test electrode material was fitted into the equipment described and the results came out as shown. This type of system can be arranged to measure whatever is required – the paper presents what the author feels is the most powerful way of analysing the data. If one takes data of currents, potentials, time and impedance, then every piece of data must be interpreted using the same reaction mechanism and it must all predict one from the other. So in the figures presented in the paper, the current/potential characteristics are predicted from the impedance and vice versa.

2. Dr Warren, Chemetics, asked whether this really fingerprinted the system at all. For an electrode system for example, for chlorate or chlorine production, does not the real interest lie in the Tafel equations for the evolution of, say, chlorine and oxygen gas. What are presented in the paper are the gross plots so to characterize an electrode one must characterize the products as well as the current/potential curve.

   In reply, Dr Harrison explained that there was obvioulsy some preparatory work to be done in any electrochemical investigation – for example analysis of products. But once that is done, the details of the kinetics, e.g. the Tafel slope, ohmic loss corrections, etc., are all included. The graphs presented in the paper had all the electrochemical characteristics built in and they were built in, in such a way that any discrepancy appears in, say, the rate constant and in fact it is so easy to build in any property of the reaction that is required. The paper presents what he felt was the best way.

   Dr Warren of Chemetics extended the discussion. He was very interested in the HOCl mechanism in considering the evolution of chlorine and the simultaneous evolution of oxygen on these electrodes. It is clear from experimental observation that the amount of oxygen evolved under the same conditions at different electrodes varies and this can be related to the Russian work evolution. It can also be shown in conformity with that, that a given electrode evolves a different proportion of oxygen to chlorine as the current density changes. Now with the HOCl mechanism as proposed, one can postulate that the different electrodes are exerting some catalytic mechanism in decomposing the HOCl to oxygen on the electrode surface.

   The possibility of a common intermediate between the chlorine evolution reaction and the oxygen evolution reaction has been considered by Dr Harrison but he was not sure whether it was possible or not.

It may be that if such a common intermediate exists then it will be difficult to design electrodes to eliminate the oxygen evolution reaction which may be an inherent feature of the chlorine evolving reactions.

An anonymous questioner asked whether the decomposition of hypochlorous acid was fast or slow because if it were fast it should be kinetically equivalent to the Volmer–Heyrovsky reaction postulated in Dr Janssen's paper (see Chapter 16) and if it were slow then one should get supersaturated hypochlorous acid which would then decompose to chlorine which could perhaps be observed.

Dr Harrison replied that he was not too certain about that conclusion. There is a lot of detail about the reaction in the literature and it is a very complex reaction. Most of the work on the kinetics of the hypochlorous acid equilibrium was done about 10 years ago by Eigen who found the rate is such that when the reaction is being driven forwards, i.e. the anodic reaction from chloride, there is no kinetic hindrance due to that step. So the reaction in the direction hypochlorous acid to chlorine must be fast enough such that it does not appear in the forward rate.

When the reduction reaction is carried out, however, the reduction of HOCl and chlorine gas can be distinguished separately so this rate must be relatively slow.

LIST OF SYMBOLS

a, b	anodic and cathodic Tafel slopes in mV
C_{dl}	differential capacity
C_j^s	surface concentrations of species j
C_j^b	bulk concentrations of species j
$\overline{\Delta E}$	Laplace transform of ΔE
D_j	diffusion coefficient of species j
E	potential
E_0'	standard potential, against standard hydrogen electrode, as in Tables
E_0	standard potential, against S.C.E., for $1\,mol\,cm^{-3}$ concentrations
F	Faraday
i	current density
j	square root of -1
k	equilibrium constant
$[Cl^{\cdot}]$	surface concentration of $Cl^{\cdot}$
k_f	forward electrochemical rate
k_b	backward electrochemical rate
kSH	standard rate constant
p	Laplace variable

R_{ct}	charge transfer resistance
R_w	ohmic resistance
$Z(jw)$	impedance
σ	Warburg coefficient
δ	diffusion layer thickness
α, β	constants which control the potential dependence of electrochemical rate
w	frequency of a sinusoidal signal

CHAPTER 16

Kinetics and Mechanism of Chlorine Evolution on Metaloxide Electrodes

L. J. J. JANSSEN, Department of Chemical Technology, Laboratory for Electrochemistry, Eindhoven University of Technology, Eindhoven, The Netherlands

The mechanisms and the data of kinetic parameters given in literature for the chlorine evolution on various types of RuO_2/TiO_2 electrodes are discussed. Moreover, some new experimental results are given. It has been concluded that the chlorine evolution on this type of electrode occurs according to the Volmer–Heyrowsky mechanism, where the Volmer reaction is in quasi equilibrium and the Heyrowsky reaction and eventually the diffusion of molecular chlorine away from the electrode surface into the bulk of the solution determine the potential–current density relation.

16.1 INTRODUCTION

The chlorine evolution on dimensionally stable anodes if of great practical importance. The active mass of these anodes consists of a mixture of metal oxides. Particularly, ruthenium dioxide/titanium dioxide anodes have great advantages over other anode materials which have been used for chlorine production. Hence, many electrochemists have extensively investigated the mechanism and the kinetics of chlorine evolution on RuO_2/TiO_2 anodes.

The aim of this study is to discuss and compare the mechansims and the data of kinetic parameters given in literature for the chlorine evolution reactions on various types of RuO_2/TiO_2 anodes. To explain the experimental Tafel slopes, it has been found desirable to include the effect of the diffusion of dissolved molecular chlorine away from the electrode surface on theoretical relations between the potential and current density according to the two likely mechanisms, viz. the Volmer–Tafel and the Volmer–Heyrowsky mechanism. The results of these calculations are published separately.

16.2 THEORY

The overall

The overall reaction of the chlorine formation by oxidation of chloride ions is:

$$2Cl^- \rightarrow Cl_2 + 2e^- \quad (1)$$

In general, the two mechanisms proposed in literature are the Volmer–Tafel mechanism with two steps:

(a) the Volmer reaction $Cl^- \rightarrow Cl_{ad} + e^-$ (2)

(b) the Tafel reaction $2Cl_{ad} \rightarrow Cl_2$ (3)

and the Volmer–Heyrowsky mechanism with two steps: the Volmer reaction eq. (2) and the Heyrowsky reaction

$$Cl^- + Cl_{ad} \rightarrow Cl_2 + e^- \quad (4)$$

In particular, electrochemists from the USSR, e.g. Krishtalik [1], propose a mechanism with three reaction steps: the Volmer reaction followed by the reactions (Krishtalik mechanism):

$$Cl_{ad} \rightarrow Cl^+_{ad} + e^- \quad (5)$$

and

$$Cl^+_{ad} + Cl^- \rightarrow Cl_2 \quad (6)$$

For the three mechanisms mentioned, the theoretical Tafel slope b and reaction orders z with respect to Cl^--ions and Cl_2 molecules are represented in Table 16.1 [1,2], assuming a Langmuir-type isotherm for the adsorption of atomic chlorine and a transfer coefficient of 0.5 for the charge transfer reactions. Moreover, it is assumed that the diffusion of both Cl^- and Cl_2 does not affect the E/i relation. The parameters for the Krishtalik mechanisms are only partly obtained from [1]; see for the basic equations for the various reactions [3,4].

16.3 EXPERIMENTAL RESULTS

Three types of ruthenium oxide electrodes can be distinguished, viz. the RuO_2 electrode consisting of pure RuO_2, the RTO electrode consisting of RuO_2, TiO_2 and eventually other metal oxides with an RuO_2 content higher than 20 mole-%, and the MO electrode consisting of one or more metal oxides doped with a relatively small amount of RuO_2.

Moreover, the RuO_2 electrode type is subdivided into two types, viz. the electrochemical RuO_2 electrode prepared by electrochemical oxidation of a ruthenium substrate, and the chemical RuO_2 electrode prepared by chemical oxidation of a ruthenium salt coating on a metal substrate in air or oxygen gas.

Table 16.1
Theoretical parameters for the chlorine evolution at 25°C [1,2]

Mechanism	Rate-determining step	Tafel slope b (mV)		z_{Cl^-}		z_{Cl_2}	
		$\theta \to 0$	$\theta \to 1$	$\theta \to 0$	$\theta \to 1$	$\theta \to 0$	$\theta \to 1$
Volmer–Tafel	$Cl^- \to Cl_{ad} + e^-$	118.4	118.4	1	1	0	0
	$2Cl_{ad} \to Cl_2$	29.6	∞	2	0	0	0
Volmer–Heyrowsky	$Cl^- \to Cl_{ad} + e^-$	118.4	–	1	–	0	–
	$Cl_{ad} + Cl^- \to Cl_2 + e^-$	39.5	118.4	2	1	0	0
Krishtalik	$Cl^- \to Cl_{ad} + e^-$	118.4	118.4	1	1	0	–
	$Cl_{ad} \to Cl^+_{ad} + e^-$	39.5	118.4	1	0	0	0
	$Cl^+_{ad} + Cl^- \to Cl_2$	29.6	∞	2	0	0	0

16.3.1 Electrochemical RuO_2 electrode

The electrochemical RuO_2 electrode used by Janssen *et al.* [5] was prepared by anodic oxidation of a ruthenium metal electrode in a solution of 1M HCl + 4M NaCl solution at 20°C and 25 mA/cm^2 during 24 h. The electrode surface became rough and black.

The relation between potential and current was determined by scanning the potential between 1050 and 1300 mV versus S.C.E. with a sweeprate of 1.25 mV/s. All potentials are referred to a saturated calomel electrode. The experiments were carried out at 20°C for chlorine saturated solutions with an [NaCl]/[HCl] ratio of 4 and with various [NaCl] and [HCl].

It has been found that at the lowest chloride concentration, i.e. 0.5M, hysteresis clearly occurs when the solution is not stirred. The hysteresis disappears almost completely on stirring the solution.

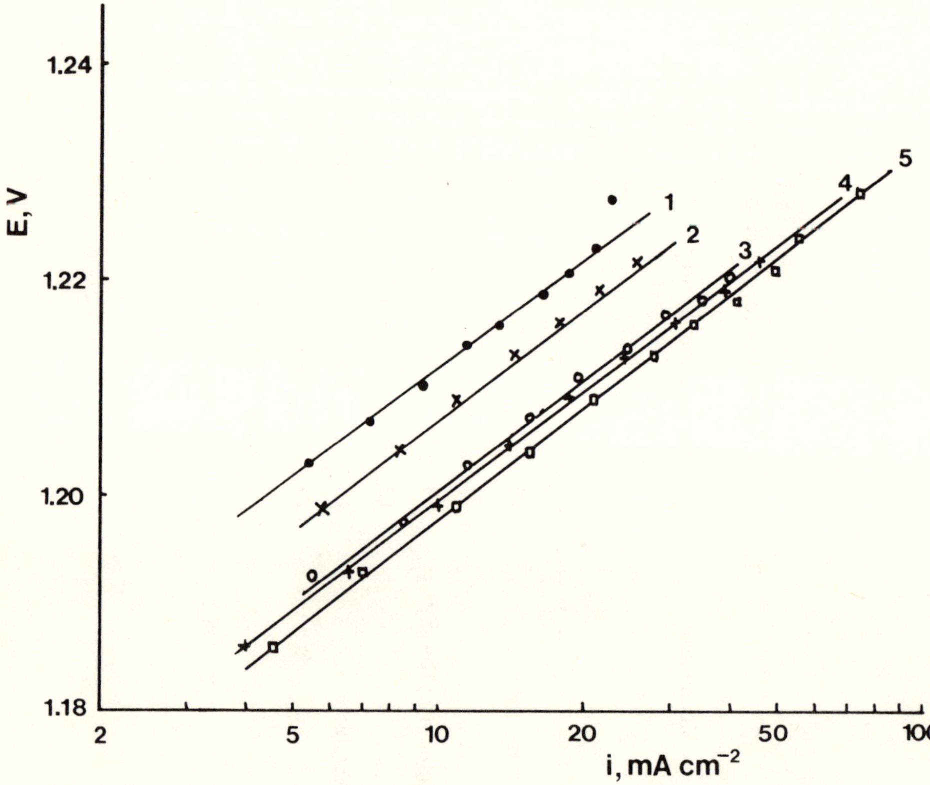

Fig. 16.1 – Log i versus E relation for an electrochemical RuO_2 electrode at 20°C in chlorine-saturated chloride solutions of various composition (1:0.4M NaCl + 0.1M HCl, E_r = 1190 mV; 2:0.8M NaCl + 0.2M HCl, E_r = 1180 mV; 3:1.6M NaCl + 0.4M HCl, E_r = 1150 mV; 4:3.2M NaCl + 0.8M HCl, E_r = 1110 mV; 5:4M NaCl + 1.0M HCl, E_r = 1100 mV).

New results for stirring the solution and cathodic scanning are given in Fig. 16.1 together with the reversible electrode potentials, E_r, for the various solutions. This figure shows that the slope b of E/log i curve does not depend on the composition of the solution; $b = 36$ mV. In the literature [5] $b = 40$ mV at 25°C is given.

Plotting the current density at $E = 1.22$V versus the Cl^--concentration on a logarithmic scale, Fig. 16.2 shows that the reaction order for Cl^- is about 0.5 at [Cl^-] from 0.5 to 2.0M and about 0.1 at [Cl^-] from 2.0 to 5M. The geometrical exchange current density, obtained by linear extrapolation of the E/log i curve, depends strongly on the composition of solution and is about 0.02 mA/cm^2 at 20°C for a 4M NaCl + 1M NaCl solution saturated with chlorine at atmospheric pressure.

To elucidate the reaction mechanism, the potential decay curves (η_n − log t_n curves) were determined [5]. The slope of the straight η_n − log t_n lines is about

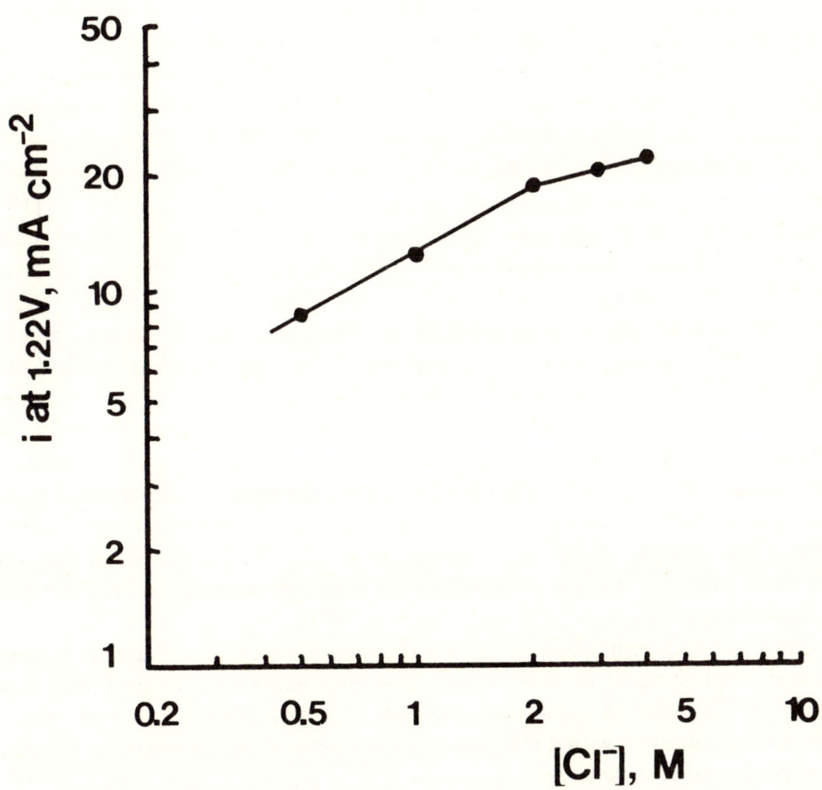

Fig. 16.2 – Plot of current density at $E = 1220$ mV and chloride concentration on a double logarithmic scale for an electrochemical RuO_2 electrode at 20°C and in a chlorine-saturated chloride solution where [Cl^-]/[H^+] = 4.

22 mV. Assuming a thickness of Nernst diffusion layer in the solution at the electrode surface of 10^{-3}cm and a diffusion coefficient of chlorine in the solution of 10^{-5}cm, it can be calculated that the transition time for the chlorine diffusion in the solution is about 0.05 s [3]. Consequently, the potential decay curve from $t_n > 0.05$ s does not depend on chlorine diffusion in the solution.

16.3.2 Chemical RuO_2 electrode

The chemical RuO_2 electrode were gradually prepared by coating with $RuCl_3$ dissolved in concentrated hydrochloric acid and then fired mostly in air at temperatures from 350 to 600°C. [Some investigations [6–9], however, used an organic solvent such as propanol to prepare a paste of $RuCl_3$. For chlorine evolution on chemical RuO_2 electrodes the Tafel slopes (about 110 mV) found by Kuhn and Mortimer [6] are much higher than those (20–40 mV) found by others [10–14]. These differences in Tafel slope may be caused by the different nature of the solvents used in electrode preparation. In the following discussion the results of Kuhn and Mortimer [6] and of Burke *er al.* [7–9] are left out of consideration due to the strongly deviating nature of their electrodes.]

The low Tafel slopes given in [10–14] can be divided into two groups, one of about 20 mV and the other of about 40 mV. The lowest Tafel slopes have been found at 80°C for a flowing and concentrated NaCl solution containing no chlorine (0.3 g $NaCl/cm^3$), pH 3–3.5 [11 and 13]. For comparison the Tafel slopes are calculated at 30°C. Using the experimental dependence given in [13], the average Tafel slopes are now 29 mV and 41 mV. The first value (29 mV) was found for a flowing and concentrated NaCl solution containing no chlorine (0.3 g $NaCl/cm^3$) having a pH of 2–3.5 [11 and 13]. For a 2.5 M NaCl + 1.5 M HCl solution saturated with chlorine at chlorine pressures of 101 and 5.7 kPa [10] and for a 5 M NaCl solution through which nitrogen gas was bubbled [14], however, a Tafel slope of about 40 mV was found.

The Tafel slopes decrease with an increase in the temperature from 34 mV at 30°C to 25 mV at 80°C for chlorine evolution with a flowing concentrated NaCl solution containing no chlorine [13]. This temperature dependence completely differs from the one for an RTO electrode in 4 M NaCl + 1 M HCl solution saturated by chlorine obtained by Janssen *et al.* [5]. They have found that the Tafel slope increases linearly with increasing absolute temperature, for instance the Tafel slope is 41 mV at 25°C and 40 mV at 70°C.

Only a few experimental results for the reaction order of Cl^--ions have been obtained. Ehrenburg *et al.* [10] determined this reaction order from the slope of $\log i/\log [Cl^-]$ plot at a constant electrode potential. They concluded that in sulphate-free solutions with a chloride concentration from 0.5 to 4 M the reaction order of Cl^--ions is close to zero, but at lower Cl^--concentrations it tends to 1 [10]. For solutions containing 2 M H_2SO_4 and various but low $[Cl^-]$ concentrations the reaction order of Cl^--ions is 1 [10, 15], however, its experimental support is rather insufficient. Arikado *et al.* [14] measured the relationship between equilibrium electrode potential and the exchange current density from polarization resistance measurements.

The current density at 1.15 V and 30°C is about 40 mA/cm^2 for a 2.5 M NaCl + 1.5 M HCl solution [10] and 4 mA/cm^2 for a 5 M NaCl solution by extrapolation [13]. The activity of chemical RuO_2 electrodes decreases with increasing pH [14].

Moreover, the shape of the $E/\log i$ curve at potentials near E_r of the Cl^-/Cl_2 redox couple is strongly affected by the pH of solution.

The effect of the thickness of the RuO_2 coating layer on the activity of chemical RuO_2 electrodes is shown in Fig. 16.3. The results of Fig. 16.3 are deduced from [12]. This figure shows that the activity of the chemical RuO_2 electrode increases at a declining rate with increasing thickness of the coating layer. As shown, the Tafel slope is practically constant at RuO_2 amounts above 0.5 g/m^2.

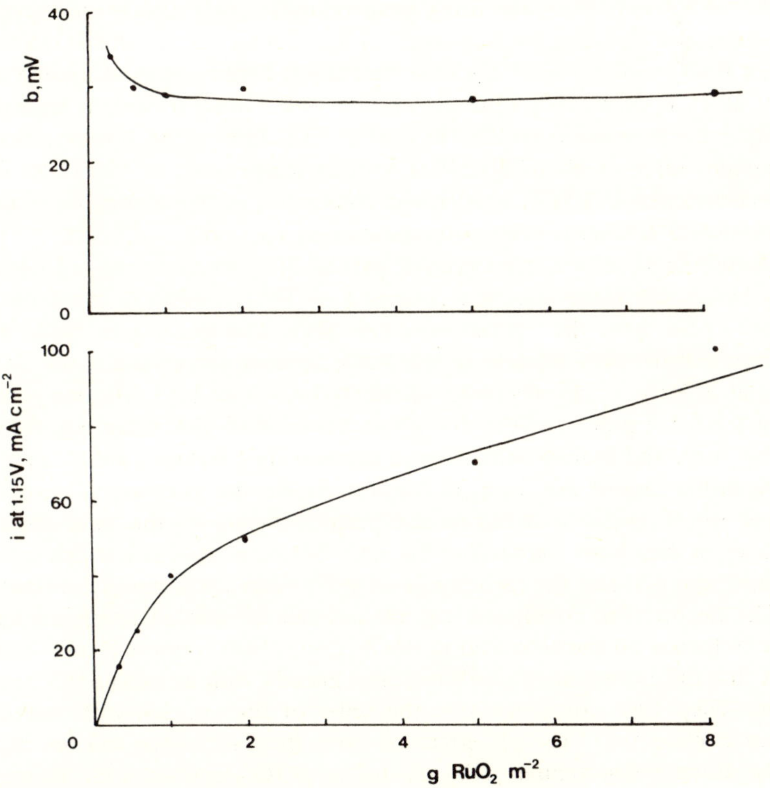

Fig. 16.3 – Plot of current density at 1.15 V and Tafel slope as a function of the amount of RuO_2 present on a titanium substrate for a 5 M NaCl solution at 80°C and a pH of 2 (see Bystrov and Romashin, *Soviet Electrochemistry*, **11**, 1141 (1975)).

16.3.3 RTO electrode

RTO electrodes with an RuO_2 content of more than 20 mole-% are frequently used [5, 11–13 and 15–19] for studying the mechanism of chlorine evolution. The RTO electrodes prepared with an organic solvent (e.g. isopropanol) are left out of consideration due to their deviating behaviour, as was the case for the chemical RuO_2 electrodes.

In Fig. 16.4 results are given for electrodes with various contents of RuO_2 and TiO_2. This figure shows clearly that b and i at 1150 mV, and thus also the behaviour of the electrode, remain almost constant at an RuO_2 content greater than about 20 mole-%. Generally, both the Tafel slope and the current density at a constant electrode potential for chlorine formation on RTO electrodes decrease gradually with increasing RuO_2 content [11–13, 20]. Obviously, the RTO electrode behaves similarly to the chemical RuO_2 electrode.

The activity of RTO electrodes depends on conditions of its preparation. Besides the solvent effect, the firing temperature strongly affects the activity of RTO electrodes [5, 8, 21]. Janssen *et al.* [5] have found that for an RTO electrode with an RuO_2 content of 30 mole-%, the current density at an overpotential of 90 mV at 25°C decreases gradually from 2.5 to 0.3 mA/cm^2 with an increase of the firing temperature from 350 to 600°C. The Tafel slope, however, reaches a minimum value of about 50 mV at a firing temperature of 500°C. At firing temperatures below 300°C, insufficient conversion of the coating layer occurs and above 600°C there is significant oxidation of the titanium substrate [21].

During firing of the coating layer, part of TiO_2 forms mixed crystals with RuO_2 having the rutile structure, and part of TiO_2 remains in the form of a separate phase with the anatase structure [11]. The amount of rutile phase increases sharply with increase in the RuO_2 content and enhances the electrochemical activity of an electrode on RuO_2/TiO_2 base [11]. The temperature during the firing process affects the properties of RTO electrodes, e.g. the ratio between rutile and anatase phase, crystal size and electrical resistance [21].

As in the case of the chemical RuO_2 electrode, the thickness of the coating layer of the RTO electrode has no significant influence on the Tafel slope [5]. Moreover, it has been found that the ratio between the current density at a constant potential and the capacity of an RTO electrode is almost constant [5].

The elecotrolysis conditions, e.g. pH and chloride concentration also have a strong influence on the behaviour of the RTO electrode. Denton *et al.* [18] have found that the current at 1150 mV increases linearly with increasing HCl concentration. They have concluded that the order of the reaction with respect to Cl^--ions is close to 1. The pH effect on both the Tafel slope and the current density at an overpotential of 80 mV for an RTO electrode with 30 mole-% RuO_2 in a chlorine saturated NaCl + HCl solution, with a $[Cl^-]$ of 5 M and at 25°C, is given in Fig. 16.5. This figure shows that the Tafel slope increases at an increasing rate and that the current density at $\eta = 80$ mV decreases with increasing pH. The pH effect is clearest at pH > 3. This is a remarkable result since the

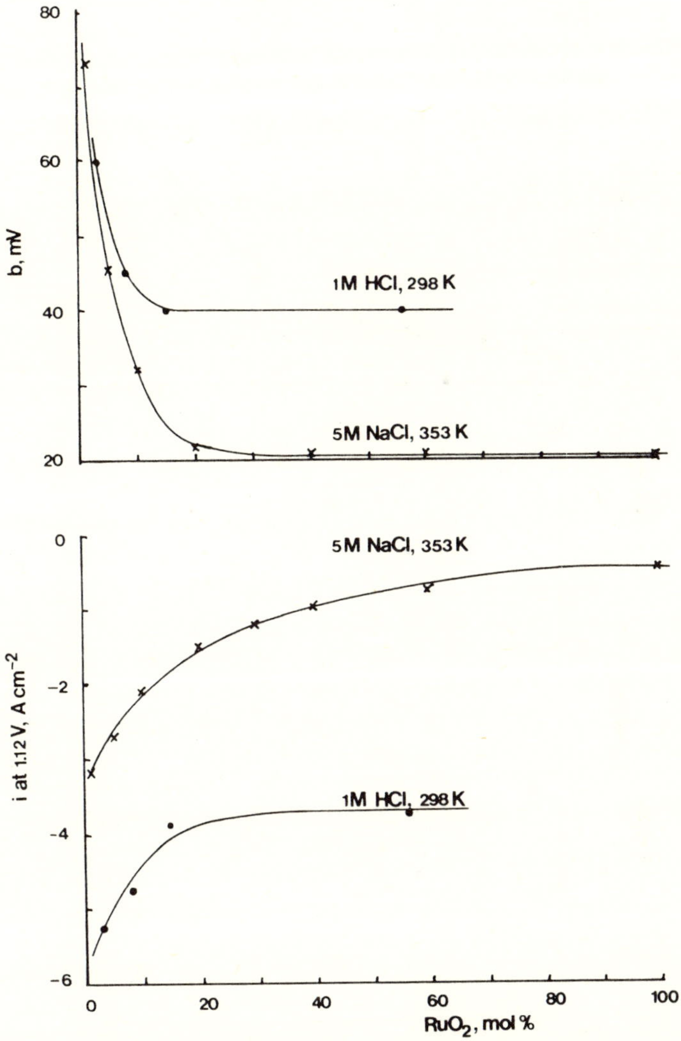

Fig. 16.4 – Plot of current density at 1.12 V and Tafel slope as a function of the mole-% RuO_2 present in the active mass of electrode for a 5 M NaCl solution of pH = 3–3.5 and at 80°C (Veselovskaya, I. E., *et al.*, *Soviet Electrochemistry*, **10**, 62 (1974)) and for a 1M HCl solution at 25°C (Burrows, I. R., Denton, D. A., and Harrison, J. A., *Electrochim. Acta*, **23**, 493 (1978).

concentration of Cl^--ions and likely also the activity of Cl^--ions is almost constant at pH > 3. It is likely that the nature of RTO electrodes depends very much on the pH of the solution.

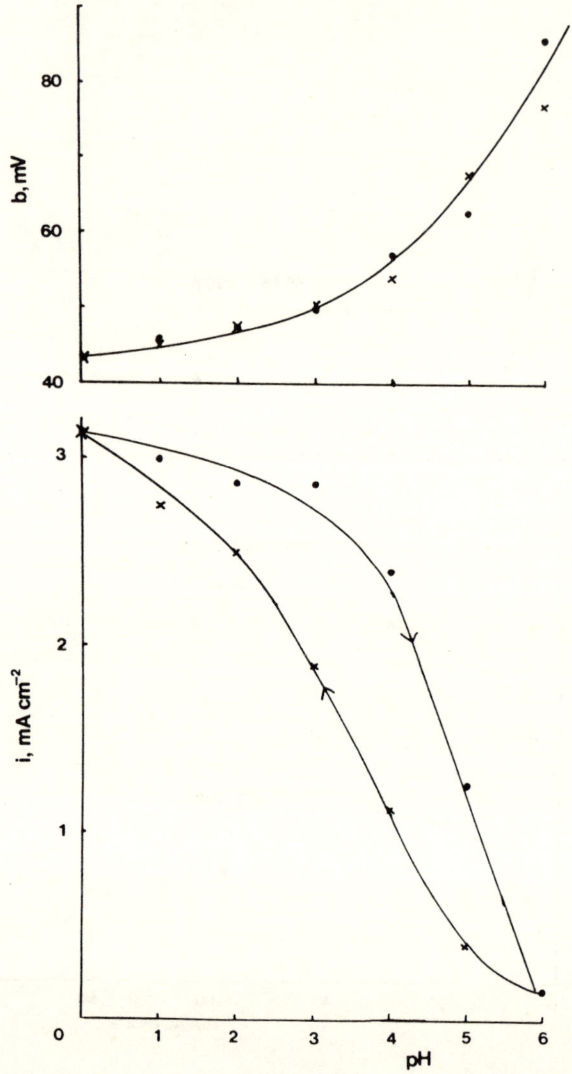

Fig. 16.5 – Plot of current density at η = 80 mV and Tafel slope as a function of pH for an RTO electrode of 30 mole-% RuO_2 in a chlorine-saturated NaCl + HCl solution with $[Cl^-]$ = 5 M and at 25°C. (The arrow indicates the direction of the pH change of solution during a series of experiments.)

The reaction order of Cl^--ions was investigated by Ehrenburg *et al.* [10, 15] and by Krishtalik [1]. From the dependence of the anodic current at 1.21 V on the Cl^--concentration at a $[Cl^-]$ from 0.1 to 1.0 M with 2 M H_2SO_4 as supporting electrolyte, the [1, 10, 15] concluded that the reaction order of Cl^--ions at low

Cl^--concentrations is equal to 1. Their conclusion is dubious since the cathodic reaction cannot be neglected at the lowest Cl^--concentration (E_r is only about 20 mV less positive than 1.21 V).

For an RTO electrode the $E/\log i$ curves are shown in Fig. 16.6 for chlorine saturated chloride solutions of various composition. Equal Tafel slopes were found. From the results of Fig. 16.6 the relation between the current density at $E = 1180$ mV and the chloride concentration is obtained and shown in Fig. 16.7. The slope of the curve in Fig. 16.7 depends on the Cl^--concentration assuming the effect of the variation in H^+-concentration (0.1–1M) can be neglected. The experimental reaction order of Cl^- (being usually equal to the slope of the $\log i/\log [Cl^-]$ curve at a constant potential) is similar to that of an electrochemical RuO_2 electrode.

From the experimental results given in the preceding discussion, it is concluded that the nature of the electrode surface depends on both the pH and the

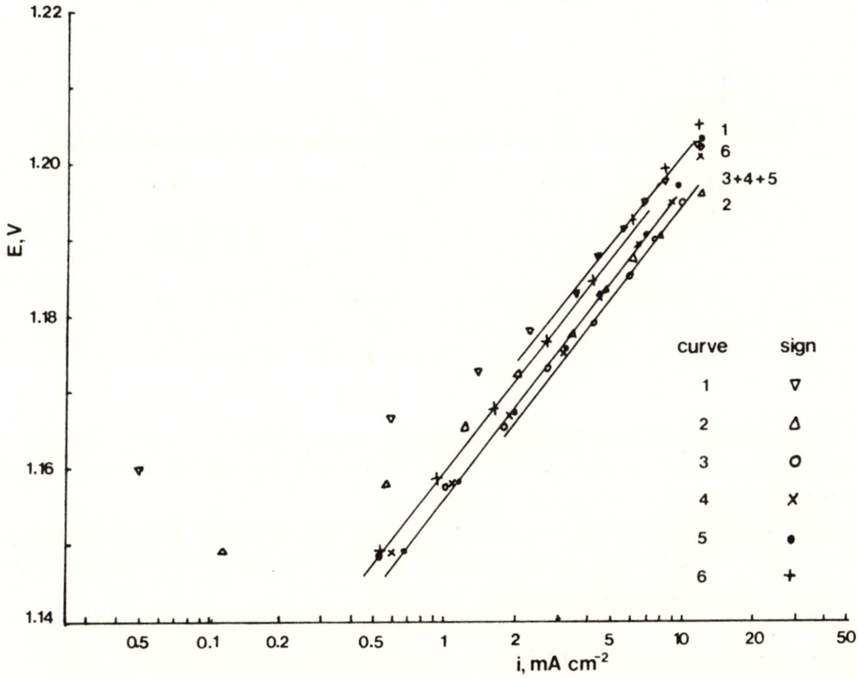

Fig. 16.6 – Plot of potential versus the logarithmic of current density for an RTO electrode with 30 mole-% RuO_2, at 25°C and in chlorine-saturated chloride solutions of various compositon (1:0.4M NaCl + 0.1M HCl, E_r = 1158 mV; 2:0.6M NaCl + 0.15M HCl, E_r = 1149 mV; 3:1M NaCl + 0.25M HCl, E_r = 1133 mV; 4:2M NaCl + 0.5M HCl, E_r = 1114 mV; 5:3M NaCl + 0.75M HCl, E_r = 1098 mV; 6:4M NaCl + 1M HCl, E_r = 1073 mV.

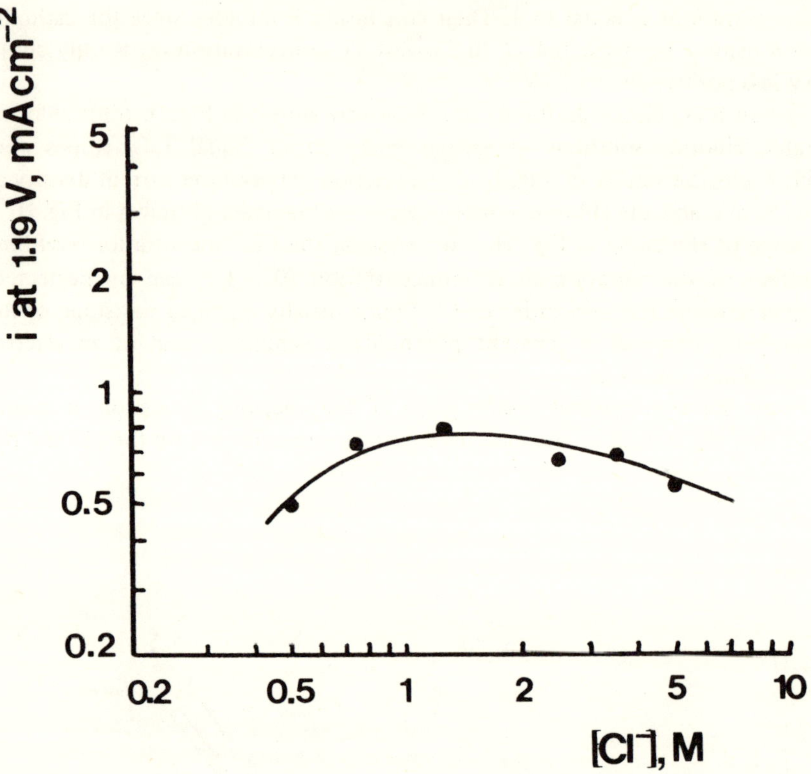

Fig. 16.7 – Plot of current density at E = 1190 mV and chloride concentration on a double logarithmic scale for an RTO electrode with 30 mole-% RuO_2 at 25°C in chlorine-saturated chloride solutions where $[Cl^-]/[H^+] = 4$.

Cl^--concentration. Consequently, the already determined experimental reaction orders with respect to chloride are not useful in elucidating the mechanism of chlorine evolution.

The literature data on the current density at a constant potential have a wide spread. Nevertheless, some data at E = 1150 mV are given to illustrate the activity of electrodes and its spread.

For an RTO electrode with about 35 mole-% RuO_2 in a 1.5 M HCl solution saturated with chlorine, at 30°C, i = 1.8 mA/cm^2 [17] and 3 mA/cm^2 (deduced from [18, 20]); in a 1.5 M HCl + 2.5 M NaCl solution saturated with chlorine, i = 0.6 mA/cm^2 [17] and 18 mA/cm^2 [15] were found. Nernst's equation was used to calculate E_r assuming the activity and the concentration of Cl^--ions as being equal.

The potential decay curves for an RTO electrode with 30 mole-% RuO_2 are similar to that for an electrochemical RuO_2 electrode [5]. The slope of the linear section of the $\eta_n/\log t_n$ curve does not depend on the current density which was used during the anodic polarization before switching off the current. This slope is about 20 mV [5].

16.3.4 The MO-electrode

The MO electrode consists of an active mass with only a small percentage of RuO_2. The behaviour of this type of electrode depends greatly on the RuO_2 content and the nature and content of other metals or metal oxides. A discussion on MO electrodes is omitted since no useful or additional information is obtained to elucidate the reaction mechanism of chlorine evolution on electrodes based on RuO_2 and TiO_2 and used successfully for industrial chlorine production.

16.4 DISCUSSION AND CONCLUSIONS

To elucidate the mechanisms of chlorine formation on RuO_2 and RTO electrodes, the Tafel slopes, the slopes of potential decay curves and the reaction orders with respect to chloride ions have been used by several authors. The preceding elucidation of the results shows that the results obtained for the reaction order with respect to chloride ions are very dubious, since the nature of the electrode surface depends on $[Cl^-]$ and pH. Consequently, the observed reaction orders for Cl^-, varying from -0.2 to 1.0, cannot be used for elucidating the reaction mechanism. Generally, the minimum Tafel slope for chlorine evolution on electrochemical and chemical RuO_2 electrodes as well as on RTO electrodes, at 30°C and in concentrated chloride solutions saturated with chlorine, is about 40 mV. A Tafel slope of about 30 mV has been found for chlorine-free concentrated chloride solutions at 30°C [11, 13]. From theoretical $E/\log i$ curves for both the Volmer–Tafel and the Volmer–Heyrowsky mechanism [22] it follows that at 30°C a Tafel slope of about 30 mV can occur if the diffusion of chlorine away from the electrode surface into the solution if the rate-determining factor. The temperature dependence of the Tafel slope, found for flowing concentrated NaCl solutions containing no chlorine [13] can well be explained assuming that both the rate of chlorine diffusion and the rate of chlorine evolution affect the Tafel slope. It is likely to assume that the activation energy for chlorine diffusion is smaller than the one for chlorine evolution. Hence, the ratio between the diffusion coefficient of chlorine and the rate constant of chlorine evolution decreases with increasing temperature. So, at increasing temperature the Tafel slope is determined (in increasing rate) by chlorine diffusion. Consequently, both the slopes, viz. 30 and 40 mV can be explained with the Volmer–Heyrowsky mechanism where the Volmer reaction is in quasi equilibrium. The Volmer–Tafel mechanism must be excluded.

This conclusion is supported by the potential decay curves for chlorine evolution on RuO_2 and RTO electrodes [5]. Also the high Tafel slope, 110 mV at 30°C, obtained by Kuhn and Mortimer [6] can be explained very well with the Volmer–Heyrowsky mechanism. In this case, both the Volmer and the Heyrowsky mechanisms determine the rate of chlorine formation. This reaction mechanism was based on a Tafel slope of 30 mV, the reaction order of Cl^--ions being 1.

The mechanisms proposed by Krishtalik [1], and in particular the barrierless reaction $Cl_{ad} \rightarrow Cl^+_{ad} + e^-$, is very unlikely. As far as we know, a barrierless reaction was proposed in literature for only very complicated systems, e.g. chlorine evolution on graphite anodes.

The activity of electrodes on an RuO_2/TiO_2 base depends on the conditions of their preparation (e.g. the RuO_2/TiO_2 ratio, the firing temperature, the firing time and the thickness of coating layer) and on electrolytic conditions (e.g. the Cl^--concentration, pH and temperature). In particular, the effect of the firing process on the activity of the electrode has to be investigated extensively. Moreover, more attention must be paid to the effect of chloride concentration and pH upon the nature of ruthenium oxide electrodes.

16.5 REFERENCES

[1] Krishtalik, L. J., *Electrochim. Acta*, **16**, 329 (1981).

[2] Hine, F., and Jasuda, M., *J. Electrochem. Soc.*, **121**, 1289 (1974).

[3] Vetter, K. J., *Elektrochemische Kinetik*, Springer Verlag, Berlin, 1961.

[4] Janssen, L. J. J., *Electrochim. Acta*, **15**, 941 (1970).

[5] Janssen, L. J. J., Starmans, L. C. M., Visser, J. G., and Barendrecht, E., *Electrochim. Acta*, **22**, 1093 (1977).

[6] Kuhn, A. T., and Mortimer, C. J., *J. Electrochem. Soc.*, **120**, 231 (1973).

[7] Burke, L. D., and O'Neill, J. F., *J. Electroanal. Chem.*, **101**, 341 (1979).

[8] Burke, L. D., and Murphy, O. J., *J. Electroanal. Chem.*, **96**, 19 (1979).

[9] Burke, L. D., and Murphy, O. J., and O'Neill, J. F., *J. Electroanal. Chem.*, **81**, 391 (1977).

[10] Ehrenburg, R. G., Krishtalik, L. J., and Yaroshevskaya, I. P., *Soviet Electrochemistry*, **11**, 993 (1975).

[11] Veselovskaya, I. E., Spasskaya, E. K., Sokolov, V. A., Tkachenko, V. I., and Yakimenko, L. M., *Elektrokhimiya*, **10**, 70 (1974).

[12] Bystrov, V. I., and Romashin, O. P., *Soviet Electrochemistry*, **11**, 1141 (1975).

[13] Ivanter, I. A., Pechnikova, G. P., and Kubasov, V. L., *Soviet Electrochemistry*, **12**, 740 (1976).

[14] Arikado, T., Iwakura, C., and Tamura, H., *Electrochim. Acta*, **23**, 9 (1978).

[15] Ehrenburg, R. G., Krishtalik, L. I., and Yaroshevskaya, I. P., *Soviet Electrochemistry*, **11**, 989 (1975).

[16] Ehrenburg, R. G., Krishtalik, L. I., and Bystrov, V. I., *Soviet Electrochemistry*, **8**, 1690 (1972).

[17] Burrows, I. R., Entwisle, J. H., and Harrison, J. A., *J. Electroanal. Chem.*, **77**, 211 (1977).

[18] Denton, D. A., Harrison, J. A., and Knowles R. I., *Electrochim. Acta*, **24**, 517 (1979).
[19] Tilak, B. V., *J. Electrochem. Soc.*, **126**, 1343 (1979).
[20] Burrows, I. R., Denton, D. A., and Harrison, J. A., *Electrochim. Acta*, **23**, 493 (1978).
[21] Hine, F., Yosuda, M., and Yoshida, T., *J. Electrochem. Soc.*, **124**, 500 (1977).
[22] Janssen, L. J. J., Visser, G. J., and Barendrecht, E., *Electrochim. Acta*, **28**, 155 (1983).

16.6 OPEN DISCUSSION

1. Dr Warren of Chemetics, made the point that Russian workers had studied the evolution of chlorine from a chlorine solution at a given electrode and also the evolution of oxygen on the same electrode in a sulphuric acid solution and had evolved the Tafel slopes for both evolution reactions which could then be compared to characterize the respective reactions.

   Dr Janssen replied that he was aware of this work but felt that the results were virtually the same with or without sulphuric acid. Also the experimental support for their results – derived at a very limiting chloride ion concentration – was, in his opinion rather weak.

2. Dr Harrison of the University of Newcastle upon Tyne, expanded the discussion by contrasting this paper with his own (see Chapter 15). There were clearly two different opinions as to the mechanism for chlorine evolution. In support of the mechanism proposed in his paper, he went on to say that they had studied a large number of ruthenium dioxide/titanium oxide electrodes and have found that using the proposed mechanism for the evolution of chlorine via hydrochlorite and also assuming that chlorine gas bubbles interfere with that mechanism, they could account for every phenomenon that had been observed in their laboratory.

   In reply to a question from Prof. Fleischman in the Chair, Dr Harrison stated that they had investigated all the parameter changes that they could, for example, concentration, pH, the catalytic structure and composition of the electrode so he felt that the discrepancy was not due to pH differences between the two sets of results.

   Prof. Fleischman continued by saving that there is a formal similarlity between the two proposed mechanisms in the sense that there is a discharge step following by some subsequent step. In Dr Harrison's case, this is a chemical step, in Dr Janssen's case it is an electrochemical step coupled to diffusion.

The mechanism of electron transfer is included by a 40 mV Tafel slope but apart from that, all the aspects of electron transfer are accounted for was the reply from Dr Harrison.

Dr Janssen then commented that Dr Harrison had considered the effect of hypochlorite on the electrode reaction and it is well known that the concentration of hypochlorite increases with increasing pH. This means that at a pH of 1 there are no HOCl molecules. In Dr Janssen's work it had been found that the Tafel slope depends on the pH. At a pH of 3, the Tafel slope increases very quickly. According to his theory the presence of hypochlorite ions affects the nature of the electrode surface and he believed that is the difference between Dr Harrison's opinion and his own. He considered the effect of hypochlorite, chlorate, etc. on the nature of the electrode surface and Dr Harrison considered a constant nature and ascribed the observed phenomena to the reaction of hypochlorite and chlorate.

NOMENCLATURE

b slope of $E/\log i$ curve
E electrode potential
E_r reversible electrode potential
i current density
t_n time after switching off the polarization current
z_{Cl^-} order of electrode reaction for Cl^--ions
z_{Cl_2} order of electrode reaction for Cl_2 molecules
η_n overpoential on open circuit; $\eta_n = E - E_r$
θ degree of coverage with atomic chlorine

SECTION E

Chlorate Manufacture

CHAPTER 17

Energy Saving in Chlorate Production with the Use of a Fuel Cell

I. H. WARREN, Chemetics International Company, Vancouver, B.C., V6J 1C7

17.1 INTRODUCTION

In the production of sodium chlorate by the well known electrolysis of a solution of sodium chloride in a diaphragmless cell, 630 m^3 of hydrogen are theoretically produced per tonne of sodium chlorate, according to the reaction stoichiometry:

$$NaCl + 3H_2O \xrightarrow{e} NaClO_3 + 3H_2 \qquad E_0 = 1.45V .$$

The evolved hydrogen is, in present practice, either vented to atmosphere and lost, burned in a boiler for steam production, or collected and purified for sale as commodity hydrogen.

With the development of fuel cell technology two additional alternatives can be considered for recovery of the energy in the hydrogen. The first alternative would utilize fuel cell technology in a so-called air depolarized cathode inserted in a chlorate cell. The overall reaction for chlorate production would then become:

$$NaCl + 3H_2O \xrightarrow[1.5O_2]{e} NaClO_3 + 3H_2O \qquad E_0 = 0.22V .$$

This alternative, although providing an obvious opportunity to decrease power consumption in the chlorate cells, would have the process disadvantage of essentially adding unwanted water to the chlorate system and, in any event, would require completely redesigned electrolyzers. The second alternative, which is the subject of the present study, would utilize the hydrogen from the chlorate circuit in a conventional fuel cell and would feed the electrical and thermal energy developed back to the chlorate production system. The simplicity of this alternative would allow the most advanced commercially available fuel cell to be added to any existing or future chlorate circuit.

17.2 THE FUEL CELL AND ITS OPERATING CHARACTERISTICS

The phosphoric acid electrolyte fuel cell [1,2] has been developed by United Technologies to the stage where two, approximately 5 MW, fuel cell power stations are about to commence operation. Alkaline fuel cells are being developed primarily for vehicular applications but none are available with the power output even approaching that of the United Technologies acid electrolyte cells, each module of which is rated at approximately 0.5 MW. In addition to their present unavailability in large sizes, the alkaline fuel cells operate at low temperatures (70°C) and thus the quality of the waste heat is significantly inferior to that produced by acid cells which operate in the temperature range 150–200°C.

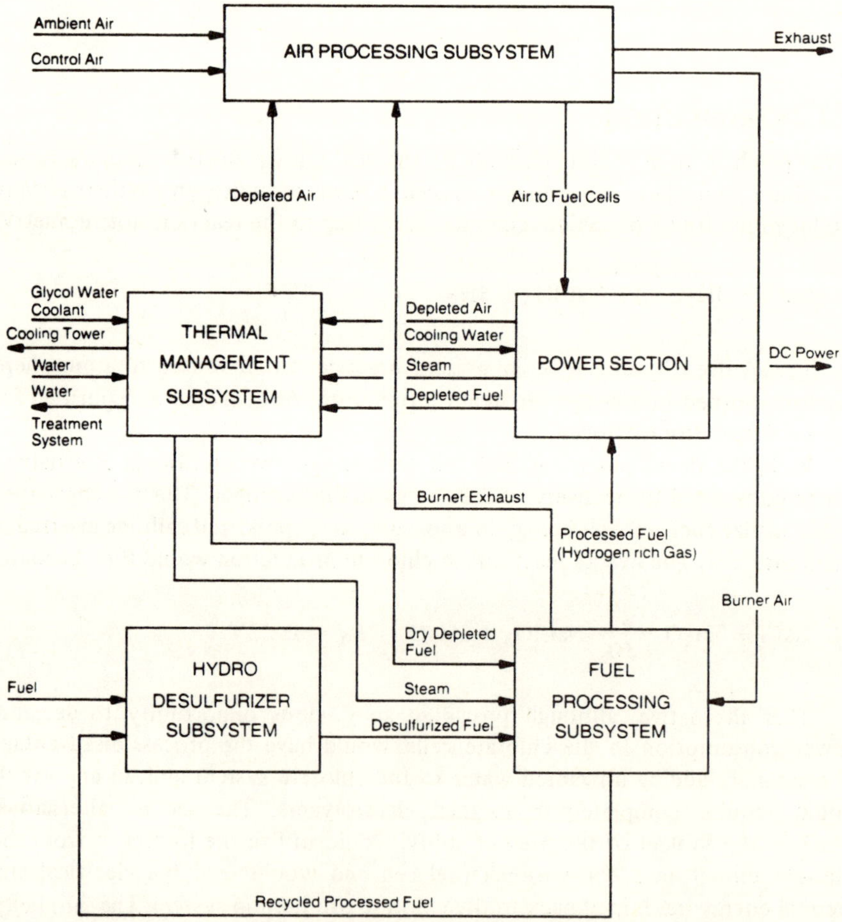

Fig. 17.1 – Functional block diagram of fuel cell utility application.

In the United Technologies free-standing utility power plant application the fuel cells are fed with hydrogen at 345 kPa produced from the refining of natural gas (Fig. 17.1). Air is fed to the cells, also at 345kPa, and the fuel cell reactions produce electricity, heat and water:

Anodic reaction $2H_2 \longrightarrow 4H^+ + 4e$

Cathodic reaction $O_2 + 4H^+ + 4e \longrightarrow 2H_2O$.

The maximum amount of electrical work (W_m) that can be derived from a fuel cell is related to the Gibbs free energy (ΔG) of the overall cell reactions:

$$W_m = \Delta G = -nFE ,$$

where n is the number of electrons in the overall reaction, F is the Faraday constant, and E is the reversible voltage of the reaction. For the hydrogen oxygen fuel cell operating at 25°C E is 1.22V. In practice the cells operate at between 0.65V and 0.8V corresponding to electrical generation efficiencies of between 44% and 54% based on the higher heat of combustion and 53% and 65% for the lower heat of combustion of the hydrogen. The typical instantaneous performance characteristic of a phosphoric acid fuel cell is that of a decreasing cell voltage with increasing power density (Fig. 17.2). During the lifetime of the fuel stack (presently 5 years) the output at a given power density falls at a rate of approximately 2% per year.

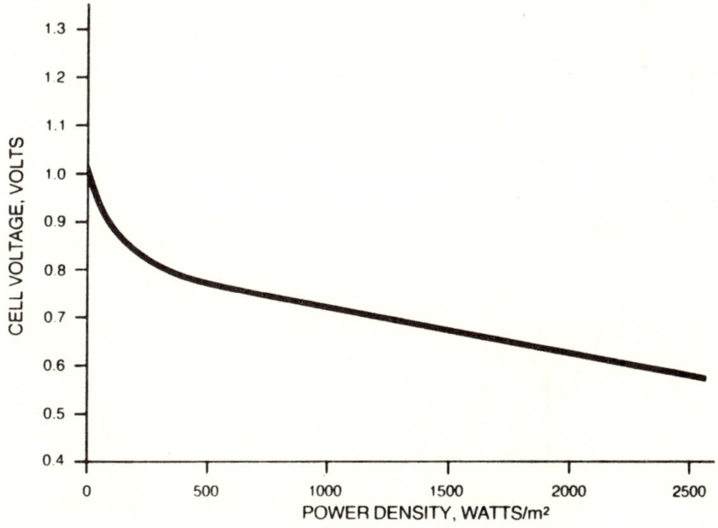

Fig. 17.2 – Fuel cell voltage–power characteristics.

Since the enthalpy change of a hydrogen oxygen fuel cell is more negative than the Gibbs free energy, the part of the total energy change of the reaction which cannot be converted to electrical energy is given out as heat. The majority of the (~90%) heat is removed from the United Technologies fuel cell via water circulated through the coolant channels in the fuel cell stack. The balance of the heat is removed by the water produced in the cell reactions, which is evaporated to produce a hot, moist cathode effluent stream. Design for operation of the fuel

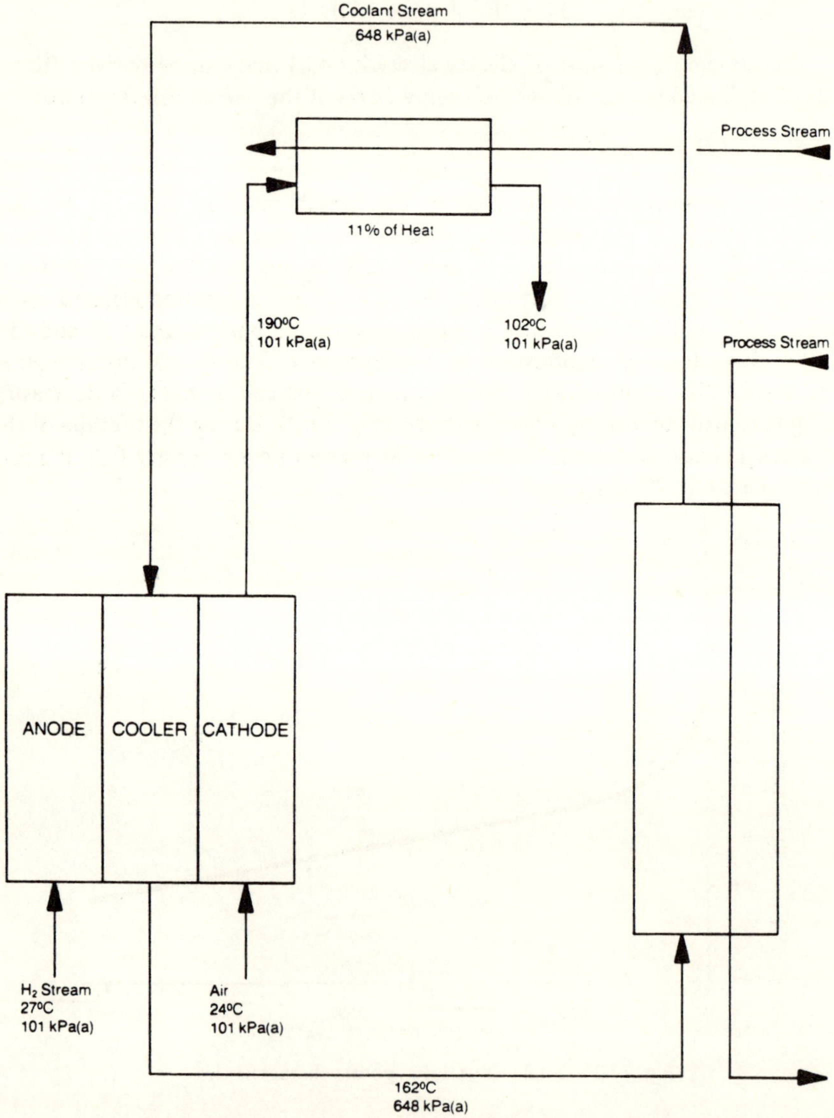

Fig. 17.3 – Thermal outputs from fuel cell when operated at atmospheric pressure.

cells at pressures other than 345 kPa is possible. Typically for atmospheric pressure operation 89% of the thermal energy would be available as 162°C, 648 kPa steam and 11% would be available at a maximum temperture of 190°C (Fig. 17.3).

17.3 THE CHLORATE CELL AND ITS OPERATING CHARACTERISTICS

Sodium chlorate is produced by the electrolytic production of sodium hypochlorite which is subsequently transformed in a separate reactor to sodium chlorate. The reaction sequences are:

Electrolyzer $NaCl + H_2O \xrightarrow{e} NaOCl + H_2$

Reactor $3NaOCl \longrightarrow NaClO_3 + 2NaCl$.

Parasitic reactions produce oxygen in amounts which depend inter alia upon operating conditions. Typically the evolved hydrogen contains 2% oxygen. This oxygen, together with water vapour and small amounts of chlorine are present in the evolved hydrogen. The chlorine content of the hydrogen can be reduced by caustic scrubbing to less than 1 p.p.m.

Table 17.1

Total energy requirements in steam for various operating modes of chlorate plants.

Output	Total heat requirement kcal/h/tonne $NaClO_3$/day ($\times 10^2$)	Minimum practical steam Supply pressure (kPa)
Liquor (Fig. 17.4) 640 g/l $NaClO_3$ 110 g/l NaCl		
60°C operation	194	Reboiler – 240
80°C operation	137	Ejector – 444
Crystal with booster ejector (Fig. 17.5)		
60°C operation	577 (Reboiler 347)	Reboiler – 240
80°C operation	554 (Reboiler 324)	Booster ejector – 444
Crystal with reboiler		
60°C operation	527 (Reboiler 347)	Reboiler – 240
80°C operation	504 (Reboiler 324)	Ejector – 444

In the electrolyzers the electrical resistance of the electrolyte together with the overpotentials at the electrodes causes heat to be generated by the electrolyzing current in excess of that required to elevate the electrolyte to the working temperature of 60–80°C. This excess heat is removed coventionally by heat exchangers in the chlorate reactor circuit. However, the overall chlorate production system operates in a thermal deficit situation because heat is required to evaporate water from the make-up brine solutions and also to effect crystallization of the sodium chlorate, if crystal, rather than solution, is the plant product (Table 17.1 and Fig. 17.4 and 17.5).

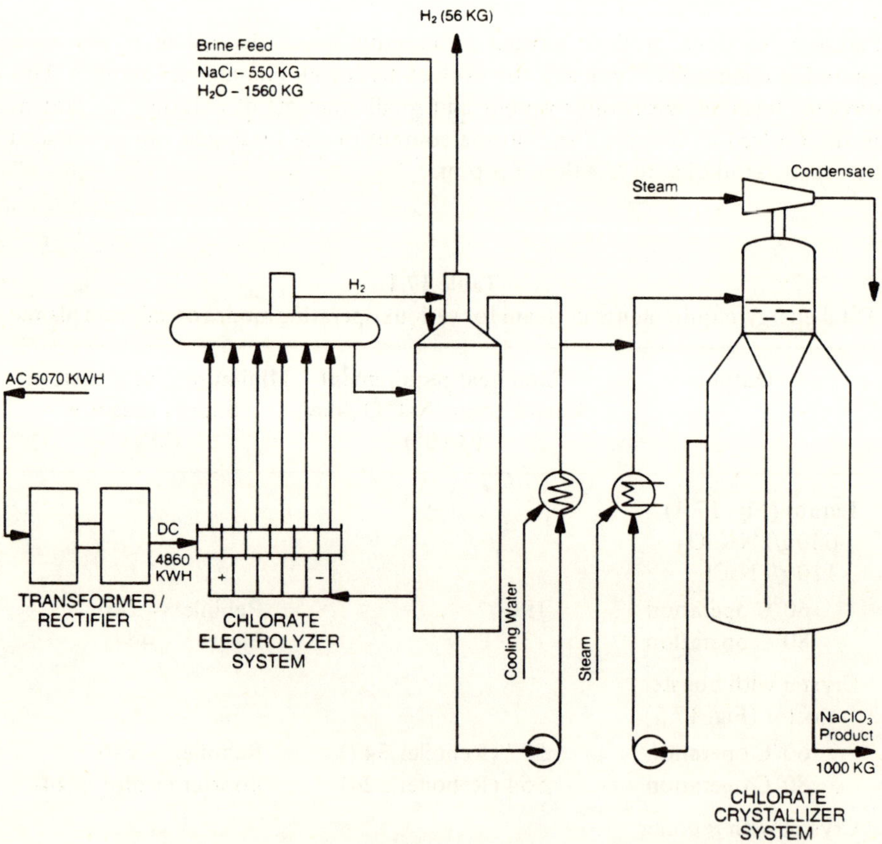

Fig. 17.4 – Chlorate–crystal product system.

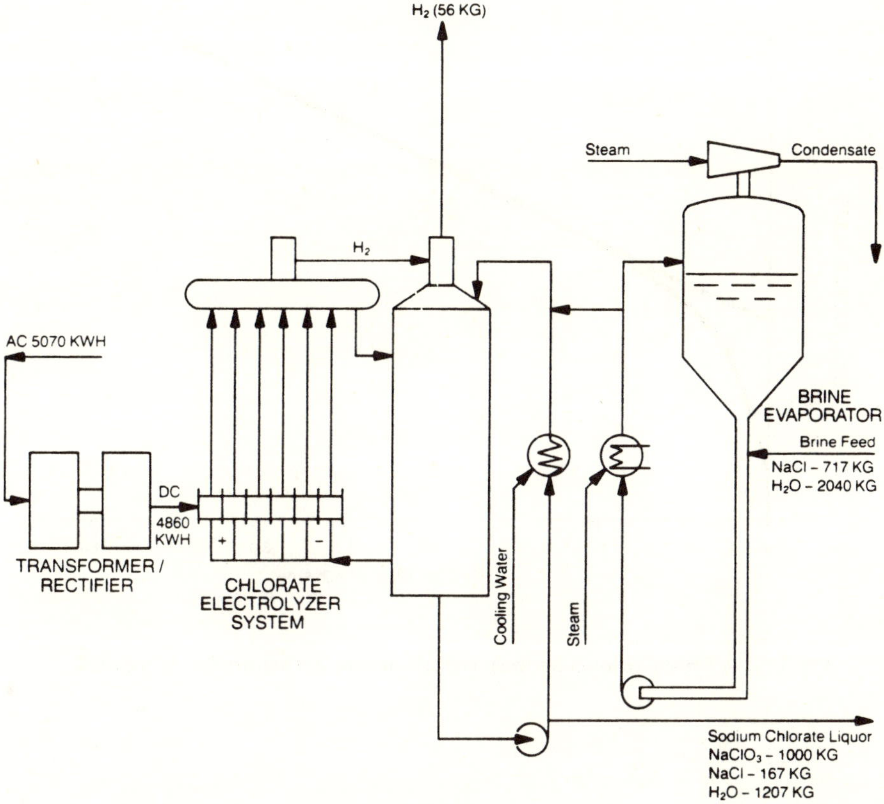

Fig. 17.5 – Chlorate–liquid product system.

The general instantaneous electrical characteristics of chlorate cells are typical of those of any electrolysis system and show a rising input voltage requirement for increasing current density (Fig. 17.6). Due to precipitation of alkaline earth hydroxides on the cathode surfaces the voltage required for a given current density may increase with time at a rate of approximately 2% per month.

17.4 INTEGRATION OF A FUEL CELL WITH THE CHLORATE SYSTEM

The ideal integration of a fuel cell with the chlorate system would be achieved if the hydrogen from the chlorate cells could be fed directly to the fuel cells and the generated d.c. electric power then fed back directly to the chlorate cells. Waste heat developed in the fuel cell would also, ideally, match the thermal requirements of the chlorate system. However, it is evident upon comparison of the characteristics of the fuel cell and the chlorate cell that simple electrical mating of the two overall systems cannot be achieved. The essentially opposite

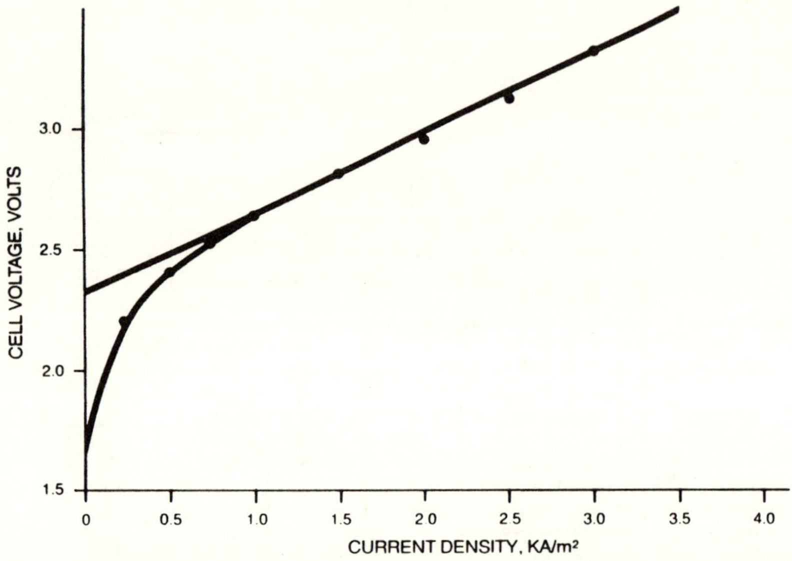

Fig. 17.6 – Typical relation between voltage and current density for chlorate cell.

instantaneous current-voltage response of the two systems would preclude any possibility, for example, of load following by the fuel cell as the total chlorate plant output was varied by changes in the current to the cells.

Long term and opposite variations in the voltage–current relationships of the two systems also appear to preclude the possibility of achieving integration for example; by feeding the fuel cell electrical output to an isolated cell bank, which is maintained at constant instantaneous chlorate output by supplementing the hydrogen supply to the fuel cell with hydrogen from a reformer even as chlorate and hydrogen production in the mains fed chlorate cells is varied (Fig. 17.7). Various other possibilities can be considered for electrical integration. Direct integration can be obtained by connecting the fuel cell stacks in parallel with the line rectifier and achieving voltage control with a buck regulator. This technique has the advantages that the circuitry is similar to that used in some electrical transportation systems, electrical efficiency in the high 90s is possible and the control of voltage or current can be established by essentially production-type hardware. The general electrical characteristics for operation of a fuel cell system with a buck regulator in conjunction with a nominal 5 MW capacity chlorate plant are shown in (Fig. 17.8).

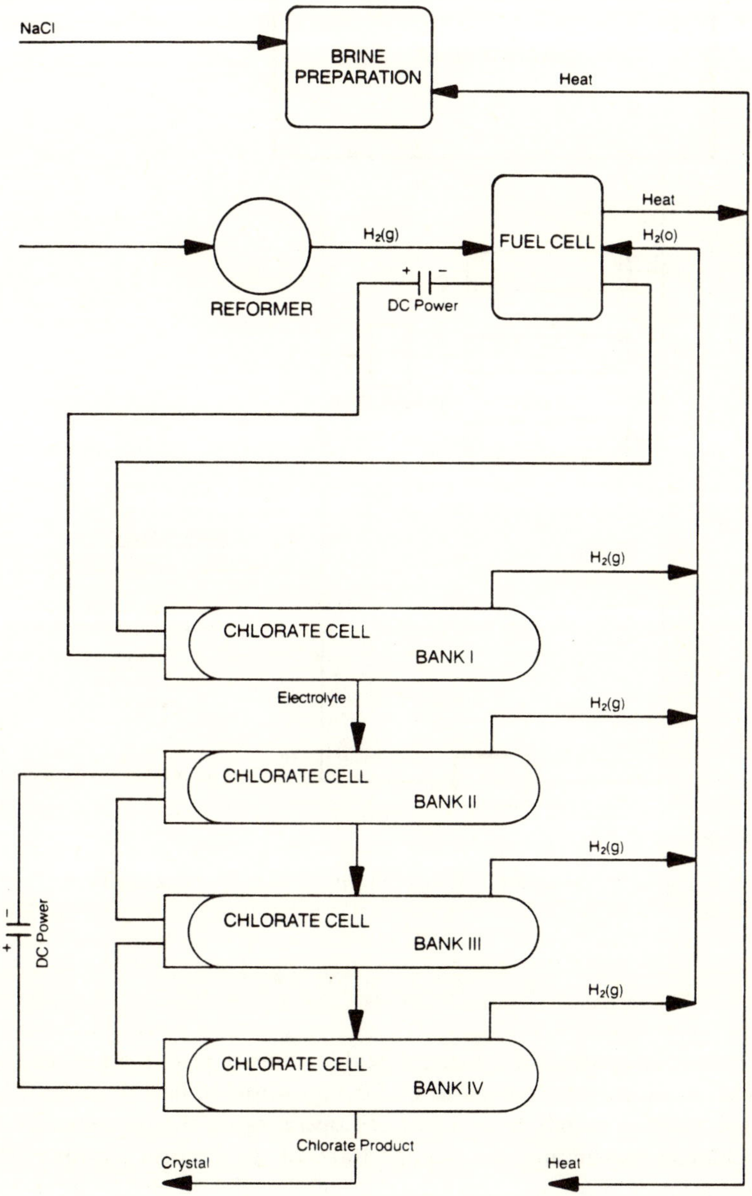

Fig. 17.7 – Fuel cell–chlorate cell integration configuration for separate cell bank.

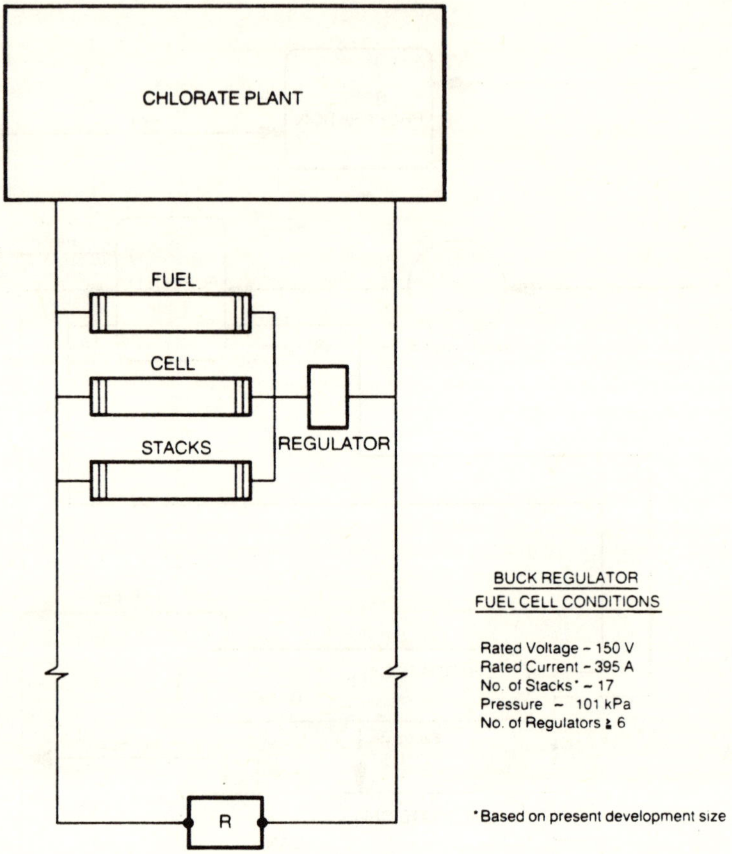

Fig. 17.8 – Integration of fuel cell with chlorate via buck regulator.

Thermal integration of a fuel cell with a chlorate plant is essentially simply achieved. As shown from comparison of the quantities of heat in Tables 17.1 and 17.2 the thermal output from a fuel cell operating on the recovered hydrogen from chlorate production can supply 100% of the thermal requirement of a solution chlorate plant and approximately 60% of the thermal requirement of a crystal chlorate plant – sufficient for operation of the brine evaporator reboiler.

It can be shown that operation of the fuel cell at other pressures and current densities will change the relative amounts of electricity and heat developed but not so as to significantly affect the pattern of integration outlined above.

Table 17.2
Thermal and electrical integration of fuel cell with chlorate system

Basis

Chlorate plant operating at 80°C produces sodium chlorate for expenditure of 4860 d.c.-kWh/tonne $NaClO_3$ and produces hydrogen at the rate of 658 m^3/tonne $NaClO_3$.

Fuel cell operation at atmospheric pressure and current density corresponding to Design Point 1 (Figs. 17.9 and 17.10) produces steam at the rate of 1239 kcal/m^3 of hydrogen and electrical energy 1.4 kWh/m^3 of hydrogen.

Fuel cell operation at 345 kPa and current density corresponding to Design Point 2 produces steam at the rate of 1318 kcal/m^3 of hydrogen and electrical energy 1.3 kWh/m^3 of hydrogen.

Design Point 1 operation provides 919 kWh of power/tonne $NaClO_3$ and 340×10^2 kcal/h/tonne $NaClO_3$/day.

Design Point 2 operation provides 854 kWh of power/tonne $NaClO_3$ and 361×10^2 kcal/h/tonne $NaClO_3$/day.

17.5 THE ECONOMICS OF INTEGRATION OF A FUEL CELL WITH THE CHLORATE SYSTEM

The projected cost of a fuel cell of the typical capacity required for chlorate system integration i.e. approximately 1MW, depends upon such units being in production for a utility power program, since this represents a very much larger potential market than the chlorate market. However, on the assumption that a fuel-cell manufacturing facility is supplying the utility market it has been estimated that the installed cost of a completely integrated fuel cell would be in the range of \$435/kW–\$485/kW of power output.

On the basis of two interest rates and two write-off periods (Table 17.3) it appears that, on the basis of value of electric power generated alone, the fuel cell should, under the most unfavourable write-off period and interest selected, be economic when power costs reach 2.4¢/kWh. It can be shown that by operation of the fuel cell at higher power densities equivalent to Design Point 2 (Figs. 17.9 and 17.10) the cost of the fuel cell would be lowered and the equivalency point, under the same most unfavourable economic conditions selected, reduced to 2.1¢/kWh.

If the value of steam generated is assumed at \$5/tonne the power cost for the equivalence point can be decreased to 1.55¢/kWh for Design Point 1 and 1.45¢/kWh for Design Point 2.

Table 17.3

Estimation of value of power saved with integrated fuel cell and comparison with capital expenditure via capital recovery factor

Basis

Nominal 5 MW sodium chlorate plant and 0.88 MW fuel cell with fuel cell cost 485$/kW.

Write off period and rate of interest	Power cost for equivalency of savings with annual payments (¢/kWh)
3 y at 10%	2.3
3 y at 15%	2.4
5 y at 10%	1.5
5 y at 15%	1.7

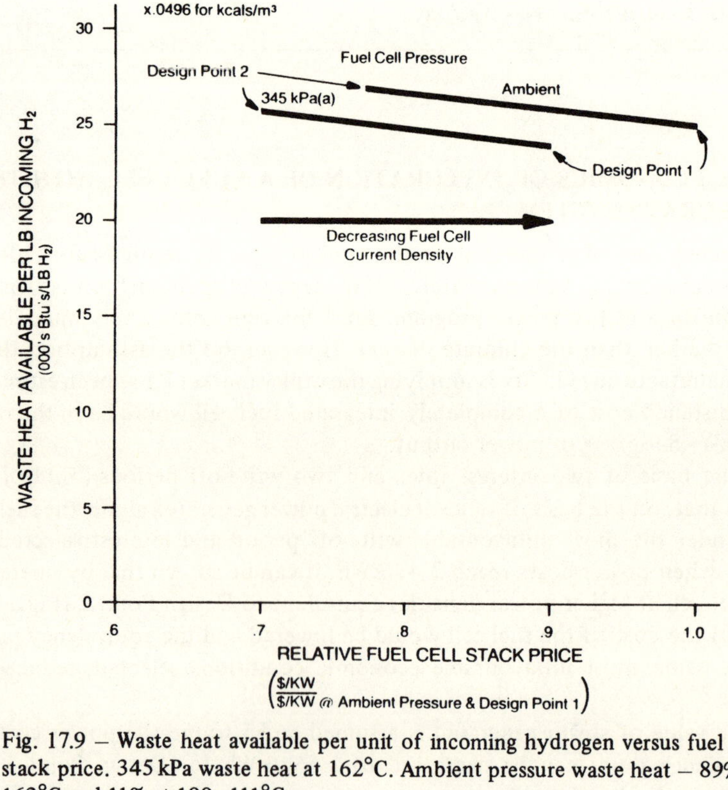

Fig. 17.9 – Waste heat available per unit of incoming hydrogen versus fuel cell stack price. 345 kPa waste heat at 162°C. Ambient pressure waste heat – 89% at 162°C and 11% at 190–111°C.

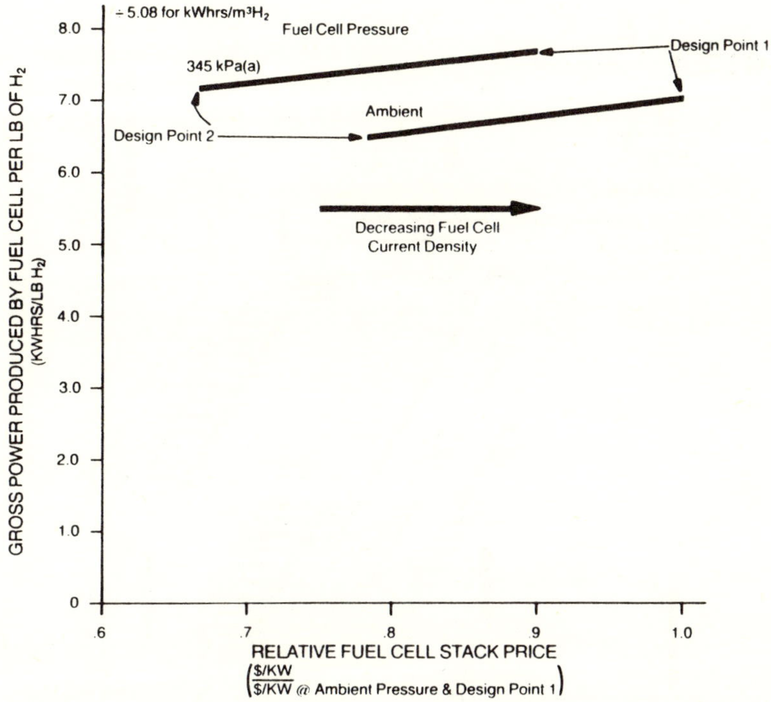

Fig. 17.10 – Electrical energy available per unit of incoming hydrogen versus cell cell stack price. (kWh not adjusted for parasitics except for turbocompressor used to pump process air in 345 kPa case. Hydrogen assumed delivered at fuel cell pressure.)

17.6 DISCUSSION AND CONCLUSIONS

The preliminary data developed suggest that an integrated fuel cell–chlorate cell could be, in the future, an economically interesting system for energy recovery. The energy produced by the fuel cell would decrease the electrical requirement for chlorate production by 18–20% as well as supplying significant thermal energy for the process. The electrical generation efficiency of the fuel cell operating at atmospheric pressure in this application would be 40%. For comparative purposes, combustion of hydrogen in a gas turbine, having approximately the same capital cost as a fuel cell would generate electricity at only 22% efficiency. A boiler linked with a back-pressure turbine would also have a similar capital cost but would generate electricity with only approximately 10–12% efficiency.

The advantages attached to the use of a phosphoric acid fuel cell for integration with a chlorate production system are:

(a) The fuel cell technology is developed and commercial-sized units are available.

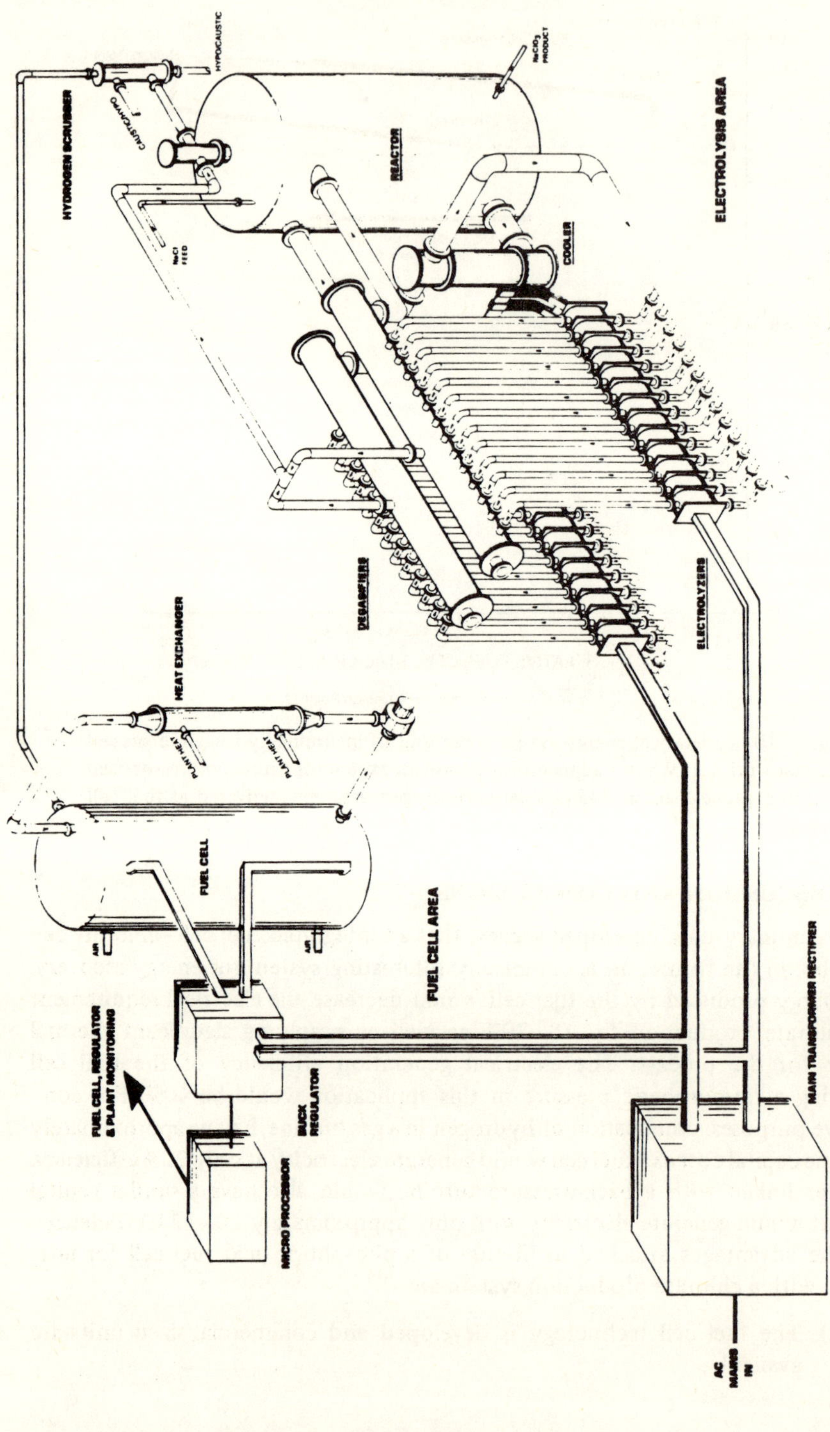

Fig. 17.11 – Schematic of fuel cell integration with sodium chlorate plant.

(b) Electrical and thermal integration of the fuel cell and chlorate cell can be simply achieved with no alteration of existing chlorate cells.
(c) The fuel cell has no moving parts and recovers proportionately the largest amount of energy as electric power of any presently available system.

The disadvantages of fuel cell integration are:

(a) Uncertainty with respect to the rate of growth of electric utility utilization of fuel cells and thus uncertainty of cost/time relationship of fuel cell units.
(b) Continued dependence of a boiler for additional steam supply if integration is with a chlorate plant producing crystal product.
(c) Lack of presently demonstrated capability of fuel cell to operate on a continuous and reliable basis when integrated with a chemical plant.

The advantages of the application of fuel cell technology to the chlorate system, together with its future economic potential, suggest that a demonstration of the integrated system should be undertaken.

17.7 ACKNOWLEDGEMENTS

The author acknowledges the receipt of a British Columbia Science Council grant which made this study possible and also the assistance received in the study by United Technologies Power Systems Division.

17.8 REFERENCES

[1] Eklund, L. G., and Handley, L. M., *Fuel Cell Power Plants for Utility Applications*, Proc. IEEE, 6, Conf., pp. 143–7, Feb. (1980).
[2] *Advanced Technology Fuel Cell Program*, Summary Report, March (1981), EPRI EM-1730-SY Project 114.

17.9 OPEN DISCUSSION

1. Mr J. L. Thakkar of Dharamsi Morarji Chemicals, India, asked for the capital cost per kilowatt for the fuel cell installation.

   Dr Warren replied that the complete installed capital cost, including the back regulator was about 450 $/kW.

2. The session chairman (Mr M. N. Dudley of IMI Marston) asked whether the quoted power savings included the presence of a boiler which it was mentioned still had to be there to take up the imbalance i.e. did the overall savings include the fuel burned in the boiler?

In all cases hydrogen was given zero value so that any system could be compared on a like with like basis. The power savings shown were purely and simply the generated power offset against the cost of the fuel cell amortized over 3 years with a capital recovery factor and the second case, which was about 1.55¢/kWh equivalency assumed the thermal energy was fed back to the system. With a liquor plant you do not require any more heat; for a crystallization plant you still need a boiler but it is there now anyway, unless a grass roots plant is being considered, so should not be regarded as additional.

An interesting fact is that the fuel cell operating to give an output of about 900 kWh/t represents about 40% recovery of energy as electric power. Burning hydrogen in a boiler and operating a conventional power generation system, which would be very small from a chlorate cell system, would at best give between 10 to 15% electrical generation efficiency. This would increase to between 20 and 22% using a gas turbine and the capital cost of these units would be about equivalent to the cost of the fuel cell unit. Thus it can be seen that the fuel cell is a very attractive way in comparison with other present systems, of recovering electrical energy.

3. Mr C. H. K. Rao of the Titanium Equipment and Anode Manufacturing Co., India, asked whether there were any restrictions on hydrogen purity for the fuel cell.

   The oxygen content is combusted in the electrode. If the content is greater than about 2% problems may arise due to heat generation on the catalyst. We are assured that 1 p.p.m. chlorine would not be a problem but greater concentrations might lead to attack of the catalyst. Fuel cell electrodes are graphite with platinum metal alloy deposited and the accumulation of large amounts of chloride would be a problem. However, carbon dioxide is no problem at all, that's the beauty of the acid phosphoric fuel cell which in fact can run on a hydrogen/carbon dioxide mixture when the feedstock is naphtha or natural gas. There is no data available on the effect of mercury, which might be present in the hydrogen from mercury cells, but it would be expected to be a problem.

CHAPTER 18

LCD Mark II A New Sodium Chlorate Cell Design

C. G. RADER, E. M. SPORE, J. C. DELONG, Occidental Chemical Corporation

Occidental Chemical Corporation (OCC), formerly Hooker Chemical Company, a wholly owned subsidiary of Occidental Petroleum Corporation, has historically maintained a high degree of commitment to the development of process technology for the manufacture of sodium chlorate. This technology commitment realized startup of the first metal anode chlorate plant in the United States in 1973 at OCC's Taft, Louisiana, facility. The chlorate cell developed for this application, designated the DCB-100, represented a radical departure from the graphite electrode cell concepts previously employed for sodium chlorate manufacture.

A second generation metal anode chlorate cell was developed by OCC in 1977, to address the ever-increasing energy costs associated with electrolytic sodium chlorate manufacture. This electrolyzer, designated the LCD (low current density) chlorate cell, replaced the first generation DCB-100 cells at the Taft, Louisiana, commercial facility in 1980.

Concurrent to the development of the LCD chlorate cell, a computer simulation of the process occurring within a brine electrolyzer was developed to model the micro and macro phenomena. Optimization techniques were employed to develop criteria for a new cell design. Based on the computer simulation, equipment design ensued, predicated on a potential energy reduction of 20% if design optimization could be achieved on a practical scale. A 10 000 A prototype electrolyzer was installed and operated in a stand-alone pilot plant facility during 1981. Subsequently, a 100 000 A commercial-scale prototype was fabricated and installed in an operating circuit at OCC's Taft, Louisiana, chlorate plant. OCC's third generation metal anode chlorate cell, the LCD MK-II, is now available for commercial applications.

18.1 OVERALL CONFIGURATION

Figure 18.1 shows a schematic drawing of the LCD MK-II chlorate cell. The design combines the electrolyzer and the chemical reactor in a single unit. Sodium chlorate is produced by the electrolysis of NaCl and can be described by the overall chemical reaction

$$NaCl + 3H_2O \longrightarrow NaClO_3 + 3H_2 \quad . \qquad (1)$$

This reaction requires 6 F to produce one gram mole of chlorate at 100% current efficiency. The constitutive electrochemical reactions appear:

$$6Cl^- \longrightarrow 3Cl_2(aq) + 6e^- \quad \text{(at the anode)} \qquad (2)$$

$$6H_2O + 6e^- \longrightarrow 3H_2 + 6OH^- \quad \text{(at the cathode)} \qquad (3)$$

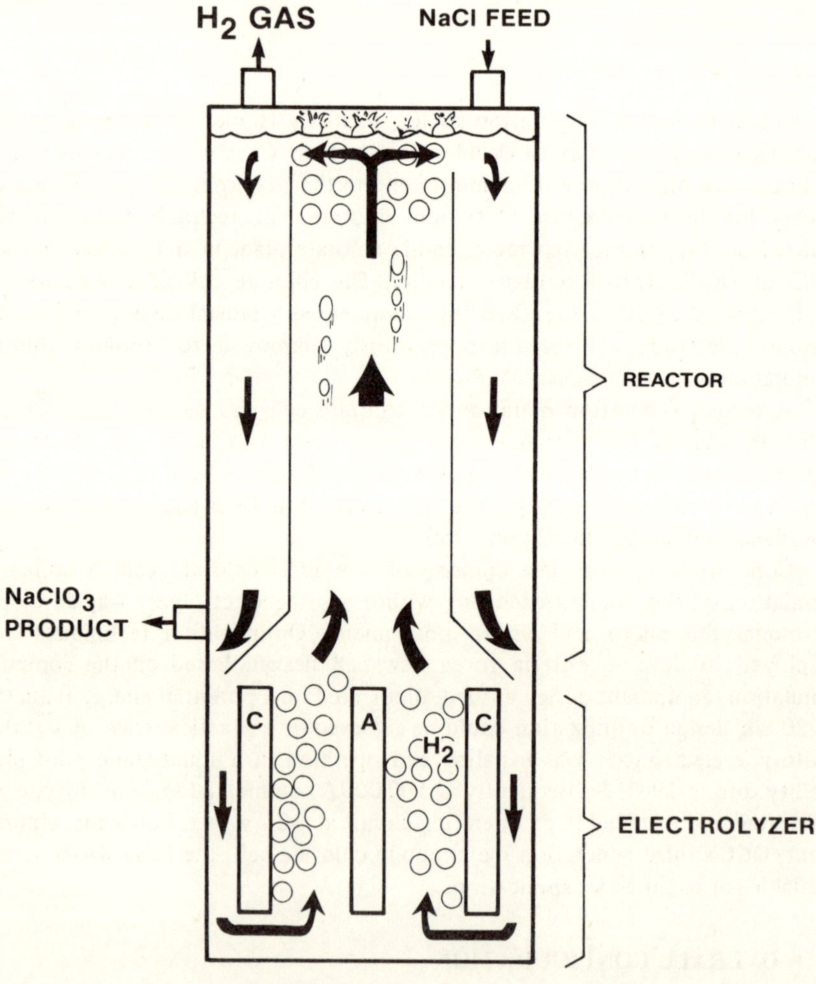

Fig. 18.1 – OCC LCD MK-II cell. Internal recirculation.

The anodically-formed Cl_2 is subsequently hydrolyzed to yield HOCl and OCl^- which, in turn, react in the bulk to form chlorate. This may be represented:

$$3Cl_2(aq) + 3H_2O \rightleftarrows 3HCl + 3HOCl \quad (4)$$

$$HOCl \rightleftarrows H^+ + OCl^- \quad (5)$$

$$2HOCl + OCl^- \longrightarrow ClO_3^- + 2HCl \quad (6)$$

Because reaction (6) proceeds very slowly, at controlled conditions of pH and temperature, volume of the reactor top is a key design parameter.

The present design configuration effectively utilizes the natural flow of solution induced by the H_2 gas evolved to control flow rates and velocities at various locations throughout the cell. The electrolysis of brine depletes NaCl in solution and forms H_2 gas which rises upward, 'lifting' the solution. The mixture of H_2 gas bubbles, NaCl, NaClO and $NaClO_3$ flows upward through the electrolyzer into the reactor top through a chimney structure contained within the reactor top. The hydrogen gas disengages from the solution and is removed from the reactor for purification and recovery.

Saturated brine solution is added at the top of the reactor to enrich the NaCl in solution. The NaCl, $NaClO_3$, and NaClO solution, overflowing the top of the chimney structure, is recirculated to the electrolysis zone. At this point, a portion of the solution of NaCl and $NaClO_3$ is removed as product.

Flanged openings are provided on the top of the reactor assembly for purge nitrogen, brine feed, acid addition, and gas release. A flanged opening is provided on one side of the electrolyzer for a temperature proble and level sight glass. The cell rests on a simple 'I' beam base, electrically insulated from the floor. A flange located beneath the electrolyzer allows for cell draining.

18.2 THE ANODE ASSEMBLY

The anode portion of the electrolyzer is comprised of thin, solid, rectangular titanium sheets extending horizontally in to the electrolyzer from one side of the electrolyzer body. These sheets are welded to a backplate using a proprietary welding technique which reduces labor, fabrication time, weld shrinkage and machining while ensuring dimensional stability and extremely tight fabrication tolerances.

The short horizontal current path inherent in this design affords the opportunity to use solid titanium sheets for the anode without supplemental titanium-clad copper current conductors. Anode thickness has been optimized with respect to cost of material and voltage drop.

The anode assembly has been designed as an independent unit for convenient servicing and easy access to the electrodes. The anode backplate is equipped with stainless steel lugs for receiving and connecting the copper busbar assembly. The busbar assembly electrically connects the anode with the intercell bus work. The anode assembly is bolted to the electrolyzer shell with a single gasket maintaining electrical insulation.

The anode coating can be RuO_2, Pt/Ir or any of the mixed oxide coatings now offered commercially. The initial cost, the cost of recoating and anode performance characteristics influence anode selection. For recoating, the anode sheets are simply removed from the backplate by flexing each sheet near the weld. All anode assembly components are suitable for reuse.

This simple design of the anode assembly is most economical with respect to both capital and operating costs.

18.3 CATHODE ASSEMBLY

Like the anode, the cathode portion of the electrolyzer is comprised of solid, rectangular plates extending horizontally into the electrolyzer from the side of the electrolyzer body. These steel sheets are welded to a steel backplate which is bolted to the electrolyzer shell to complete assembly. Following cell assembly, a copper busbar is bolted to the outside of the backplate using stainless steel lugs. Both the copper and steel surfaces are silver-plated to prevent oxide formation and minimize electrical resistance.

The simple design of both cathode and anode assemblies offer substantial cost savings and easy access for field maintenance and cleaning.

18.4 REACTOR TOP

The LCD MK-II chlorate cell reactor top is made from prefabricated titanium pipe. This design feature offers a savings in fabrication cost from both availability and material strength considerations. The tall vertical dimension also permits achievement of the required reactor volume while minimizing floor space utilization.

The reactor top as an integral part of the cell, eliminates extraneous titanium piping, expansion joints and electrical insulation inherent in designs which physically separate electrolyzer and chemical reactor. As demonstrated, the reactor top also provides efficient circulation of electrolyte by maximizing hydrogen 'gas lift' effects.

The top of the reactor is flanged. A gasketed cover plate is bolted to the flange. The bolted assembly is not only cost effective, but also provides a pressure relief point for the cell in the event of a system upset. The reactor assembly is attached to the electrolyzer by a bolted flange, with electrical insulation maintained with a fluoropolymer envelope gasket.

The chimney structure within the reactor top is also fabricated out of titanium. The lower section serves to collect the gas-filled electrolyte rising from the electrolyzer. At the point of contraction, mixing is increased by increasing velocity without generating severe back pressure. This increased mixing serves to facilitate the chemical formation of chlorate.

18.5 INTERNAL DESIGN

As stated earlier, a computer simulation of the internal material and energy balance was developed for the cell. The design parameters incorporated in the model include:

Anode : cathode gap.
Electrode height, thickness and width.
Operating temperature.
Fluid dynamic design of the electrolyzer/reactor assembly.
Current density and total current.
Brine feed conditions.
Chlorate product concentration.
Solution pH.
Physical property relationships.
Electrochemical relationships.

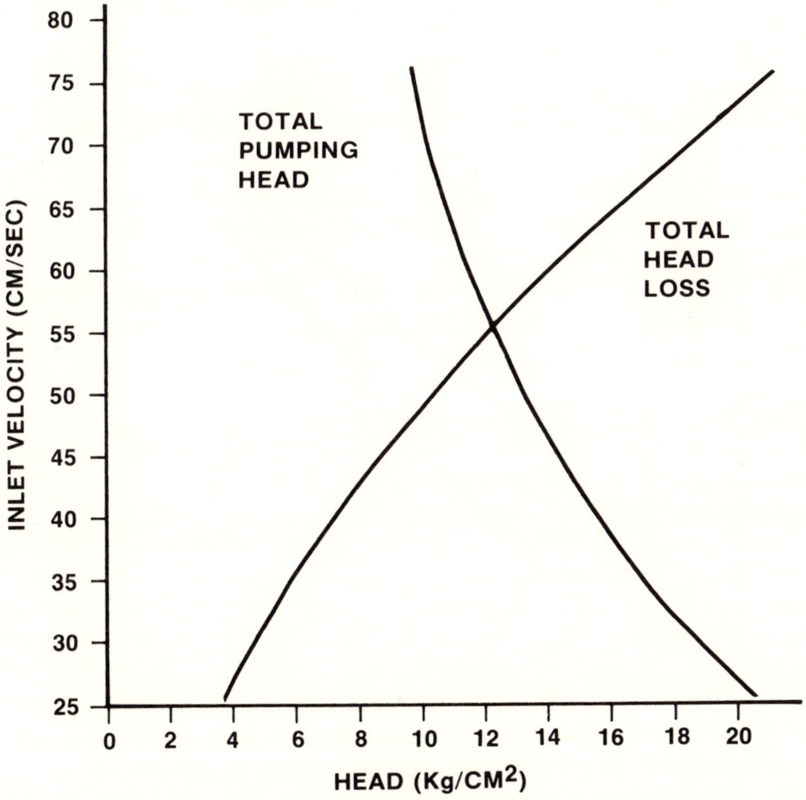

Fig. 18.2 – OCC LCD MK-II cell. Inlet velocity versus head loss.

The computer simulation permits calculation of material and energy balances for the cell using an iterative procedure wherein design parameters can be systematically varied. More specifically, the computer simulation calculates pumping head due to the H_2 evolution, iteratively, over the height of the cell for a range of inlet velocities or flow rates at the bottom of the cell. This flow component is 'matched' with the friction head loss due to electrolyte flow over the height of the cell for the same range of inlet velocities or flowrates based on hydraulic analogs for the cell configuration. Figure 18.2 presents a plot of inlet velocity versus head loss for a typical set of operating conditions for the LCD MK-II. Intersection of the curves represents the operating point at which all material and energy balance constraints are met.

The computer simulation then derives a voltage distribution for the electrolyzer based on local current distribution and conductivity, along with ion concentrations, flow rates, velocities, etc., for the various parts of the cell, subject to maintaining internally consistent energy and material balances.

Figure 18.3 presents a plot of electrode gap voltage drop versus anode height for various cell temperatures at constant current density. Figure 18.4 presents a

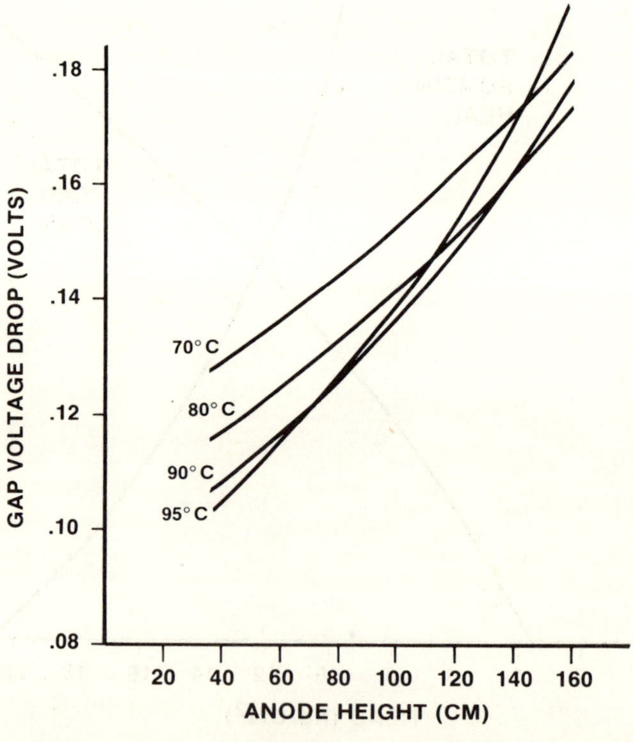

Fig. 18.3 – OCC LCD MK-II cell. Electrode gap voltage versus anode height.

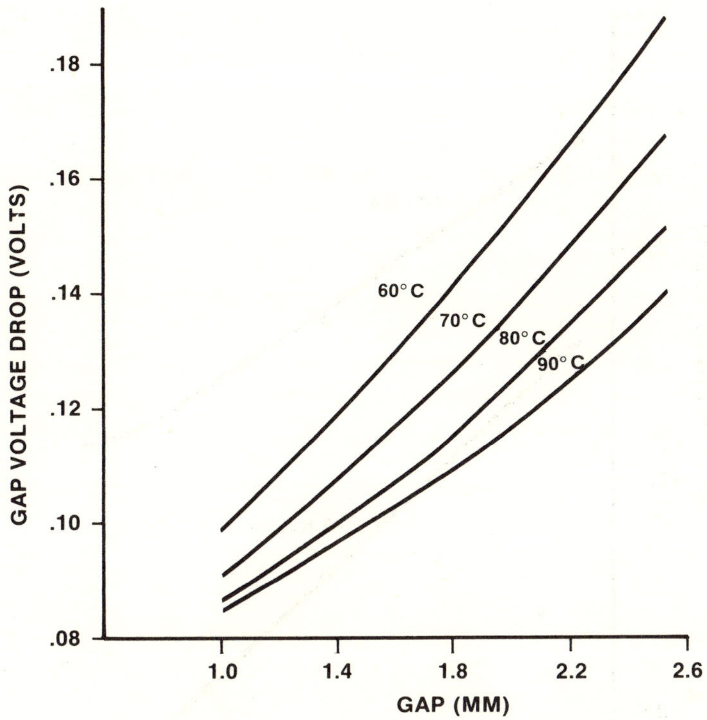

Fig. 18.4 – OCC LCD MK-II cell. Electrode gap voltage drop versus electrode gap.

plot of electrode gap voltage drop versus distance between electrodes for various cell temperatures at constant current density. The following operating conditions were used for the illustrative examples contained in Figs. 18.2, 18.3 and 18.4:

Product	130 g/l NaCl
	450 g/l $NaClO_3$
Brine feed	300 g/l NaCl
Load	100 kA
Current density	2.3 kA/m^2

The methodology depicted above was used in conjunction with capital and operating costs to arrive at optimum design and operating conditions.

18.6 PROTOTYPE CELL

On June 4, 1982, a 100 000 A prototype LCD MK-II chlorate cell was installed in an operating circuit of cells at OCC's Taft, Louisiana, plant. Figure 18.5 shows a plot of overall cell voltage versus operating temperature for the commercial

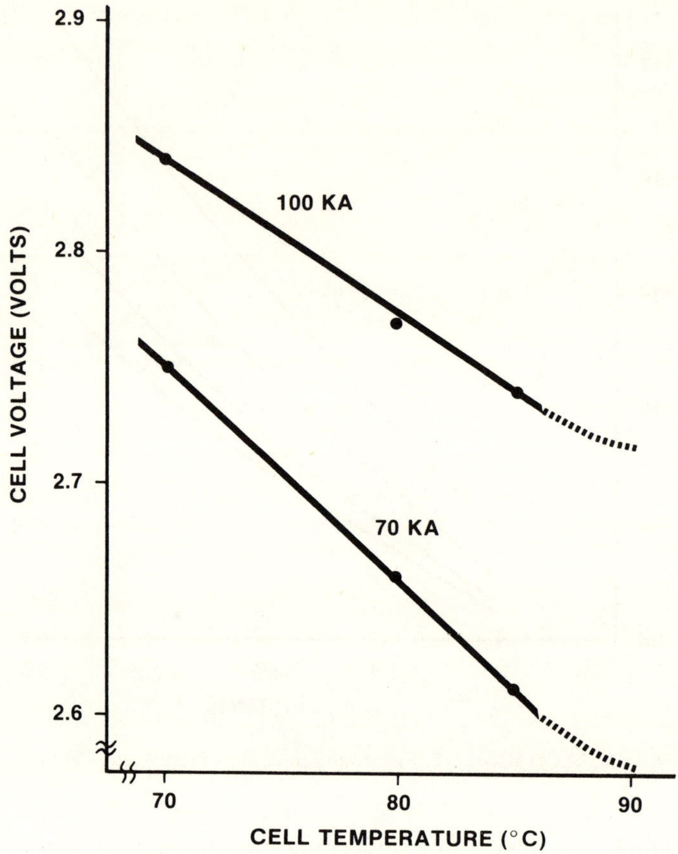

Fig. 18.5 – OCC LCD MK-II cell. Cell voltage versus cell temperture.

prototype. Operating temperature was systematically varied to confirm the heat transfer requirements of the cell and to verify energy consumption predictions. Figure 18.6 shows a plot of cell energy consumption versus operating temperature. The data confirms that the cell operating at 85°C will consume less than 4400 d.c.-kWh/metric ton of sodium chlorate.

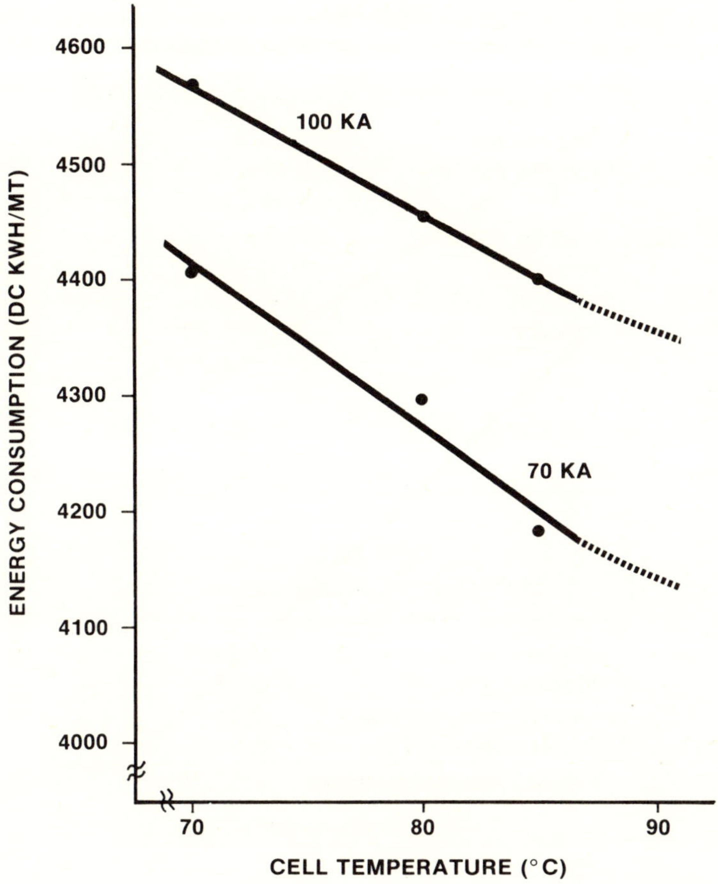

Fig. 18.6 – OCC LCD MK-II cell. Energy consumption versus cell temperature.

18.7 CELL OPERATION

The energy consumption of the LCD MK-II cell (less than 4400 d.c.-kWh/metric ton) is the lowest of any commercial cell operating at comparable current density. The low energy consumption is attributed to voltage drop minimization which results from design optimization. The internal cell design was based on maximizing the utility of the H_2 to circulate electrolyte and thus maximize the flow of gas-free, conductive solution through the electrode gap. Table 18.1 presents typical operating parameters for the LCD MK-II chlorate cell which have been confirmed by full-scale prototype cell operation. Table 18.2 presents a voltage component analysis for the cell.

Table 18.1
LCD MK-II cell typical operating conditions

Cell load	100 000 A
Cell voltage	2.760 V
Current density	1.49 A/in^2
	2.3 kA/in^2
Current efficiency	95.0%
Brine feed	310 g/l NaCl
	25°C
Product	120 g/l NaCl
	450 g/l $NaClO_3$
	90°C
Gas analysis	2.2% O_2
	0.8% Cl_2

Table 18.2
LCD MK-II cell voltage analysis

Hardware (IR)	0.080 V
Electrode gap (IR)	0.13
Cathode overpotential	0.84
Anode overpotential	0.035
Cell decomposition voltage	1.673 V
	2.76 V

18.8 CONCLUSION

The LCD MK-II has the lowest energy consumption of any commercial cell of comparable current density. The completely optimized design effects substantial savings in both capital and operating cost due to the simplicity of design.

SECTION F

Energy and the Chlor-alkali Business Scene

CHAPTER 19

Factors Affecting the Future and the Chlorine Industry

D. S. DAVIES, Duncan Davies Associates, London; Tosco Corporation, Los Angeles (formerly Chief Scientist and Engineer, Department of Industry)

Editor's note: This text is based on a transcript of an informal talk.

19.1 INTRODUCTION

There are many ways in which patterns of resource and raw material availability can affect an industry. First, they may affect product design and choice; oil prices are massively affecting the kind of cars bought by Americans. Secondly, they may affect process choice; in the 1950s, chlorine prices fell less rapidly than ethylene prices, which caused the chlorohydrin process for ethylene oxide to be replaced by direct oxidation. Thirdly, they may affect overall demand. The 1973 and 1979 oil price rises have affected product and process choices, but they have also caused people to manage with less energy; conversely, steadily falling oil prices over the period 1950–1973 were a considerable factor in causing growth in consumption – with substantial political consequences. Fourthly, they may affect geographical distribution of product and technology: for example, current petrochemical developments in the Middle East. And fifthly, all of these factors can have political repercussions – on employment, on balances of payment, on exchange rate, and on military power. Dr Rudi Bloch has urged that human history has been more affected by the availability of salt than by any other factor. Figure 19.1 lists all of these factors.

The symposium is mainly concerned with the internal and important matter of process choice – bipolar cells, diaphragm cells, membrane cells, mercury cells – and to this matter I can make little contribution, for I am out of date. Accordingly, I would like to tackle the other matters which will be very important in determining investment patterns, and hence how, when, and where the entry fees for new process technology may be made available.

ENERGY AND CHLORINE

MATTERS FOR ATTENTION NOW THAT ENERGY IS DEAR

1. PROCESS CHOICE – HOW TO MAKE?
   Membranes, Mercury, Diaphragms.
2. PRODUCT CHOICE – WHAT TO MAKE?
   PVC, Paper, Aluminium, Rayon, Soap, Solvents, Germicides.
3. DEMAND AND RECYCLE – HOW MUCH TO MAKE?
   Economy options, events to data (package changes, smaller cars, being careful).
4. GEOGRAPHY – WHERE TO MAKE?
   Everywhere is different – what do you do if you start afresh?
5. POLITICS – WHAT GOVERNMENTS ALLOW YOU TO MAKE.
   Balancing trade, preventing disaffection.
6. TIME – WHEN TO MAKE?
   Substitution rates, speed of change, effect of habits.

Fig. 19.1 – Matters for attention now that energy is dear.

19.2 POLICYMAKING – FORECASTING – DISCONTINUITY – TIMESCALE – THE HEADLINES

If we attempt such broadly-based policymaking, and to escape from the standard error of always preparing for the last war, we have many influential people whom we must persuade. This is not easy; in a big company, it is less risky to one's career (though not to the company!) to assume that the future will be very much like the past, only bigger and better. Forecasters and scenario-writers, who aim at better future policies, face two fundamental difficulties. One is that they are certain to make wrong forecasts: success does not require the virtual impossibility of being 100% right, but merely somewhat less wrong than one's competitors. But the critics will see the mistakes the forecaster did make much more clearly than the mistakes he helped people to avoid. So he must be brave, and say at all times that he knows he will be proved wrong on all sorts of occasions, and hopes to be judged on his average performance.

Secondly, the forecaster-planner must win committed support from management colleagues. Thus, he must try and make sure that those responsible for decisions pay attention to the forecast. There is only one way to do this and that is to get them to pay for it. If no costs have been paid, it is easier to ignore the forecast. However, if some of the cost is being paid by users, this limits what you can say because users do not believe that the world can change quite as rapidly as it sometimes does. So you must be user-friendly and omit some of the things which might happen but which users are not yet ready to accept. As an example of this – in about 1970, the number of people doing research and development in ICI had increased by 3% compound annually for 40 years. Now the UK population in the UK at that time was going up by only about 0.5% obviously this research and development expansion could not go on for ever and a re-think was needed in ICI, or the whole population of the UK would be doing research and development for ICI, in due course! But ICI researchers were not ready for this message until economic difficulty forced attention, in 1970–71.

Another example is provided by the glass industry. This industry did not want to handle scrap glass or cullet, when energy was cheap and it was said in the industry that the maximum amount of cullet which could be handled was about 20% of the charge to the furnace. When energy prices increased, cullet became valuable since putting more of it into the furnace put in more fixed energy and also reduced the furnace time cycle. It was now found that up to 40% of cullet was acceptable and campaigns for collecting more scrap glass were started.

When we go through a discontinuity like we are doing now, the new prices not only have an immediate effect upon the things which you are doing; they affect all sorts of people all around, all of whom are doing things which affect you with varying degrees of directness and only when you look at what everyone is doing are you beginning to be able to see what might happen next. Fortunately, many inexorable changes are slow. The pace of change based on social behaviour occurs in decades rather than years and therefore providing one takes the trouble to understand what is going on, there is time to adjust, which is important and helpful.

So Fig. 19.1 lists the important factors to discuss. Elsewhere in this book there is a lot of detail about process choice and about process detail. Behind these there are product choices; the things that are made from chlorine are competing with things that are not made from chlorine. Then, what happens to demand, over time, with high energy prices. We need to discuss the options for this whole economy: for a full understanding of the market for chlorine (and everything else).

19.3 CONTROLLING FACTORS

19.3.1 The product pattern of the chlorine industry: substitutes

The study of the future geographical and economic pattern of the chlorine industry must begin with the consideration of the final products and their markets. The industry began with the conversion of an unwanted by-product of the old LeBlanc Alkali process – dilute HCl gas – into dilute chlorine gas (by the Deacon process) and then into bleaching powder. This product can be thought of as a 'maintenance chemical', used to inhibit biodegradation and sweeten foul effluent. The Castner–Kellner process then added 100% chlorine gas (or liquid chlorine) as a further byproduct of alkali manufacture: this could also be sold as maintenance chemical, in cylinders. It could also be converted to trichlorethylene and perchlorethylene which, as dry cleaning or degreasing fluids, again were sold as maintenance/service chemicals. Growth of such markets is strongly linked with the pattern of other activities – engineering, clothing fashions and maintenance practice, etc.

The rapid growth of the chlorine industry has been associated with more independently promotable materials, most especially PVC. These can be called 'construction chemicals' (the word 'construction' embracing packaging film, auto components, etc., as well as guttering, pipes, roofing and cladding). The strength of the big PVC market is now associated with users accustomed to its properties and limitations. Being well established, it will be with us for a long time. It is a long way down its experience curve.

However, other thermoplastics may well prove to be preferable in the long run. Just as polyester fibre is replacing the nylon fibre that opened up melt-spinning, the polyolefins, polyurethanes, and perhaps polyacetals and regular polyesters are likely to erode the demand for PVC polymer that opened up injection moulding and extrusion. You will be told, by traditionalists, that the polyolefin and PVC markets are quite separate, with little overlap. Be very suspicious of this argument: cheap, distressed price, high quality polypropylene or HD polythene can lure many people away from PVC.

It may therefore turn out that PVC becomes something of a speciality polymer, with a long term market considerably below the present level. This argument in no way depends on fears about divinyl chloride toxicity which has probably been dealt with adequately. It simply recognizes that other polymers, less well established with users, seem likely to have property and price advantages that will tell in the long run. They may also prove easier to recycle, when this becomes important. PVC is an adequate rather than an excellent polymer, which fails at moderate temperatures and is somewhat degradable chemically. Users have learnt to live with its defects, partly by using additives. Their loyalty will have limits.

Accordingly, chlorine's use as a 'maintenance chemical', somewhat eclipsed in recent decades by the glittering growth of PVC, may again become more important. If so, this will impinge considerably on your work and its geography and politics. It may be exaggeration to speak of 'clogs back to clogs in three generations', but balances are likely to move back towards old patterns, and there are many important 'construction chemicals' like polychloroprene, that may be more sustainable than PVC.

19.3.2 Toxicology and environment

Organic chlorine compounds tend to be biologically active and therefore need careful handling, which costs money and involves regulation. I have mentioned the toxicology of vinyl chloride monomer as a problem that is probably contained. The continued use of chlorinated solvents depends on benefits to the public being great enough to justify expense and care in ensuring that safe atmospheric concentrations are not exceeded. Again, I believe that the game is well worth the candle and that the problems are adequately understood and handled.

There are some products where the balance of toxicity is less favourable, and where markets will continue to be restricted and may become more so. I refer to the chlorinated insecticides (important though these have been, and are), and more particularly the PCBs, which may vanish completely.

Volatile chloro compounds are also atmospherically active in the stratosphere; they are amongst a variety of substances that can catalyse the photodecomposition of ozone and thus modify the atmospheric protective action against ultra-violet radiation. This is a much more complex problem, on which work will be needed for many years yet, and there are natural as well as man-made components. Again, my own belief is that the problem is containable.

In your thinking about the future of your industry, there will be those who say that environmental problems are figments of the imagination of irresponsible academics who can safely be disregarded and caricatured as 'econuts'. Reject this argument. Environmental problems are real, but manageable at a definable cost. Their impact varies with climate, population density, and other factors. You need to know about them for your planning. And I do not need to say a single word to you about mercury or asbestos toxicology.

19.3.3 Demand – overall price elasticity –market saturation and overcapacity

Since 1973 the escalating oil price has increased the incentive to use oil-based products more carefully and cleverly. In the case of plastics, this is promoting better design, and filling, foaming, or orientation as means for enhancing the duty or utility achieved from each kg of resin. While oil was cheap, the crude procedure of making a moulded item thicker in order to make it stronger, was adequate. Not any more. It now pays to add more value by spending money on capital assets, development, or software. Further, consumers change their habits, and now make things last longer. Price elasticity takes time to have an effect, but over and above the substitution of PVC by other polymers, we have to consider the long term total markets for all polymers, and for materials in general. So let us consider four questions; what is the scope for economy? has anyone seriously economized yet? and are there any historical cases that are illuminating? and are governments going to do anything useful?

Is there scope for economy?

Yes, there certainly is. Let us consider three basic 100:1 ratios. On average we use 100 calories of fuel for every food calorie we use. It is the food calorie we need to keep warm; the 100 fuel calories, which include industrial use and everything else, we use extremely wastefully. It seems unreasonable to try to maintain a 100:1 ratio. If this came down it would have quite an effect on energy demand.

Another 100:1 ratio relates to water. We consume, and have consumed on one behalf, roughly 100 gallons of impounded water for every gallon we drink. The gallon we drink is used to flush nitrogenous waste out of our metabolic system whereas the 100 gallons are used for moving rubbish about in one way or another, either by washing the car or washing clothes and so on. Again it seems unreasonable to suppose that the sophistication of the bowel and the kidney cannot be matched in the outside world. If we used all water impounded as efficiently as the water we drink, we would save a lot of money and energy.

The final example is hardware. Roughly we buy 100 tonnes of hardware for every tonne we like well enough to keep. So there is plenty of scope for economy here also.

Has anyone seriously economised yet? – the US experience 1973–82

The most dramatic example is what has happened in America since 1973 when oil prices increased. First, imports of oil increased. The world use of oil was about 54M bbl/day of which the Americans used about 20M bbl/day at that time. Immediately after the oil price rise, American imports moved up to 9M bbl/day. Since that time the American use of oil was cut down and with it of course the imports of oil, now running at about 4.5M bbl/day.

The first to economize have been industrial companies which can deploy resources; the results show in the profit and loss account. The saving here was of the order of 1M bbl/day. Another 1M bbl/day was due directly to recession and another 1M bbl/day was due to the use of smaller cars for fewer miles. The 'gas guzzlers' were still there, but in the cities people are driving a smaller car and leaving the big car in the garage so that, coupled with about 10% less milage, produced the saving. So the scope for economy is not just theoretical – it is actually happening in America.

Are there any illuminating historical comparisons?

Yes, there are. Steel, in 1900, was where plastics are in 1980, and the comparison with the steel industry is both useful and daunting. It is published in detail elsewhere.

Steel and plastics both started as military materials. Plastics began in Germany between the wars as part of the campaign to replace imported materials for military purposes. Polythene was invented in this country but was first used for military purposes as a radar dielectric. Steel was a military material, its production up till about 1850 being measured in thousands of tons a year only for military purposes. The first part of the industrial revolution was with iron not steel. Steel was not cheap enough. Railway lines were iron. Cast-iron was used for compression and wrought iron for members in tension (as witness many Victorian railway station buildings). Two process inventions initiated the use of steel as a serious material, the Bessemer converter and the Siemens open hearth furnace. Both came in 1860 and reduced the cost of steel to a level at which it could begin to compete with iron. This caused steel over a period of 4 decades to expand at about 15% compound (just like plastics) and in about 1900 the use of steel in the UK (which was a leader) was about 0.1 tonne/head of population. At that point demand bagan to be saturated. Energy costs did not limit demand and environmental factors certainly didn't. Nobody believed that the industry was mature and as a result, ever since 1900, the steel industry has been a place where life has been very difficult indeed.

Now the plastics industry is similar. Plastics were originally military materials and demand escalated rapidly as a result of the (1940) introduction of the ethylene cracker. All the product inventions in steel came after the process invention and the same was true for the major plastics. Plastic production (in the form of polymers) in the leading country, West Germany, has approached about 0.1 tonne/head of population relatively recently. But of course the further growth of plastics has been held back not merely by saturation of demand, but also because of higher materials and energy costs and (as already noted) and because you can get approximately five times the utility out of a ton of plastic simply by orienting it, filling, foaming it, or just by clever engineering which nobody has bothered to do so far because resin has been cheap.

So, a variety of factors, environmental, engineering and sophistication are bearing down upon plastic and would have done even if there had been no energy price rise in 1973.

The role of government, product standardization and collection costs in recycle

If recycle is to be used as a source of energy economy then governments have a part to play because products are now very diverse, and scattered about a very wide area so that more standardization is needed. The Monsanto Cycle-Safe project showed how the collection of used Cola bottles could be organized so that recycle was economic and attractive. The project only failed because of the toxicology of acrylonitrile (and in this case there probably is a case for saying that toxicological caution was excessive). Recycle will demand product standardization in which governments will almost certainly have to be involved – and systematic collection which involves local authorities. This leads a further series of matters of which those planning the chlorine industry need to be aware.

19.3.4 Geography and politics

From all of the above, we can see that the future planning of the chlorine industry is a complex, difficult and very broad task. Simplistic thinking that markets *must* expand, will not get us very far. Free trade may be desirable, but the doctrine of comparative advantage is being attacked by countries anxiously seeking methods for avoiding the export of jobs that will not provoke retaliation. Chlorine plants are going to be built in some countries while serviceable, older ones with better marginal economics have spare capacity elsewhere. The idea of secrecy in chlorine technology has gone: many more people are now prepared to license.

19.3.5 Timing

A few years ago Fisher and Pry [1] determined the take-over time for a number of substitution processes in the recent past. The take-over time was defined as the time from 10% to 90% penetration. Their results as shown in Table 19.1 which clearly illustrates that substitution is a slow process, partly determined by social factors.

Table 19.1
Substitution rates for replacement inclusion

	Δt/years	To
Synthetic/natural rubber	58	1956
Synthetic/natural fibres	58	1969
Margarine/butter	56	1957
Electric arc/open hearth spec. steels	47	1947
TiO_2/PbO–ZnO pigments	26	1949
Plastic/hardwood floors	25	1966
Organic/inorganic insecticides	19	1946
Detergents/natural soap	8.75	1951

19.4 COUNTRIES AND PRODUCTS

19.4.1 Products

19.4.1.1 *Paper*

The processes for treatment of pulp and for bleaching have been improved a great deal by collaboration between the chlorine industry and the paper makers but of course who says we want paper? For a long time VDUs and head up displays have been there as alternatives but that change, even if it occurs will occur very slowly. We are now just at the beginning of the use of LCDs for screen and projection so as to have microelectronic display equipment including a radio receiver in your pocket – just pick it up, open it out and there is your newspaper. Is this science fiction or real possibility? Further, as the price of paper increases, as it will, the importance of recycle is likely to increase.

19.4.1.2 *Maintenance chemicals*

It is quite obvious that cleaning and looking after equipment so that it will last longer is something that will grow gradually. While equipment was cheap it was natural to buy it and then to throw it away. That will continue but gradually, as equipment and spares become more expensive, the operation of maintenance will become more important.

19.4.1.3 *Biocidal products*

The use of chlorine-based products to kill bacteria is likely to increase slowly especially where the products reduce corrosion thus extending equipment lifetime.

19.4.2 Places

OECD countries

The OECD countries will obviously be the most careful about hazards and about mercury. When they try they will be best at recycle and they have the most material to recycle. They will also be better at standardization. They are the countries who make the most of co-production processes and manage them best. They are also the most wasteful at the present time but they could become the biggest 'maintenance market'. So you could find that the OECD countries are the ones where the 'clogs back to clogs' syndrome first appears, with a re-emergence of chlorine and derived products as maintenance chemicals.

Comecon

Comecon countries will change most slowly, and will, therefore remain with PVC. They will probably be less sensitive to hazards, and may be inefficient about co-product systems and recycle.

OPEC

OPEC countries will probably eschew manufacures involving difficult co-production (for example chlorine and alkali) and will tend to go for the relatively simple manufactures. They will also be dependent on export and transport costs, an initial capital costs will be higher. But they have methane available at zero value.

The Third World

Third World countries will not have much money for some decades yet but they will have cheap trees and plant products and many bugs to deal with which will all need chlorine. They will lack experience relevant to maintenance and recycle.

General

Investment in industry will be best in countries determined to plan technology well but free trade will be difficult to maintain with everyone trying to export unemploym,ent. The politicians have a whole armoury of weapons at their command in the form of, for example, taxes and grants, energy price intervention, tariffs and quotas, standards, financial policy and by no means least, social consensus.

19.5 CONCLUSIONS

The conclusions to be drawn from this presentation relevant to the chlorine business are as follows:

1. Chlorine started as a maintenance chemical, has become a construction chemical and may be driven back to a maintenance chemical by energy prices especially in OECD countries.
2. Recent and continuing large changes have major but slow effects on politics, geography and product choice that will slowly swamp the minor changes that can be studied by conventional economics.
3. The trends will differ greatly in different countries. In Comecon change will be slower and chlorine will remain a construction chemical and OPEC countries may not bother with chlorine at all while the Third World countries, although not having much money for a long time yet, may become the major areas where chlorine is a construction chemical.
4. Survival and prosperity in the chlorine industry will depend on a very broad market, political and social skills as well as technical ability.

19.6 REFERENCE

[1] Fisher, J.C., and Pry, R. H., *Industrial Applications of Technology Forecasting* (Ed. M. J. Cetron) John Wiley & Sons (New York) 1971.

19.7 DISCUSSION

1. In response to a question from Mr Saranathan of Titanium Equipment and Anode Manufacturing Co. Ltd., India, about whether psychological and social factors play a greater role than technical and economic factors in decision making, Dr Davies replied:

   When people have psychological perceptions – things they would like to do, or when the Friends of the Earth have things they would like to do, action results more readily if economic and technical factors are in favour than if it is merely 'a good idea'. There have been cases where things have happened just because a sufficient number of people said it would be a good idea but I feel for the most part, that psychological factors play a part where there are economic and technical factors pointing in the same direction.

2. Mr C. A. Hansen (Linden Chemicals & Plastics Inc.) referring to Dr Davies's remarks about the ability of PVC to compete the polypropylene, asked whether Dr Davies saw anything on the horizon in the way of developments to PVC which might improve its ability to compete.

Dr Davies replied: Although obviously PVC has been improved out of all recognition over the past 30 years, mainly by attention to purity and polymerization conditions, there appears to be more scope for changing the properties of the other polymers rather than PVC. On the other hand, engineers, designers and other general users are so well accustomed to PVC that clearly it will stay with us for a while.

Off hand, I find it very difficult to see a way in which demand variation can make an economic difference other than in a very small way. Clearly the various plastic companies are looking at the question but I think that over the next 20–30 years the whole structure of the plastic industry will change particularly by the resin makers moving into engineering, moving nearer the market moving into fabrication and other more specialized subjects. If that happens, perhaps a combination of chemists and engineers might find new combinations and fabrications but I would be included to think that the polyolefins would stand a better chance.

CHAPTER 20

Impact of Rising Energy Costs on the Worldwide Structure of the Chlor-alkali Industry

GEORGE L. INNES, Charles H. Kline & Co. Inc., Fairfield, New Jersey, USA

20.1 INTRODUCTION

The chlorine industry today looks much different than it did even as recently as 10 years ago. At that time it was considered – perhaps not a growth industry – but certainly one with better than average prospects. Chlorine consumption had been growing at an annual rate of about 7% and seemed likely to continue at something close to this rate for some years to come. Forecasts made in the mid-1970s by ourselves in our first world survey of the chlor-alkali industry and also by some US producers predicted increases of 5 to 6% a year for chlorine into the early 1980s. The big problem then was what to do with the surplus caustic soda that everyone expected to be produced.

The oil crises of 1974 and 1979, particularly the latter, changed all that. Although world chlorine production did recover from the recession of 1975, reaching an all-time high in 1979, it then declined again in 1980. In the US it dropped still further in 1981, to a level below that of 1974. So there has been no growth at all in the US since 1974, as shown in Fig. 20.1. Western Europe and Japan show similar patterns.

All chemicals, like other products, pass through stages in their commerical life. An introductory period is followed by one of rapid growth, then a gradual levelling off and finally a decline. Chlorine has apparently reached the levelling off or mature stage sooner than expected, at least in the industrial countries of the world. In the less developed countries of Asia, Africa and Latin America, chlorine and caustic soda are still growth products as these countries begin to industrialize and develop their own chemical industries.

There is only a limited market for merchant chlorine. Most of the larger producers with world-scale plants have integrated downstream into chlorinated derivatives such as vinyl chloride monomer, chorinated hydrocarbons, or other chlorinated compounds.

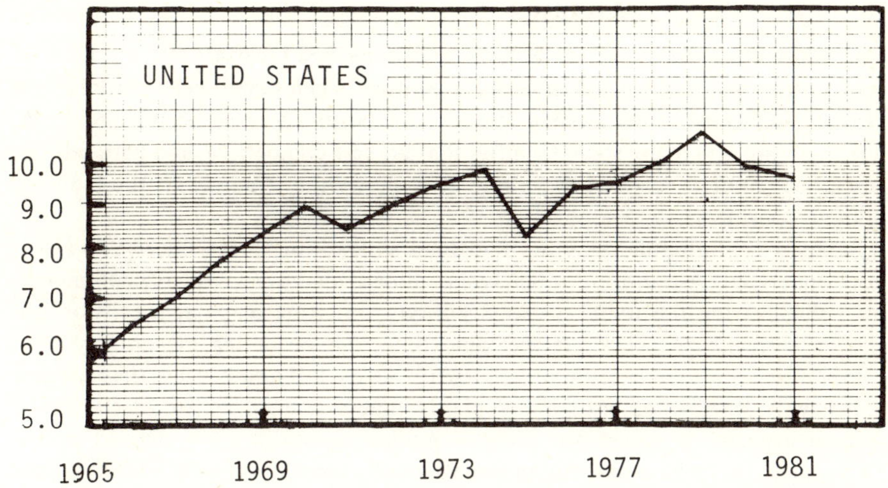

Fig. 20.1 – US chlorine production 1965–1981 (million tonnes).

That the market for chlorine would eventually mature was inevitable but the process has apparently been hastened in the industrial countries by such developments as restrictive legislation on chlorine and some of its derivatives, technological changes, including the use of other bleaching agents in pulp, and more particularly, the rapid increase in energy costs.

During the last few years regulations have been enacted in a number of industrial countries on chlorinated derivatives: vinyl chloride monomer and trichloroethylene as possible carcinogens, and the fluorocarbons for their effect on ozone in the atmosphere. Even the use of chlorine as a sanitizer in drinking water has been under attack.

Energy costs escalated to the point where they now represent about 50% of total manufacturing costs of chlorine. They are expected to go still higher; some forecast as high as 75% of manufacturing costs by 1990. Profitability has declined, since these included costs cannot be passed on entirely through higher selling prices.

Overcapacity has intensified the cost problem still further. Because of lower operating rates and the cost-price squeeze, some producers are unable to generate enough capital to invest in the expansions and modernizations that will be required in the future. Some smaller plants, particularly the older ones, are being forced to terminate operations. The very large producers, like Dow and ICI, can afford to expand overseas in areas where energy costs are lower, but smaller ones cannot.

20.2 INDUSTRIAL COUNTRIES

The chlorine industries of the US, Western Europe, and Japan are old; a number of plants were first built in the nineteenth century when commercial chlorine and caustic soda were in their infancy. Many of the larger ones have been modernized, but there are still many small facilities that are less competitive and have felt the cost-price squeeze very keenly. Table 20.1 shows the structure of the industry in these countries.

Japan's chlorine industry has remained almost unchanged since 1975 and is now undergoing a transition which will eliminate mercury cell plants completely by 1984, leaving the industry based entirely on either diaphragm cells or ion-exchange membrane cells. At the same time it is becoming increasingly non-competitive, not only because of high energy costs, but also because the Japanese industry is fragmented into many plants that are small by world standards. Annual capacity in Japan was about 4 million tonnes in 1980, divided among 34 suppliers with 53 plants for an average yearly capacity of only 75 000 tonnes per plant. With traditional Japansese loyalty to the customer, the smaller producers have preferred to continue operating, even at a loss, to keep their customers supplied. However, the industry will probably be forced to rationalize itself into a smaller number of more efficient producers, and look increasingly to invest overseas for future expansions.

Table 20.1

Average plant size industrial countries

	US	Western Europe	Japan
Total capacity 1980†	13 400	11 600	4 000
Producers	32	72	34
Plants	67	102	53
Average per plant†	200	114	75

† Thousand tonnes.

By constrast, the US chlorine industry in 1980 has 67 plants with an average capacity of 200 000 tonnes a year, and Western Europe, even with 15 producing countries, had 102 plants with an average yearly capacity of 114 000 tonnes. But even in these countries there are many small units: the US, for example, had 32 in 1980 with capacities under 100 000 tonnes a year and Western Europe 57.

The world industry appears to be due for a restructuring as producers look for ways to cut their energy consumption, especially feedstock costs. A gradual shift away from the industrial countries is evident, as shown in Table 20.2. In the US, Western Europe, and Japan, chlorine capacity increased by 13.3%

from 1975 to 1980 and based on current plans, is expected to expand by another 8.3% by 1985.

Table 20.2
Increases in chlorine capacity 1975–1985 industrial countries
(million tonnes)

Region/country	1975	1980	1985
US	12.3	13.4	14.5
Western Europe	9.3	11.6	12.9
Japan	4.0	4.0	4.0
Total	25.6	29.0	31.4
Percentage increase	–	+13.3%	+8.3%
Percentage of total free world capacity	88.0%	84.8%	82.8%

The expansions in Europe and the US may be even smaller than indicated here since some plants now scheduled may be postponed or even cancelled altogether if the present recession continues for long.

The recession has of course slowed the growth of expansion all over the world, but in the countries of low-cost energy capacity increased by nearly 50% from 1975 to 1980 and will probably grow by another 25% by 1985, as shown in Table 20.3.

Table 20.3
Increases in chlorine capacity 1975–1985 low-cost energy countries.
(million tonnes)

Region/country	1975	1980	1985
Asia and Africa	1.5	2.2	2.7
Canada	1.3	1.5	1.7
Latin America	0.7	1.5	2.1
Total	3.5	5.2	6.5
Percentage increase	–	+48.6%	+25%
Percentage of total free world capacity	12.0%	15.2%	17.2%

The less developed parts of the world are growing more rapidly and would in any event need more capacity, but in such countries as Canada and Saudi

Arabia and other Middle East locations, the new plants are not for domestic consumption but for export. Caustic soda is easily shipped but chlorine cannot be conveniently transported in its elemental form so that it is customarily exported as vinyl chloride monomer or ethylene dichloride for use in the production of polyvinyl chloride.

This introduces another energy-related factor – the cost of ethylene, which is directly related to the cost of feedstocks. The cost of producing ethylene (and therefore ethylene dichloride) is cheaper in the countries around the Arabian Gulf, the Far East, Mexico, Canada, and Australia. Economics recently developed by the United Nations Industrial Development Organization demonstrate this (Table 20.4). These are the regions to which the chlor-alkali industry is shifting. A number of facilities are planned and a few are already under construction. Some of these plants, particularly those in the Middle East, are likely to move further downstream, beyond ethylene dichloride to vinyl chloride monomer and PVC itself to serve local markets. The Saudis have in fact already begun to talk about this. Such a development will affect the export markets of the established producers, who will then have reduced outlets for their chlorine.

Table 20.4
Ethylene costs at selected locations 1980 and 1985.
(US dollars per tonne)

Region/country	1980	1985
US Gulf	529	927
West Germany	754	1103
Japan	864	828
Indonesia	640	672
Mexico	440	443
Arabian Gulf	480	545

Source: UNIDO

These new plants can be divided roughly into two categories: those intended primarily to satisfy domestic consumption and those for export. The distinction is not clearcut, because some countries wish to do both; however, Latin America is building mostly to satisfy local demand.

20.3 LATIN AMERICA

Table 20.5 lists scheduled expansions to 1985 which will add 576 000 tonnes or about 38% to the 1980 total for the region. Demand here is growing more rapidly than in any other part of the world, and new capacity is needed to keep up with local consumption and to hold down imports. Mexico, as we have already pointed out, is one of the areas of low-cost energy, but none of the current new capacity is intended for export. Pemex has talked about new production of ethylene dichloride amounting to as much as 400 000 tonnes a year, but it is unlikely that any such plant will be built until the late 1980s.

Table 20.5

Proposed Chlor-alkali projects to 1985 in Latin America.

Company	Country	Capacity (thousand tonnes)	Startup
Salgema	Brazil	228	1985
Carbocloro	Brazil	108	1981
Induclor	Argentina	90	1984
Alcalis de Colombia	Colombia	40	1983
Pennwalt	Mexico	40	1985
All other		70	
Total		576	

20.4 MIDDLE EAST AND MEDITERRANEAN

Plants now under construction or in the planning stage are shown in Table 20.6. The one nearest completion is the Shell/SABIC joint venture at Al-Jubail in Saudi Arabia. Targeted to be on stream in late 1984 or early 1985, it will have a capacity of 330 000 tonnes of chlorine and 400 000 tonnes of ethylene dichloride.

Table 20.6

Scheduled chlor-alkali projects in low-cost energy countries

Organization	Country	Capacity (thousand tonnes)	Startup
Shell/SABIC	Saudi Arabia	330	1985
Dow/INA	Yugoslavia	280	1984
Dow Canada	Alberta, Canada	230	1984
Petkim Petrokimya	Turkey	75	1981
Government	Iraq	60	1981
Government	Libya	43	1981
Total		1018	

Smaller chlorine plants in Iran and Iraq are probably already completed, but the one in Iraq is understood to be on stand-by pending the end of the war in Iran.

Another complex at Ras Lanuf in Libya is scheduled to produce 40 000 tonnes of chlorine with corresponding quantities of vinyl chloride. Other chlorine–caustic soda–VCM complexes have been proposed in Israel, Egypt and Algeria.

Another project that should be included here, although it is not in the Middle East, is the Dow/INA complex on the Island of Krk in Yugoslavia. Units for the production of 200 000 tonnes of vinyl chloride and 150 000 tonnes of ethylene dichloride were scheduled to be in operation by the end of this year, but there are reported to have been deferred until at least 1984. This production is destined primarily for export, since it cannot be supported by local consumption.

Another Middle East plant that cannot be ignored in spite of recent events is the Iran-Japan Petrochemical Co. in Bandar Khomeini, Iran (Table 20.7). Originally scheduled to be on stream last year, and nearly complete, it was damaged in the Iran–Iraq war and now lies idle while the owners, the National Iranian Oil Co. and a Japanese consortium led by Mitsui are discussing who should bear the cost of completion. When it will be finished is a matter for speculation, but it is likely that someone will eventually operate the plant, probably before the end of the decade.

Table 20.7

Pending chlor-alkali projects in low-cost energy contries.

Organisation	Country	Capacity (thousand tonnes)	Startup
Iran-Japan Petrochemical	Iran	225	1985–1990
Several	Australia	400†	1985–1990
Exxon	Indonesia	70	1986
Total		695	

† Estimated total.

20.5 SOUTHEAST ASIA

A plant in Indonesia by Exxon and a Japanese consortium is in the planning stage. It will produce ethylene, ethylene dichloride, and vinyl chloride monomer. Scheduled for completion in 1986, the plant is reportedly intended to satisfy the growing domestic demand for PVC in Indonesia.

20.6 AUSTRALIA

From time to time over the last 10 years chlor-alkali–petrochemical complexes have been proposed in Australia. The country's consumption of caustic soda for alumina refining, nearly all imported, makes it an obvious choice, but the obstacle has always been the disposition of the chlorine. There are several projects now under consideration, but the one that appears to be most active at the moment is in South Australia by Conzinc Rio Tinto and another Japanese consortium, which is reported to be willing to guarantee to take part of the output of ethylene dichloride.

Others that have been proposed include Dow at Stony Point, South Australia; ICI at Point Wilson, Victoria and Occidental in Queensland.

20.7 CANADA

Canada does not have a large population and domestic consumption cannot support a large production of chemicals, but it does have relatively low-cost energy, since its crude oil and natural gas have been priced by law at levels below world prices. Canada is already exporting substantial quantities of ethylene dichloride to the Far East, mostly Japan, where domestically produced material is more expensive.

Dow Canada has agreed in principle with Kanegafuchi Chemical and Mitsui Toatsu of Japan to form a joint venture which will add another 272 000 tonnes a year of ethylene dichloride capacity to Dow's plant at Fort Saskatchewan in Alberta. Other joint ventures with Japanese consortiums have been proposed for other locations in Western Canada, but so far have not resulted in any definite plans.

However, the price in Canada is a political consideration and can be changed. In recent months, declining demand and rising Canadian gas and oil prices may make Canada less competitive in world markets, and a much less attractive proposition.

20.8 RISKS

Notwithstanding all their cost advantages, investments in the countries of low-cost energy are not without political risks, especially in the Middle East. The revolution in Iran; Iran's subsequent war with Iraq, and their effect on the new chlorine installations in both these countries have shown us how much the success of these ventures depends on stability. Another military or political upheaval in the Middle East would create an entirely different scenario – one that is extremely difficult to predict at this time.

Canada and Australia, on the other hand, have the advantage of relatively stable governments and would represent fairly reliable sources of supply.

20.9 DISTRIBUTION OF NEW PRODUCTION

Where will this new production be sold? If it is designed for export, it will obviously go to those parts of the world that have a large and ready market for ethylene dichloride or vinyl chloride monomer, or have a growing demand for them. The most likely areas are Japan, Western Europe, the US, and Southeast Asia. Some of the more populous countries of Southeast Asia, such as Indonesia, will absorb some of the output, but the bulk will probably go to the industrial countries where most vinyl chloride is produced, as shown in Table. 20.8.

Table 20.8
Free world production of vinyl chloride 1980.

Country/region	Million tonnes
Western Europe	4.1
US and Canada	3.2
Japan	1.5
Other Asia and Africa	0.8
Latin America	0.4
Total	10.0

As we have already pointed out, Japan is even now importing large quantities of ethylene dichloride from Canada and seems willing to import more. These imports are certain to have a devastating effect on the domestic Japanese chlor-alkali industry; imports in 1981 approached 20% of the total Jananese production of vinyl chloride monomer. Western Europe and the US will also feel the effects. However, the full impact will not begin to be felt until the latter part of the 1980s since most of the plants we have discussed will not be on-stream until after 1985.

20.10 SURVIVING THE GLOBAL SHAKEOUT

What should the chlorine producers in these countries do to remain competitive? There are already too many plants; the US, Western Europe and Japan alone have 220 plants, many of them old and small-scale. Some will no doubt decide to withdraw from the business, and in fact a few already have. However, for those companies that have substantial investments in chlorine plants it may be impossible or undesirable to withdraw. There is still time to take whatever defensive measures may be decide upon; the critical period will not come until the late 1980s or early 1990s when the new capacity begins to come on-stream. Each producer will have to realign its activities, re-examine its competitive position, and decide, sooner or later, whether it wants to stay in the business, or quit.

Those that decide to survive must make use of one or more of the following strategies:

(1) Achieve the lowest cost delivered to the customer.
(2) Differentiate the total offering to the customer – products, service, system.
(3) Shift sales from products to knowledge and service.

20.10.1 Reducing costs

The most effective strategy for chlorine producers to remain competitive is undoubtedly the reduction of costs. Since energy now represents at least half of total manufacturing costs, in recent years there has been an almost frantic search for new energy-saving technology. Many producers already have programs underway to reduce their energy consumption, for example by the use of dimensionally stable anodes, ion-exchange membranes, or bipolar electrolyzers.

Another cost-saving method that the industrial countries will have to face at some point is the need to rationalize production into plants of optimum size. Japan, as we have already pointed out, must consolidate its industry into a smaller number of plants of larger size. This need has already been recognized in the number of Japanese consortiums that have been formed for overseas investments.

Other methods can also be useful:

- Debottlenecking plants for maximum efficiency and yield.
- Maintaining equipment and buildings in excellent condition for optimum time on stream.
- Minimizing charges for:
    - Depreciation
    - Raw materials
    - Labor
    - Environmental protection
    - Sales expense
    - Terms of payment

20.10.2 Differentiating the product

In an old and well-established industry like chlor-alkalis differentiating the product is not easy. However, it is not impossible. A few methods of differentiation are summarized below:

(1) Seek diffuse, not concentrated markets, where unique distribution is required. For example, emphasize derivatives outside the mainstream of the chlorine market. About 30% of the consumption of chlorine is

in a variety of chlorinated chemicals, some of which, like agrochemicals, are growing more rapidly.

(2) Incorporate the product in a system. For example, sell chlorinated solvents as part of a complete metal degreasing system.

(3) Seek markets or segments with few competitors.

20.10.3 Selling knowledge and services

Companies in the chlorine industry are used to selling products of their physical assets, such as manufacturing plants, distribution systems and stock points. However, they also have intellectual assets to exploit – scientific and engineering knowledge, patents, end-use knowhow, and marketing and financial expertise.

Such assets include:

- Consulting on original plant design and subsequent operation to minimize costs.
- Managing plants owned by the state.
- Licensing technology.
- Marketing products of developing countries outside their boarder.

Some of the larger producers are already doing this. Diamond Shamrock, ICI, Hooker and Olin are active in cell technology; Asahi Chemical, Du Pont, and Asahi Glass have invested heavily in membrane technology which they are now promoting.

20.10.4 Redeploying assets

In aging industries the weaker competitors must withdraw from the business if they cannot reduce costs or retrench to remain competitive. For large diversified producers this means milking the business as much as possible by reducing overhead and other related costs to a minimum and pricing as high as the market will bear to get the greatest return possible. The assets can then be redeployed to:

- More favorable locations.
- Acquisitions of compatible firms.
- Investments in downstream products.
- Innovative products of new technologies.

For small undiversified producers there are fewer options; some may simply have to go out of business.

In summary, low-cost energy is an important requirement for a successful chlor-alkali operation, but not the only one. Efficient management, good marketing, and technical knowhow are also important. Chlorine has no real substitute; it is the most available and least expensive halogen and there will always be a market for it, no matter how old. The world is not going to depend entirely on faraway and often politically unstable sources for chlorine, whatever its price and there will always be room for an efficient and well-managed producer.

Of the 220 chlorine plants in the industrial world today, I venture to say that not more than 75% will still be in operation by 1990, but gentlemen, I trust you will all be among the survivors and will look back on the 1980s as a time of challenge and stimulation.

20.11 DISCUSSION

1. Mr C. W. Swenson, Borregaard Ind. Ltd., Norway, opened the discussion. It was mentioned that the size of the plant should be optimized but in Norway two smaller membrane plants have recently been built – one for the bleaching of wood pulp, the other for metallurgical reasons which doesn't appear to fit the suggested philosophy.

   Mr Innes replied that he did not want to imply that all small plants would be non-competitive but those based on the older technology will be. Certainly ion-exchange membrane technology appears a new flexibility which may be very applicable to small plants. This may lead to more small plants being built which may even lead to satellite plants in pulp and paper mills or other consumer industries.

2. Dr A. T. Churchman of the Electricity Research Council (as session chairman) developed the theme of in line plants with captive use of chlorine and thus no transporting problems suggesting that this removed the political risk of chlorine supplies to a consumer possibly being interrupted by political action and cited the case of the Middle East.

   Mr Innes replied that he did not think Middle East producers were going to stop at just making both chlorine and ethylene but they were eventually going to integrate downstream (as they have said) but this may be 10 years away.

CHAPTER 21

Past and future trends in the European Chlor-alkali Industry and their impact on Chlorine Technology

S. M. ANDREW, J. P. H. SHAW, ICI Mond Division, Runcorn

21.1 INTRODUCTION

Chlorine technology and the chlor-alkali business both appear to have entered a period of radical change in the past decade. In particular, membrane cells look poised to take off as the principal technology of the future, while at the same time the business has entered a period of low growth and irregular behaviour unprecedented in recent years. This makes forecasting unusually difficult, but it nevertheless remains essential for those responsible for developing and financing the technology to take a clear view of the future. Such a forward view must be primarily a synthesis of past experience, and this paper reviews the history of the chlor-alkali business in Europe, and especially in the UK, in an attempt to provide some pointers to the future.

Discussion is mainly in the context of four principal factors which appear to have determined the development of the industry and technology, namely:

- the market for the products,
- the availability and cost of 'raw materials' (including energy),
- the availability of technology suitable for industrial scale use,
- the availability of finance and the perceived prospects for a profitable business operation.

Other factors, most notably social and environmental ones, have from time to time played a significant role, and in fact environmental effects (HCl emissions in the nineteenth century and mercury in the nineteen-seventies) have supplied the catalyst for important technological changes at least twice. However, in the long term these are secondary to the four basic factors listed above.

21.2 INDUSTRIAL BEGINNINGS

As far as the UK is concerned the production and technological records are well documented over many years and it is even possible to trace some of the aspects back to the beginning of the nineteenth century. Much of this paper is based on this UK data, but it also draws on parallels for the more recent past in Europe and North America. An attempt is made to summarize each 20-year time-span from 1800 to the present day (see Table 21.1).

In 1800 chlorine had been isolated and could be used for bleaching of textiles. Caustic soda, however, was a much more important industrial chemical and at that stage production of the two basic chemicals was not linked as they were obtained by two quite different routes. The industrial manufacture of chlorine used the original chemical route

$$2NaCl + 3H_2SO_4 + MnO_2 \longrightarrow Cl_2 + 2NaHSO_4 + MnSO_4 + 2H_2O$$

while the world supply of alkali was dependent upon natural materials of limited availability such as seaweed and crude Egyptian natural trona.

The textile industry, the origin of the industrial revoluton in England, provided a ready and expanding market for chlorine as a bleach, and there were ample outlets for alkali, but both these potentially abundant markets were limited by the absence of any technology capable of supplying them economically.

The first major technological breakthrough came with the exploitation of the Leblanc process, originally invented in France in 1790, in locations where the basic raw materials of salt, sulphur, limestone and coal could be made available. Leblanc soda manufacture by the two stage process

$$2NaCl + H_2SO_4 \longrightarrow Na_2SO_4 + 2HCl$$

$$Na_2SO_4 + CaCO_3 + 2C \longrightarrow Na_2CO_3 + CaS + 2CO_2$$

began on Merseyside at almost the same time as the world's first railway system commenced operation, and for the first time alkali was available as a bulk commodity. From 1820 to 1860 this technology expanded rapidly, satisfying much of the world demand for alkali from Merseyside and a number of other European sites. At this stage environmental problems due to the direct discharge of the byproduct HCl to atmosphere brought about the enactment of the world's first national environmental control legislation – the UK Alkali Act of 1863. The ensuing development of techniques for absorbing or recovering the HCl led to the development of technology for the separation of chlorine, and from that date the production of chlorine and caustic were inexorably linked.

Two processes for the separation of chlorine from HCl were successively developed. The first was the multistage Weldon process, based on oxidation with manganese dioxide as manganites. The process was highly efficient in recovery of valuable manganese, but extravagent on chlorine, only 30% of the chlorine enetering the process as HCl ending up as chlorine gas. However, with abundant byproduct HCl available this did not matter, and the process produced

Table 21.1

The development and application of UK chlorine and alkali technology since 1800.

Period	Chlorine	Alkali	External events
Pre–1800	First industrial process 1799–previously no commercial outlets.	Considerable demand for the product, but limited mainly by supply from natural products. No industrial manufacturing technology.	Commerce entirely dependent on waterborne transport.
1800–1820	Chemical technology available but limited to demand for textile industry.	Financial investment being considered in the Leblanc process.	Repeal of the Salt Act 1820.
1820–1840	No change.	Market for alkali very strong. Industrial production expanded rapidly but deficiencies in raw material usages and waste disposal a severe constraint.	Introduction of rail transport and rapid extension of the industrial revolution.
1840–1860	Commercial demand increasing but technology little changed.	Leblanc technology comes under severe environmental pressure but continues to expand.	First commercial undertakings for domestic water supply but no thought of chemical treatment.

Table 21.1 – *Continued.*

Period	Chlorine	Alkali	External events
1860–1880	Chemical technology changed to use byproducts from the Leblanc soda ash process. Increased rate of growth in demand for bleach products.	Improved soda ash technology developed (Solvay). Capital intensive but much improved efficiency of raw materials. UK production levels static as export markets develop their own sources of production.	Passing of first 'Alkali Acts', 1863.
1880–1900	Commercial market for bleach products saturated. Availability of raw materials for chemical process recedes with progressive demise of Leblanc process.	Demand static, Leblanc process progressively only used to supply lime soda while Solvay process meets the soda ash market.	First commercial electricity undertaking 1882. 1892/95 invention of electro-chemical routes to chlorine and caustic.
1900–1920	Chemical processes for chlorine cease, electrochemical technology developed initially most widely in diaphragm cells but rate of application of mercury cells accelerating. Overall chlorine market remains static. 'Raw material' costs reduce rapidly with development of electricity supply industry.	Cessation of Leblanc process, Domestic soda ash demand starts to increase at approx. 6% per year but caustic soda remains static.	Use of chlorine as a war gas gives the product a bad name but no noticable impact on production levels. Water treatment processes started and accelerated rapidly after Croydon typhoid epidemic 1920.

Table 21.1 – *Continued.*

Period	Chlorine	Alkali	External events
1920–1940	Chlorine demand growth (approx. 7% per year) led initially by inorganic chemicals but eventually more strongly by organic derivatives. Electricity prices continue to fall but chlorine technology static.	Soda ash production doubles in 20 years. Approx. 50% caustic demand still supplied by lime soda route, while caustic production rates increase at approx. 6% per year.	
1940–1960	Demand growth remains at approx. 7% per year, mainly led by organic derivatives with limitations on the availability of finance. Pressure to improve technology to reduce capital costs.	60% of caustic now supplied by electrolytic route and Cl_2 growth curve starting to catch up with caustic.	Fuel cost inflation, and shift towards domestic rather than industrial consumer, slows down improvements in electricity costs.
1960–1980	Chlorine growth slows abruptly in 1973/74, chlorine and caustic come into balance. Capital cost improvement in technology continues. No major new derivatives in prospect.	Recession in demand for total alkali. Lime soda production stopped.	High environmental concerns stimulate the development of new technology. 1973 oil price leap. High interest rates and inflation stifle industrial investment.

Table 21.1 – *Continued.*

Period	Chlorine	Alkali	External events
1980+	Low growth for most derivatives? Some surplus Cl_2/NaOH capacity. Improved technology available with lower capital cost but driving force in its installation will be raw material costs and efficiencies, plus plants in new producing territories.	Demand for a wide range of outlets will be similar to industrial production. Export 'surplus' from major industrial countries will balance imports for Third World and alumina. Some additional demand for oil exploration.	Cl_2 and NaOH levels of production now tied to the electrochemical equivalents. Which is the price leader and which is the by-product will vary with shorter term fluctuations. Industrial development in oil producing and other Third World countries will give opportunities in new areas.

strong chlorine gas. The Weldon process was supplanted by the much simpler direct oxidation Deacon process:

$$2HCl + O \longrightarrow Cl_2 + H_2O \ .$$

This was cheaper to build and operate, and much more efficient in HCl, but produced a weak chlorine gas (<10%) suitable only for making bleaching powder, and with a consequent influence on the market.

A rapid growth in the chlorine industry then followed, and it is notable that the four 'conditions' for growth were favourable. The markets were there, raw materials were available and cheap (indeed the HCl was of negative value), the technology was adequate, and with the wealth generated by the industrial revolution there were few financing problems.

Figures 21.1 and 21.2 show the development of chlorine production in the UK from 1800 to the present day. By 1860 the output of chlorine by the original chemical route had reached 10 000 t/y. Then, under the influence of the favourable factors discussed above, output grew to 50 000 t/y by 1880, at which level it remained for the next 40 years.

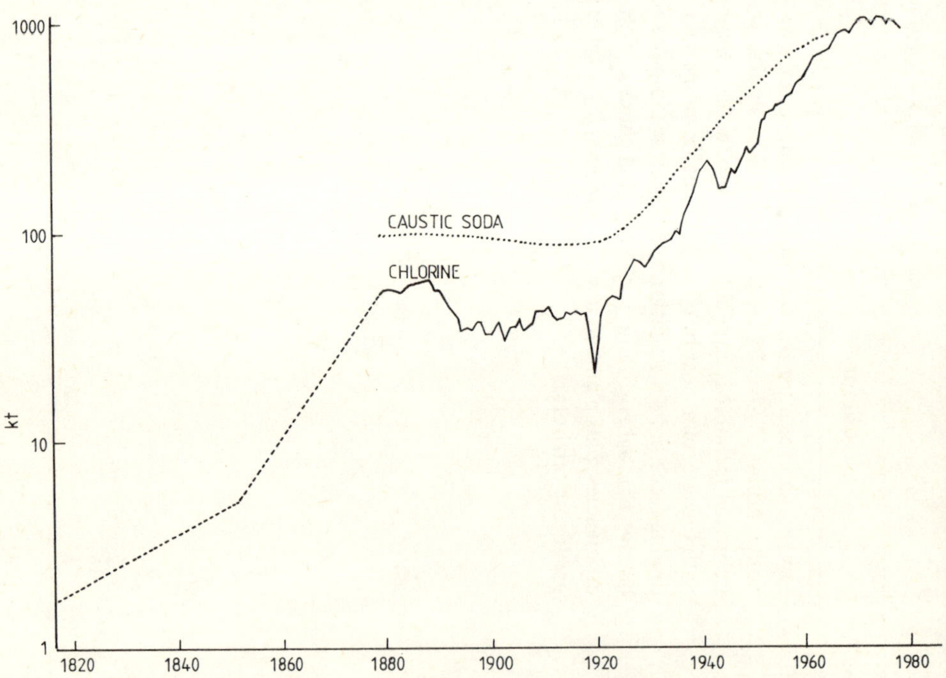

Fig. 21.1 – UK chlorine production 1820–1980 and chlorine equivalent of caustic soda production.

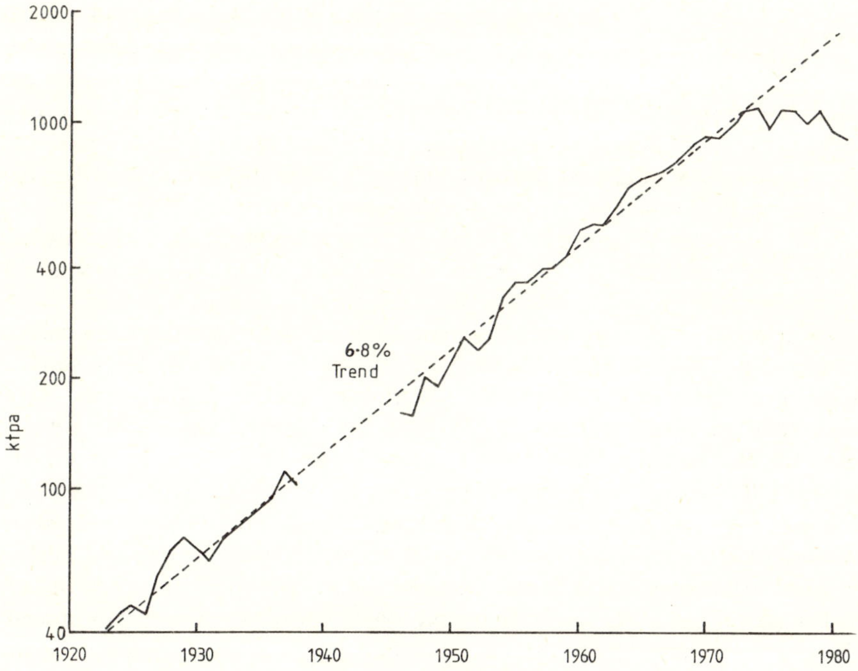

Fig. 21.2 – UK chlorine production from 1923.

At this stage, the world demand for alkali was still well in excess of the electrochemical equivalent quantity of chlorine. The Leblanc process remained the dominant provider of both carbonate (soda ash) and caustic soda (via the lime soda process). In 1880 the UK production of alkali (ash and caustic soda expressed as 100% NaOH) was 300 000 tons compared to the chlorine figure of 50 000.

The period 1860–1880, however, saw the development of the Solvay soda ash process, which had a dramatic effect on the alkali industry. Its efficiency was so much greater than Leblanc that it rapidly replaced the Leblanc process for soda ash production even in a period of largely static demand; in fact between 1880 and 1900 the Leblanc industry fell to a third of its former size, the surviving operators continuing to produce caustic soda from the lime soda reaction. A consequence of this change in technology was a fall-off in the availability of low cost HCl, and this along with the development of competitive manufacture outside the UK led to a fall-off in the production level of chlorine.

At the same time, the end of the nineteenth century saw the development of the first industrial scale electrochemical processes for caustic and chlorine. Caustic at that time had a potentially very large market which was limited by the costs and capabilities of the lime soda process; from 1880 to 1920 the UK production of caustic soda remained at roughly 100 000 tons, with almost half of it exported. In 1880 the only route employed was based on Leblanc soda ash, though by 1900 there were small inroads from both the Solvay soda ash source and from brine electrolysis. In 1920 the Leblanc process was finished, partly in favour of Solvay origin material, but 50% of the caustic soda produced was of electrolytic origin. At the same time soda ash production in the UK, after an initial small cut back, more than doubled between 1880 and 1920.

The rules were therefore set for the development and potential rapid growth of electrolytic technology. The caustic market size and price were established. The advent of improved technology for soda ash at a time of a general slowing down in world trade had made the basic commodities of soda ash and caustic more readily available, the reduced production costs helping to bring about a doubling in the size of the market. Chlorine production had remained static since 1880 at approximately 50 000 tons, with its end use almost entirely based on the oxidative properties of chlorine, bleaching and water treatment, and the old chemical routes dependent primarily on Leblanc byproduct HCl for their raw materials were inevitably phased out. The output of chlorine therefore became basically dependent on the rate at which hardware for the new electrochemical technology could be installed and the availability of money to finance it. The latter was made available by the operators of the Solvay ash process, who saw the opportunity to offer a complete portfolio of products and to sound the ultimate death knell of the Leblanc industry.

21.3 EARLY MERCURY CELL DEVELOPMENT

Though diaphragm cells were developed at about the same time, in Europe the prime interest was always in mercury based technology, and the early progress of this industry is an object lesson to present day industrialists. In 1892 Castner's chief chemist, Baker, inspired by the publications of Kellner, produced chlorine in a bell jar over a pool of mercury. In 1893 he had three prototype rocking cells working at 4.4 V, 84% current efficiency and a current density of 0.45 kA/m^2. By 1897 he was producing 500 tons of chlorine at Runcorn and by 1900, 3500 out of a total UK electrolytic production of 5000 tons of chlorine. The rocking cell was a fairly robust piece of hardware, and in fact some installations were still using this technology up to the past war.

The ability to produce the total UK demand of 40 000 tons Cl_2/year in the early years of the 1900s from rocking cells would however have required excessive amounts of capital, and it was actually the long cell which was developed in most large installations in Europe. By 1920 the UK production of chlorine from mercury cells was just over 16 000 tons, with just under 21 000 from diaphragm cells and 5000 from the vestiges of the Leblanc/Deacon processes. A point to note is that the d.c. electrolytic power consumption of the earliest long mercury cell was approximately 3500 kWh/ton Cl_2, a figure that has remained virtually unchanged through the 80 years history of the technology. The development of electrochemical technology was of course dependent on the availability of the main raw material, electricity. Electrical undertakings were established in the UK in 1880 with an initial commercial price approximately twenty times today's levels. By 1900 the effective cost of this raw material had fallen by 50%, and it continued to halve in each of the next 10 years. The relatively constant energy usage of electrochemical technology, therefore, posed no specific cost disadvantage.

From 1900 to 1960 the world supply of soda ash and free alkali was dominated by the Solvay ammonia soda process. UK soda ash production expanded from 200 to 1200kt/y and lime soda caustic production from 100 to 300kt/y, with the percentage going to export falling from 40 to 20%. This process therefore continued to determine the free caustic market position, with the electrolytic process competing on the basis of a better product (particularly suited to certain end users such as rayon manufacture) but limited by the size of the chlorine market. Throughout the period from 1920 almost up to 1940, mercury cells were still being installed as long cells with non-adjustable graphite anodes at 2 kA/m^2. Only towards the end of this period were improved materials (ebonite), better anode designs, etc., put into use. Capital costs per unit of capacity therefore only improved in respect of the benefit of scale, and variable costs with roughly constant unit power usage were dependent upon the electric supply industry for any improvement.

The rate of application of electrolytic technology in this period was thus principally dependent on the development of new chlorine end uses, as in the absence of a substantial pulp and paper industry in the UK the traditional uses of chlorine based on its oxidative properties remained essentially static at 40 000 t/y from 1880 till recent times. These new uses in fact materialized first through inorganic compounds of chlorine and latterly mainly through the development of chlorinated organic derivatives. These developments have largely led chlorine growth in the UK since the introduction of electrolytic technology.

21.4 ORGANIC CHLORINE DERIVATIVES

In 1900 the production of organic chlorine derivatives in the UK was effectively zero, but by 1920 Cl_2 use in their production had become significant and by 1940 it had increased to 35 000 tons. At the same time total chlorine production in the UK had increased from its plateau of approximately 40 000 t/y up to 1920 to almost 135 000 t/y in 1940, 77% of this from mercury cells, 19% from diaphragm cells and 4% from fused salt electrolysis.

In the period up to 1940 the main increase in demand for organic derivatives was for solvent materials. Leaving out the transient effect of the war period, chlorine demand from the forties through to the early seventies continued to be determined by the growth of organic derivatives, but latterly it was dominated by vinyl chloride for PVC which was growing at over 15% per year in the fifties and sixties. This rapid growth demanded frequent new investment, and in a period of relative financial stringency but low energy costs technology advances were channelled mainly into reducing the capital invested per unit of product. The capital intensiveness of the process at this time is illustrated by the fact that up to 1950 new capacity cost £200 per annual ton, while total operating costs were less than £25/ton. Interestingly, most of the innovations, and there were many, including individually adjustable anodes, improved gas release, more use of copper, etc., were originally introduced to increase output or reduce costs on existing plants, but they were rapidly taken up by the designers of new plants and the net result was more and more output from cells of a given size. Overall, improvements in electrical resistance were balanced by increased current density, so the power consumption per unit of production still remained constant. It should also be noted that the maintenance of an electrolytic cell at the desired temperature conditions requires a certain level of heat input to counteract the losses in water vapour and by radiation. This in itself tends to increase the level of energy consumption above that required for achieving electrolytic dissociation. Thus the figure for the electrical power consumption of the best modern mercury cell in 1980 is 3500 kWh/t chlorine at 10–12 kA/m^2, compared to the same unit consumption at 1–2 kA/m^2 in the earliest long cell. However, in 1980 a new mercury cell chlorine plant cost £300/t year (money of the year) compared with a typical operating cost of £120 – a totally different proportion from that quoted for 1950 above.

21.5 FUTURE TRENDS

In attempting to forecast the future direction of chlor-alkali technology it is the most recent data which might be expected to be most relevant; this is summarized for the UK in Fig. 21.3, Western Europe in Fig. 21.4 and the USA in Fig. 21.5. Here we come up against a difficult problem, namely the dramatic change in the business which took place in the early seventies, with trends which had been steady for many decades apparently changing quite suddenly – in Europe largely in coincidence with the 'great oil price leap' of 1973–74, though in America rather earlier. From 1960 to 1973 trends followed those of earlier years, one consequence of which was that chlorine production caught up with caustic demand bringing to an end the need to make caustic independently of chlorine by the lime soda process. Steps were taken to anticipate the forecast caustic 'surplus' as this process continued by, for example, developing export caustic markets (some of which were effectively substituting soda ash) and making provision for carbonation. However, in the event the 'surplus', manifested in weakening caustic prices, never became as serious as expected, because from the early seventies the dramatic fall in chlorine growth rates brought the two products largely into balance. A consequence of this change is that one of the attractions of diaphragm cells – the ability to run without evaporation and with low energy

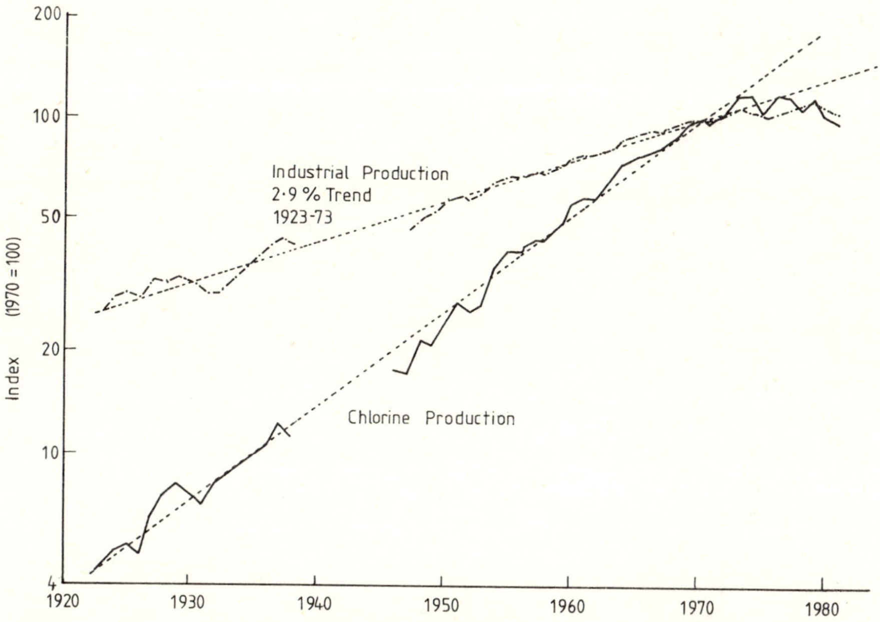

Fig. 21.3 – UK chlorine production and industrial production indices.

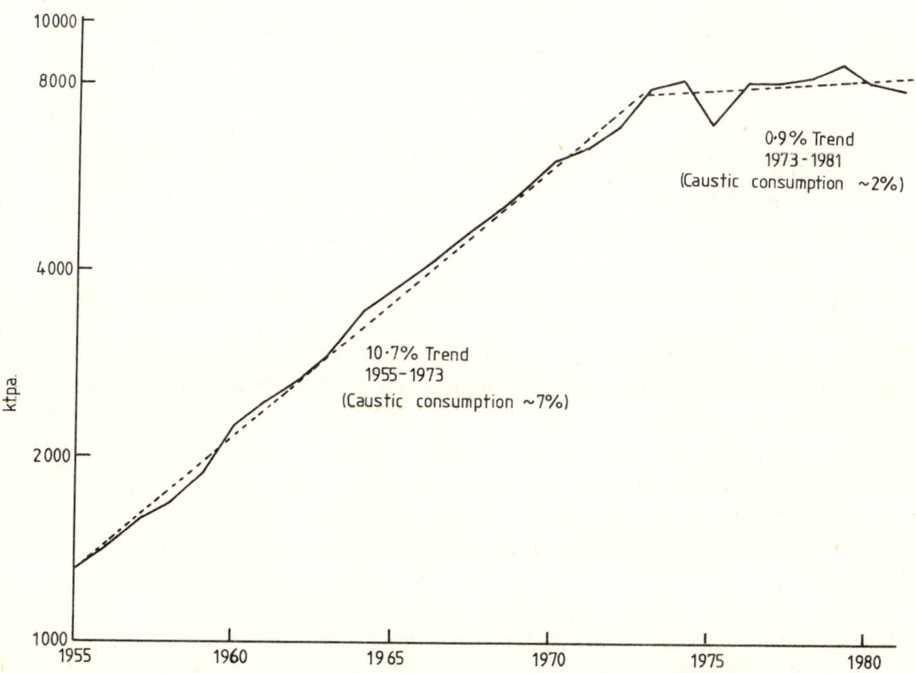

Fig. 21.4 – West European chlorine production.

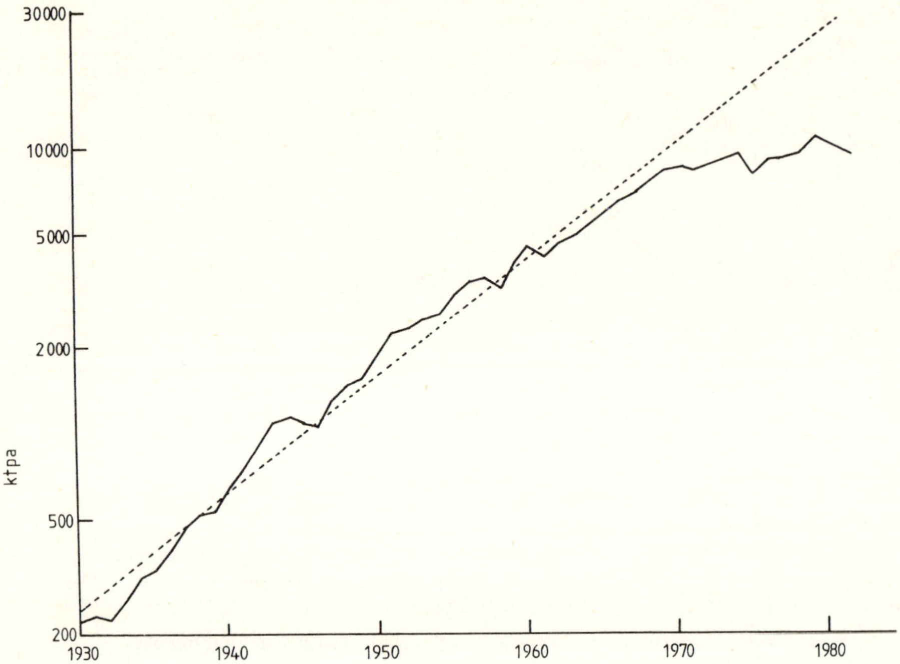

Fig. 21.5 – US chlorine production.

costs if the purified caustic is to be carbonated – has been lost. With probably roughly equal growth rates for chlorine and caustic in the future the advantage goes to processes that can produce balanced quantities of pure caustic and chlorine cheaply – for example membrane cells.

One result of the changes since the earlier seventies is that caustic and chlorine growth rates have come close together. The other, which has the more profound impact, is that both have fallen to low levels, and have indeed fallen more than have relevant economic parameters such as industrial production. Thus whether in Europe, the UK or the USA, chlorine growth traditionally exceeded industrial production growth by some three points; however, since 1973 chlorine has tended to grow at a rate similar to industrial production or somewhat below it, while industrial production growth itself, in Europe at least, has dropped sharply (see Fig. 21.3). This new, flat trend, admittedly with a good deal of 'noise' in the data, has now persisted for 8 years, and without any clear reason to expect a change looks set to persist for some years to come. The fact that the 1973–74 events had a more drastic effect on chlorine growth than caustic probably reflects the former's stronger links with the petrochemical industry which itself was very directly affected by oil price escalation.

The fall in chlorine growth can also be associated with certain identifiable specific depressing effects whose timing added to the energy crisis effect, and which have not been compensated as in earlier times by the introduction of new derivatives – the process of the invention of new products seems to have run out of steam for chlorine derivatives, as in many other fields. Examples of negative effects on chlorine growth are improved solvent recovery, both for economic and environmental reasons, better processes such as oxychlorination which avoid chlorine waste, recovery of HCl, the ozone controversy which has reduced fluorocarbon growth, and replacement of processes which use chlorine as an intermediate in their manufacture – e.g. for ethylene oxide.

Effectively, therefore, it can be said that the application of chlor-alkali technology in Europe in the period from 1960 to 1980 has been dominated by the commercial forces associated with the demand for the products, with an element of financial pressure surrounding the installation of new capacity in the middle of this period (in terms of scale and capital cost). The input cost of electricity has also ceased to improve for many years, improvements in fuel efficiency and scale of power generation being lost in the progressively increasing cost of fuels. Environmental pressures, though much publicized, have not in fact had a great impact in Europe on the type the technology in use of installed, the majority of it still being as mercury cells, although the same pressures in Japan and North America brought the installation of new mercury cells to a halt, with extensive replacement in the case of Japan. These pressures however provided a substantial spur to the development of new technology, which will have its impact in Europe in the future.

21.6 SUMMARY

To summarize, the history of the industry over the period 1800–1980 is outlined by twenty-year segments in Table 21.1; for each time period this table indicates which of the major factors of market, raw materials, technology and finance have been most important at each stage, with most rapid progress being made when all four are favourable together. What will determine the pattern from now on? The prediction of the future from past performance is only a reliable tool if the circumstances which shaped the experience gained prevail in situations to come, and here it is the trend over the last decade or so that seems predominant. As there is nothing obvious to change it, it looks as if chlorine and caustic growth may mirror that predicted for industrial production for the next few years at least, suggesting a trend growth rate in Europe of about 2% per year. Any significant increase above this for chlorine would depend on some major downstream technological break-through with associated market development comparable with the introduction of PVC or solvents in the past, such events are by their nature difficult to forecast, but none are obviously on the horizon. There will be some scope for replacement of existing plants as they wear out, but with generally plentiful capacity in the developed world the number of major new plants will be relatively few. There will be more scope in developing countries. In a continuing high energy cost regime, low operating costs are likely to be the dominant factor in the rate of application, particularly if the investment has to be justified on operating cost savings rather than to satisfy new demand. Here, membrane cell technology should have the advantage. However, with continuing financial stringency there will continue to be a signficant premium on the installation with low capital cost provided this is achieved without sacrificing efficiency.

21.7 ACKNOWLEDGEMENTS

The authors would like to thank their UK predecessors in the art of forecasting chlorine and caustic demand for their careful record of production data, and the BITC and Chlorine Institute some of whose production statistics have been included in the above analysis.

CHAPTER 22

The American Business Scene from the Perspective of the Chlor-alkali Industry

C. A. HANSEN, President, The Chlorine Institute, LCP Chemicals & Plastics Inc.

22.1 INTRODUCTION

I am very happy to be with you today to share with you my views on the American business scene from the perspective of the chlor-alkali industry. Before we look at the future of our industry in the US during the balance of the 1980s, let us first look back.

Many of the circumstances that are impacting us now had their roots in the 1970s and even the 1960s. For the US chlor-alkali industry, the 10-year period, 1970–1979, was one of great change and turmoil and was considerably different from the 1960s. Much of the change was the result of unanticipated external forces, rather than the normal problems of a maturing industry. After these changes are in perspective, we will look to our expectations for the next 2 years, and then with our crystal ball we will look at the latter half of the 1980s. More importantly however, I will present a scenario to you which you can debate and which will provide a basis for modification as actual events happen. I am certain that over the next 8 years, there will be many rapid changes of various direction and magnitude and that the US chlor-alkali industry in 1990 will be significantly different than it is today. The major challenge to the chlor-alkali industry will be the need to respond flexibly to rapid change while recovering its economic health.

22.2 THE PAST DECADE

Let's take a few minutes now to look back at the decade of the 1970s to those factors that seriously affected our business.

Figure 22.1 presents chlorine production in millions of tons per year and, as can be seen, three recessions or business cycles occurred during the 1970s. Overall production increased by 17%, abount 1.7 million tons per year. During the 1960s, production increased by 110%, almost 5 million tons per year. Capacity increased

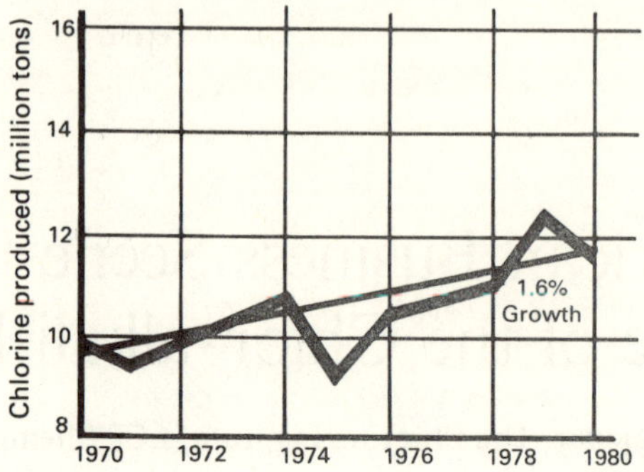

Fig. 22.1 – Chlor-alkali industry: chlorine production (tons/year).

100% in the 1960s and 40% in the 1970s. The large disparity between capacity additions and requirement during the 1970s created economic havoc in the industry.

22.2.1 Energy effects

Certainly of major impact in the 1970s are those things that occurred in the general field of energy, particularly the oil embargo early in the period, recurring energy shortages and increasing costs during the remainder caused by OPEC's pricing policies. As a nation, America learned the importance of energy efficiency and the need for conservation; in fact, we have done a better job than expected but much remains to be done. In the chlor-alkali industry, where electrical energy is a key raw material and cost, the need for efficiency was recognized very quickly and much progress has been made in reducing energy consumption per unit of product. The order of magnitude is about 25%.

Also significant is the trend that began in the middle of the period and continues today toward the equalization of electric power costs across the USA due to deregulation of oil and gas. No longer are there significant but artificial pockets of extremely low cost power available for the production of chlorine and caustic in specific regions of the country. From an energy cost standpoint, as well as others, plants located in the Southeast, Mid Atlantic, and Midwest are now competitive with those located in the heretofore low cost gulf area. Most of the cost disadvantages of the Northeast are also disappearing. These changes will have a major impact on the location of new production units, growth will be spread more evenly in the sunbelt, the Northeast will stabilize and the viability of regional plants will be enhanced.

22.2.2 New technology

A second area that will have a lasting effect on the US chlor-alkali industry, and indeed the world chlor-alkali industry, is new technology. For close to 25 years, there was little change in the basic technology used to produce chlorine and caustic soda. Three factors, although developed individually, tended to combine to bring about the major transformation in the basic technology of our business in the 1970s. These factors were the introduction of metal anodes, environmental problems concerned with mercury cell production and the aforementioned basic need for energy efficiency. This resulted in the demise of mercury cell plant construction and the installation of metal anodes in essentially all chlor-alkali plants and set the stage for membrane cell development.

I believe that membrane cell technology will ultimately satisfy the demand for energy efficiency in an environmentally safe way. We as an industry are now in the process of implementing this next major change, the introduction of membrane cells on a commercial scale. A major deterrent to this developmet is the lack of enthusiasm by companies for investing major sums in an area where economic performance has been poor and where the outlook is not particularly bright.

22.2.3 End-use patterns and process developments

While energy and technology are certainly important aspects of the changes of the 1970s, I think equally important were the rather drastic changes in the historic end use patterns, which were generated by forces outside the natural maturation process, primarily by government actions. Certainly the demise of DDT was a forerunner and an indicator of the future actions. Other major pesticides followed suit, and in short order serious restrictions were placed on several chlorinated solvents, and on the most significant products of our industry in the US, vinyl chloride monomer and polyvinyl chloride. The vinyl chloride monomer and polyvinyl chloride problems were solved, a major contribution to the health of chlor-alkali industry. But, perhaps, the one restriction that was most illustrative of the problems facing the US chlor-alkali industry and the most frustrating in that it involved a product considered environmentally sound, was the ban on the use of fluorocarbons in aerosol sprays.

Add process technological developments; such as introduction of oxychlorination to the vinyl chloride process and the use of direct oxidation for the production of ethylene oxide and propylene oxide and you have a decade in which over half of the growth potential of the US chlor-alkali business disappeared, without the introduction of new products (with the possible exception of polyurethanes) to offset these drastic changes.

There was certainly no longer any need to be concerned about a world-wide glut of caustic soda. The downward revision of chlorine demand took care of that.

Instead of the predicted 6 to 7%, chlorine growth averaged less than 2% during the 1970s. So there we were, an industry faced with serious changes in its end use pattern and with changes in key elements relating to production and yet we continued to build significant additions to our capacity.

Figure 22.2 presents major new plant capacity additions in terms of percent of production capacity added. As can be seen, several new batches of capacity were added during the 1970s and, as Fig. 22.2 overlay shows, we added more than 42% to industry capacity in the 1970–1980 period and waited for the high growth and good economics to return. Unfortunately, that was not to be in the 1970s since requirements went up only 20% and we ended the period with an 80% operating rate.

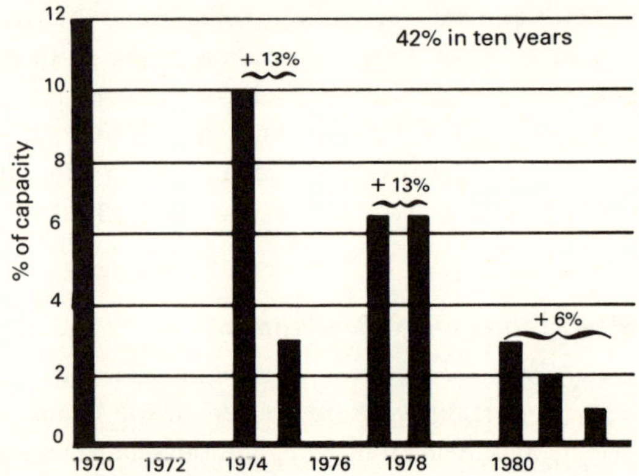

Fig. 22.2 – Chlor-alkali industry: major new plant capacity additions.

Figure 22.3 presents chlorine production rate during the 1970s and is a graphic illustration of the volatility of the period. Low growth rates and consequent low operating rates caused intense competition and thwarted recovery of rapidly increasing costs. This review of the 1970s brings our industry into the 1980s where we are faced with a new dilemma, the US economy has been stumbling, certainly for the last 3 years, and perhaps for the last 5 years.

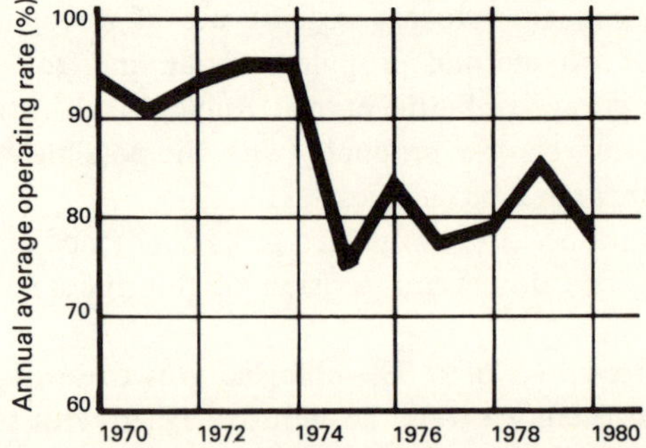

Fig. 22.3 – Chlor-alkali industry: chlorine plant operating rate.

Since 1978, the US economy has been plagued with idle manufacturing capacity and high unemployment, not caused so much by business cycles, but more by a determination to remove inflation and its serious deterimental aspects from our economy. The deliberate use of tight money policies to accomplish this end has had a repressive effect on our industrial economy. We are in 1982 with our chlorine plants operating at less than 65% of capacity. We have capacity ready for the boom, but we wonder when it's going to come. The outlook is for continued low production rates and low economic returns until the end of the 1980s.

22.3 FUTURE TRENDS

22.3.1 Production and capacity

However, the news may not be all bad. Let me suggest a scenario to you. The current recession has had its desired effect. Inflation has been controlled and is at a level that can be handled. The tight monetary policies will probably be eased so that interest rates, while still high by historic standards, can drop out of the teens. As we recover from this recession, President Reagan's economic policies will start to bear fruit so that the proposed budget deficit does not have a negative impact on either interest rates of inflation. The recovery will start during the latter half of this year or early next year and will continue for 18 months to 2 years.

The combination of Reagonomics and monetary control will restrict inflation and permit a growth economy, but at a rate substantially reduced from what we have experienced in the past. During the next few years US industry will address its serious problems as a world-wide competitor so that during the last 3 or 4 years of the decade there is new life in the US economy based on the return of industrial and technological strength. There will only be one business cycle either in 1986 or 1987.

Figure 22.4 presents our forecast of chlorine production and capacity. The

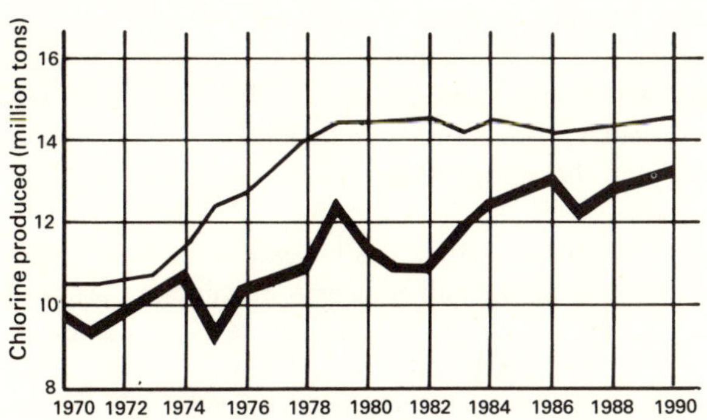

Fig. 22.4 – Chlor-alkali industry: chlorine production (tons/year).

heavy line is production in millions of tons per year. The lighter line is capacity in millions of tons per year. With this forecast in mind, let's look at several key elements of our business which are part of the basis for these projections. The first is growth. The changes if the 1970s in end use patterns were permanent. We don't anticipate any significant departure from the current end use pattern during the 1980s. There will remain continuing threats such as the removal of chlorine from the pulp bleaching process, but we don't see any significant impact until well into the 1990s. During the next 18 to 24 months of recovery, we see chlorine growing in excess of a 6% annual rate with operating rates approaching 85% of capacity by 1984. Thereafter with a business cycle and as the US economy returns to a more normal growth rate, we expect chlorine growth to be at the 1% to 2% per year level. Alkali demand, on the other hand, will continue to be at a significantly higher level than chlorine, probably around 4% per year. The result, of course, will be long term continuing demand pressure on caustic soda and caustic soda prices and continuing growth for soda ash.

22.3.2 Exports

A significant factor in both sides of the growth equation is the importance of exports. During the last years of the 1970s, the relatively favourable US position in the world energy situation created a substantial export market for chlorinated hydrocarbons. This amounted for up to 10% of chlorine production in the US and also made available significant quantities of caustic soda for export. The US petrochemical industry and the chlorinated products associated with it will no longer be as competitive on a world scale. The emergence of significant petrochemical industries in areas with lower raw material costs, such as Canada and Mexico and, of course, the Middle East will remove these chlorinated related market opportunities from the US chlor-alkali industry. Eastern Canadian developments in oil and gas production and low cost hydroelectric power will be of special interest to the European Community and Eastern US. The US has historically exported alkalies at levels in excess of 10% of US production. During the 1980s the strong cost and production capability position of the US natural soda ash industry will result in increased soda ash exports to world markets. At the same time, caustic soda exports from the US will be reduced. This is the basis for our belief that there is ample room for soda ash and caustic at satisfactory returns levels for each.

Technology, as we stated earlier, has become one of the important factors in the future of the US chlor-alkali industry. Membrane cells with their most recent advances provide the means to the high level of energy and operating efficiency that will be necessary for profitable participation in the chlor-alkali business in the future. Conversion to membrane cells will be widespread but relatively slow. They will set the competitive norm for efficiency for at least the next 10 years. But they will have an even broader influence on out industry.

22.3.3 New technology

Membrane cells will determine the nature of new productive capacity. Major world scale grass roots plant construction will be severely restricted, if not eliminated. When there are new sites they will be smaller membrane cell units located near specific markets. This is due to rising transporation costs, anticipated tightening of transportation regulations and equalization of power costs off-setting economies of scale. Most new capacity will be tied to conversion of membrane cells and will come from incremental, low cost growth of existing plants. Significant conversion will start in 1985–1986. By 1990, possibly as much as half of the US capacity could be in membrane cells.

Energy costs, industry profitability levels, and the need for capacity will determine the extent of the conversion to membranes. Overall, productive capacity will remain relatively constant as marginal plants are shut down to improve costs and return on investment.

22.3.4 Profitability

We learned during the 1970s that the profitability of the US chlor-alkali industry suffers most with rapid increases in the cost of energy, and from supply demand imbalances caused by additions of capacity rather than recessions. Pricing tends to hold steady during a recession unless capacity comes on and causes someone to push for more market share. People then react and prices tumble. We don't anticipate major new additions to capacity during the next 8 years and, with an improving economy, pressure on profitability will be reduced. Supply–demand changes due to business cycles will continue but with little change in capacity, profitability should be relatively stable. Minimum levels of profitability during business downturns in the past have been achieved because of the co-product nature of our business. When chlorine is long, caustic is short. This also works in reverse as those of you know who can remember the caustic gluts of the 1960s. This phenomenon has been tested severely during the 1980–82 period with caustic soda pricing being used to dampen the adverse chlorine volume and price impact.

We can, however, anticipate some restrictions on the ability of caustic soda to carry this burden in the future. For example, caustic soda imported from new low cost world sources will have a ceiling effect on US prices. In addition, US soda ash capability will have a similar effect as soda ash moves into traditional caustic soda markets.

To offset these restrictions we have the probability of a more positive economic climate. The result will be a period of modest profitability without strong positive or negative pressures and with more balanced pricing within the electrochemical product unit.

The changes in the chlor-alkali business will cause many producers to reconsider their position in the industry. Not everyone would be able to achieve profitability nor will the return generated be satisfactory to many present producers. There will be a strong tendency away from the merchant side of the chlorine portion of the business leaving some producers with solely a captive position in chlorine. Many others will simply question the reality of their position in the chlor-alkali business at all and will move into more specialized products. This process of rationalization in underway already with Allied Corporation and BASF Wyandotte as leading examples of total removal of once major factors from the US chlor-alkali industry. In addition, many other producers are reviewing individual plant situations, stressing the necessity of optimizing efficiency and when necessary shutting down facilities which cannot meet minimum standards.

These changes have a positive side too, by creating a need for a few efficient producers who remain dedicated to a supply position. There has already been a concentration to seven or eight significant producers, filling this need.

22.3.5 Role of government

The last point I want to consider with you today is the role of government, both local and national, in our business for the balance of this period. Certainly we would be foolish indeed if we thought government did not have a role to play in our business. Our task is to minimize the emotionalism and overreaction that frequently is part of government involvement. We can expect continuing environmental pressures. These will have their cost impacts as will the costs of clean-up of hazardous waste sites through superfund and related taxes. We can expect the US Government to increase restrictions on the transportation of hazardous materials. We need not be paranoid about government involvement, we expect the responsible corporate manager of the 1980s to take the lead in discharging his responsibilities to the environment and the general public.

22.4 CONCLUSIONS

I have reviewed with you the factors that will impact the US chlor-alkali industry in the 1980s. Let's just reflect a minute on what the end result will be. In 1990, you will find fewer producers with a compression to five or six key producers. Those remaining will be operating extremely efficient plants. Membrane cells will make up a signficiant portion of the capacity. Many of the plants will be smaller regional plants. You will see a business that is based primarily on the domestic industry with little active participation in export to the world markets.

Figure 22.5 demonstrates what might happen during the 1980s at various growth rates. In 1990 the industry will have passed through a period of modest growth. Overall growth for the decade will be the same 1.6% we experienced during the 1970s.

As can be seen, if we were to be fortunate enough to experience a growth rate of 3% during the 1980s, we would be out of capacity in 1987 – a happy but not likely problem for the industry to encounter.

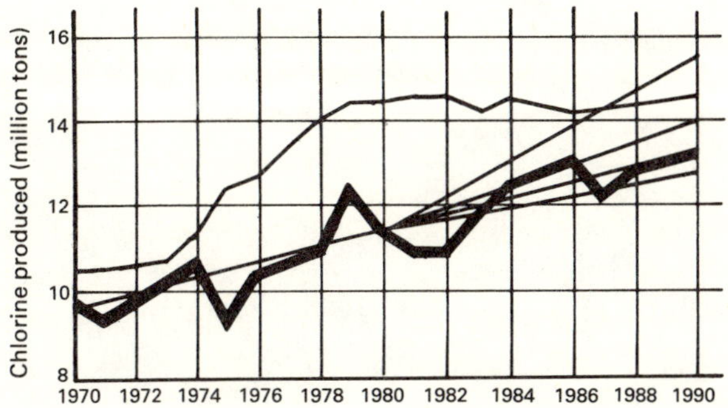

Fig. 22.5 – Chlor-alkali industry: chlorine production (tons/year).

Figure 22.6 presents the industry operating rate as percent of capacity for the 1980s. As can be seen from the chart, we will be starting to see a revitalized US economy. Chlor-alkali operating rates will finally exceed 90%.

The overlay for Fig. 22.6 illustrates what might happen to various growth rates during the period. As can be seen, even if the growth rate is only 1%, the lack of capacity additions predicted for the period would still enable the industry to operate at a rate close to 90% by the end of the period.

This is not a picture of a dynamic business, but it is certainly one of a well managed industry which has been able to survive serious adverse changes and position itself for future rewards.

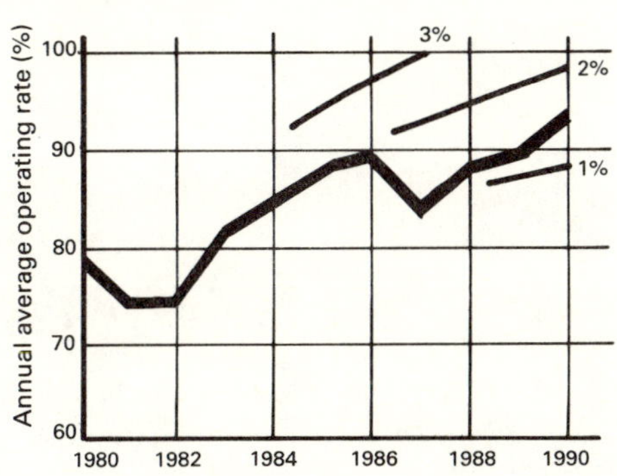

Fig. 22.6 – Chlor-alkali industry: chlorine plant operating rate.

SECTION G

Environmental/Safety

CHAPTER 23

Environmental Trends and North American Chlor-alkali Industry Challenges: Past and Projected

EDMUND J. LAUBUSCH, The Chlorine Institute Inc.

23.1 INTRODUCTION

As a young lad, I visited the 1939–40 World's Fair in New York and marvelled at the various exhibits and displays. One in particular was very impressive – the Du Pont nylon exhibit with its very inspiring slogan: 'Better Things for Better Living Through Chemistry'. Today, as well we all know, the chemical industry is nearly obsessed in its quest to quell the assaults upon it that accompanied the environmental movement.

The first real shot at the chemical industry – in the battle against DDT and other insecticides – was fired in 1962, with Rachael Carson (and her book *Silent Spring*) as champion. Thus, a new platform was launched for public protests about the uses to which science and technology were being put. Thereafter, a number of highly institutionalized public interest groups came into being. And with the proclamation and overwhelming public response to Earth Day in April 1970, came the realization to American politicians that there were votes in the environment. For by then, ecology had become entrenched as a way of life. The environmental movement blossomed as the issue of the decade.

Four major influences shaped American environmental policies of the 1970s:

- Strong, well organized environmental groups whose many causes, though well-intended, were characterized as calamitous.
- An ill-prepared, largely defensive industry (chemical, automotive, or whatever), the villains and victims of these causes.
- An eager news media, with an insatiable thirst for sensationalism.
- Zealous political leaders, and their appointed regulators, who rallied on the ecology bandwagon, hell-bent on saving society from inevitable doom at any cost, even to legislating non-existing technology.

The result was a long litany of frequently overlapping legislation and regulations that had a very profound impact – some good, some bad – on industry's *modus operandi.* Today I would like to highlight some of these, especially as they affected the US chlor-alkali industry.

By way of perspective, and in connection with the various operative laws and regulations about which I will talk, let me identify for you the commonalities that exist among them:

- Laws, of course, are enacted by elected politicians, motivated as they perceive the people's wishes – largely as reflected by inputs from industry, environmentalist and other special interest lobbying groups.
- Regulatory agencies interpret and implement the various statutes, nominally objectively but clearly influenced and biased by the administration in power. Few opportunities exist for pre-rulemaking peer inputs, for example on the *need* for specific regulations and the development–evaluation of the data base on which they are established.
- Regulatory rulemaking proceedings do occur in a public forum, with primary inputs from (1) affected industries, individually and collectively through trade association groups (for example, The Chlorine Institute), and (2) environmentalist groups. Some statutes even provide for expense reimbursements for participation by public action groups. In general, academic and states' participation are woefully deficient.
- Regulatory administration and enforcement is at best a shared federal-state's prerogative. States unwilling or unable for any reason to qualify for, or to accept, the administrative and enforcement burdens of any regulations, in effect, surrender those roles to the federal government.
- Those states that do accept this role have an obligation to accept federal regulations as minimum requirements. Each state has the option of adopting more stringent regulations should it choose, and several have vied aggressively in this direction. Even those states not accepting the federal programs often have parallel, and frequently more stringent, state rules. The result, particularly impacting multi-plant corporations, is duplicative and often divergent and frequently conflicting regulations.

  I must go on to say that even where states have accepted management of federal programs, the federal government continues to look over their shoulder, and particularly in enforcement proceedings, will bring action if the state fails in its enforcement duties.
- Once regulations are promulgated, essentially anyone – even a private citizen – can challenge them in the courts. The doctrine of presumptive applicability applies, and it is the burden of the opposition to prove the government of wrongdoing. In consequence, we have been characterized as a litiguous society, because rarely is any group totally satisfied.

– Criminal proceedings, involving heavy fines and/or jail sentences, can be initiated against those regulated and their employees for willful falsification of data, 'midnight dumping', non-compliance, etc. (We don't go so far as did King Edward II of England some 650 years ago (in 1308) who adopted the rather more straight forward strategy of executing the polluter.)

23.2 THE US EPA

The US Environmental Protection Agency – the EPA – has responsibility for developing, administering and enforcing most national environmental regulations. The bureaucracy – with its current nearly one billion dollar annual budget – is huge and extremely complex. EPA's basic organizational logic follows statutes – not necessarily media. Separate offices and mini-bureaucracies exist, for example, for developing effluent standards, for air emission standards, for water quality standards, etc. Each of these offices functions pretty much independently. Essentially no policy co-ordination exists among them; they tend not to communicate with one another. Now at the state and municipal levels, the bureaucratic structuring is equally inefficient. With some 64 chlorine producing plants operating in 26 separate states, you can begin to sense the magnitude of the problems.

For our industry, the first and nearly crippling crisis emerged at just about the same time that our Congress was readying statutes to satisfy the demands of the environmental movement. It was early in 1970 that a Scandinavian graduate student, studying at the University of Western Ontario, found mercury in walleye pike from Lake St. Clair and reported it to the Canadian authorities. Within weeks, five Canadian chlor-alkali producers were ordered to eliminate mercury discharges. Commercial fishing was suspended in the area, initially in Canada and shortly thereafter in the US. On July 22, the Justice Department sought injunctions against ten US producing plants alleged to be discharging as much as 15–20 kg mercury daily to the waterways. What is interesting, however, is that the authorities had to draw upon the 1899 Rivers and Harbor Act which banned non-permitted 'refuse' discharges that might be an impediment to navigation. 'Refuse' was defined as anything –soluble of insoluble – not present in the water taken from a stream. No better enabling legislation was then available. Our industry was an immediate and well defined target for legislation under development and shortly to be enacted.

(As an aside, it is equally curious, but not at all surprising, that the very first citation issued by the US authorities under the Occupational Safety and Health Act of 1970 was against a chlorine producer for alleged worker overexposure to mercury. After all, the mercury crisis was in full bloom. The mood of the country was action.)

23.3 WATER POLLUTION CONTROL

The genesis of water pollution control stems from the correlation of dirty water with contagious disease – pretty much a health department function that bears little resemblance to current policies and administrative responsibilities.

Now prior to the 1970s, federal programs to combat water pollution consisted mainly of supporting and funding individual state initiatives to adopt water quality standards. They were completely ineffectual. Culminating a long series of political hassles and debate a new approach to water pollution control became entrenched in 1972 with the adoption of the Clean Water Act. The old concept of dilution as a solution to pollution was dead. Instead, the thrust was to disallow discharge of treatable wastes without prior treatment. It minimized state authority over pollution control. The Act declared as *national* goals the achievement of fishable and swimmable waters by mid 1983 and the elimination of discharge of pollutants into any and all of the nation's waters by 1985. These were laudible goals indeed, but hardly either practical or necessary.

We should keep in mind that at this point in time, mercury discharges to the waterways by our plants had been reduced drastically, to no more than 50–150 g per day per plant (net basis). Some plants simply shut down and those that remained were in a tizzy, operating under the government's decree of zero discharge. We were desperately striving to tighten further our operations, to reduce our mercury consumption and, above all – hoping that emotion and hysteria would yield to reason and reality – simply to survive.

There was some hope in the new Clean Water Act and its statutory provisions. But what we did not then foresee was that the newly created regulatory agency, the EPA, would take over a decade to come to maturity. We did not foresee the painstaking, monumental industry effort that would be required to cope with the new federal bureaucracy. We had been accustomed to dealing directly and solely with autonomous state and local authorities.

One of the provisions of the Clean Water Act was for the prompt identification of 'toxic substances', and the adoption of national effluent standards for each of them. Naturally mercury was included in the original published list of nine. But the EPA did not publish the requisite effluent standards. Exasperated by the lethargy and lack of progress on this and other issues, an environmentalist group successfully sued the government and the ensuing consent decree provisions were later incorporated in amended water pollution legislation in 1977. The original list of nine toxics was expanded to 65 – now also including such materials as asbestos, a dozen other heavy metals, and a variety of chlorinated organics – and 21 major industry groups (later expanded to 34), and including chlor-alkali, were identified for which final substance-specific toxic effluent regulations were to be promulgated by mid 1980 and to be in place by mid 1984. That statutory goal also has not been reached, except for a very few of the toxics, and only one of the mandated industry-wide effluent standards has yet been promulgated.

Another provision of the 1972 Clean Water Act mandated the definition of 'Best Practicable Control Technology' standards for existing facilities, and it established mid 1977 as the deadline for compliance. With regard to chlor-alkali, the initial EPA proposals would have limited mercury discharges from a facility to 0.1 lb per day, regardless of plant size. Ultimately, when the standards were finally promulgated, we were persuasive in getting the Agency to regulate discharges on the basis of production capacity, and measured exit the central mercury wastewater treatment plant. The actual limit was established at 0.14 g/ton, 30-day average basis. Even so, the impact of this was significant. By the time Best Practicable Control Technology requirements became effective in mid 1977, eleven US mercury cell chlorine producing plants had shut down. Among the 26 plants that survivied, in excess of $40 million was spent for installation of necessary polution control systems, with about $1.7 million in annual operating expenses. Related to capacity, capital costs have been about $15 per annual ton and operating expenses approximate $6 per ton. Of course, with constantly evolving and moving regulatory targets, no new mercury cell plant has been built since 1970. And the industry's mercury cell commitment has diminished from a peak of 28.6% of total industry capacity in 1969 to the present level of 19.2%.

Yet another provision of the amended Clean Water Act is the industry-specific effluent standards, including toxic substance-specific effluent limitations which are to be in place by mid 1984. This second regulatory phase, identified as 'Best Available Technology Economically Achievable', is planned to be published in July 1982. If promulgated as proposed a year or so ago, restrictions applicable to both existing and new sources will be imposed for seven heavy metals in addition to mercury – arsenic, cadmium, copper, lead, nickel, silver and zinc – and restrictions for TSS, residual chlorine and pH. The limit for mercury would be 0.10 g/ton, 30-day daily average basis *exit treatment plant*, and based on the EPA's definition of best technology as sulfide precipitation–separation treatment. For comparative purposes, this is about five-times more restrictive than the values adopted in the recent EEC mercury directive. And whereas the EEC directive provides daily limit values of four-times the monthly values, ours would be set at 2.5-times the average monthly value. Although the US has not yet officially adopted any water quality *standards*, proposed quality *criteria* for mercury are of a similar order of magnitude different from the EEC's.

The mood of the present administration – applauded by industry – is that perhaps industry has already progressed far enough in technology-based controls of point source discharges. The EPA is inclined to seek modifying legislation either to abondon or at least to postpone best available technology requirements in favor of controlling toxics on the basis of site-specific receiving water quality objectives that are related to ultimate end-uses. The outcome is now uncertain. The opposition politicians perceive that the public demands that there be no retrenchment in environmental objectives, and already public action groups are organizing the votes for upcoming local and national elections. There may be some backlash if the offing.

23.4 AIR POLLUTION CONTROL

I wish at this time to address briefly the situation regarding air pollution. In the early part of the century, air pollution was synonymous with smoky skies. In industrial cities like Pittsburgh and Liverpool where problems were most onerous, the city fathers were reluctant to do much by way of control at the expense of the local economy. Even the Donora, London and New York smog episodes (between 1948–53) did little to muster public support for federal control action. As was the case for water pollution, early federal involvement was pretty much in the form of ineffectual advisories to local and, later, state authorities, with incremental encroachments on these authorities and jurisdictions becoming well entrenched by the 1970s.

The US adopted a Clean Air Act in 1970. From the chlor-alkali industry's viewpoint, the most relevant feature of that legislation was the provision for establishing national air emission standards for designated 'hazardous' air pollutants, applicable to both existing and new sources. (The Act defines a hazardous air pollutant as one, 'which in the judgement of the EPA Administrator, may cause, or contribute to, an increase in mortality or an increase in serious irreversible, or incapacitating reversible illness'. It provides also that the Administrator establish standards which, again in his judgment, provide an ample margin of safety to protect the public health.)

Considering that mercury was at the time the No. 1 crisis chemical, it is not at all surprising that it, along (only) with asbestos and beryllium, was designated as a hazardous pollutant in March 1971. (It is relevant to note that in the intervening 11 years only one other material has been so designated – vinyl chloride.) Two years later, following rulemaking proceedings with the customary public hearings, emission standards were promulgated. The EPA had concluded 'that in view of the present limit knowledge as to the effects of inhaled mercury in the general population ... it is prudent to consider exposures to methyl mercury (diet) and mercury vapor (air) to be *equivalent and additive*'. The agency, by fiat, set as an ambient threshold goal an average daily concentration not exceeding $1 \mu g/m^3$ – 100-times less than the permissible occupational exposure limit. The EPA determined that two point sources – mercury ore processing facilities and mercury chlor-alkali plants – emitted mercury in a manner that could exceed this ambient threshold. It established 2300 g per day per facility as the national mercury emission limit, derived from dispersion models based on worst case assumptions ... the Administrator's judgment of 'an ample margin of safety'.

Thus, this daily limit applies to the sum of mercury emitted from hydrogen systems, cell end box vent systems and the general cell room air, never mind that the last named source can not properly be measured. And it applies uniformly to all plants, whether production capacity is 100 or 1000 daily tons or whether emissions have any measurable impact on the environment. Again, the impact on the industry paralleled that for the effluent standards. Among the plants that survived, necessary capital expenditures were $20 million and annual operating expenses approximate $6 million. Or put another way, capital costs have been about $6.50 per annual ton and operating expenses are about $2 per ton of chlorine.

23.5 DRINKING WATER

As an industry, we would have comparatively little direct concern over, or involvement in drinking water regulations. But let me cite just as an example, one or two of the problems we have encountered that illustrate the inter-relationship of the various environmental media.

Our Safe Drinking Water Act of 1973, also administered by the EPA, requires the establishment of maximum contaminant levels in drinking water. Adopting the same old philosophy and safety factors as they did in connection with the Clean Air Act a few years before, the Agency continued its erroneous judgment that the toxicology of all forms of mercury is equivalent and additive. A MCL for total mercury (all forms) of 2 p.p.b. was adopted, to our everlasting chagrin. And so, yet another few years later a multiple of that level was used as the ultimate criterion for deciding whether or not the leachate from mercury-containing industry wastes presents an endangerment to public health or the environment.

Now the federal authorities are considering – as I believe also is the situation here – the issuance of maximum contaminant levels for a host of synthetic chlorinated organics in drinking water. Also under consideration is the regulation of water treatment additives. For example, it recently has been proposed to limit carbon tetrachloride in chlorine sold to water utilities to 11 p.p.m. And that is a tough specification to meet consistently under some existing chlorine production technologies, notwithstanding the deficiencies in state-of-the-art risk assessments or even that available analytical methodologies have a precision of no better than 50% at his concentration. Unlike the situation in Europe, most North American drinking water utilities use chlorine for disinfection, and thus the impact could be quite significant.

23.6 WASTE MANAGEMENT

The Resources Conservation and Recovery Act of 1976, one of the most complex and far-reaching pieces of environmental legislation devised – is designed to encourage wastes conservation and recycling, and to minimize land disposal practices. Prior to passage of the Act, nominal federal guidelines that had been issued were solely advisory. Implementation of the Act's goals is structured similar to the Clean Water and Air Acts, but in this instance – possibly reflecting a swing of the pendulum – individual states are *encouraged* to enforce federal standards on a permit basis and subject to federal approval. Five years after adoption of the Act the projected states' role has only partially materialized.

The act covers, with the principle exception of domestic sewage, essentially every conceivable waste material. In fact solid wastes are all encompassing, including solid, liquid, semi-solid or contained gaseous material resulting from industrial, commercial, mining, agricultural and community operations. The term embraces, for example, water and waste treatment plant sludge and sludges from an air pollution control facility. It prohibits open dumping and provides for 'womb to tomb' management of such wastes through a manifest system – from the point of generation, through storage, treatment, transport and final disposal.

The principal focus of attention is on wastes that are characterized as hazardous, because of their quantity, concentration, physical, chemical or infectious characteristics. The Act mandated as a minimum that those materials listed as 'hazardous' under the Clean Air Act, or 'toxic' under the Clean Water Act, be initially designated as hazardous materials, and therefore subject to the special requirements. In identifying hazardous wastes, the federal authorities also chose to list generically certain classes of wastes, and of course near the top of the list were mercury cell brine purification muds, mercury cell wastewater treatment sludges, and chlorinated hydrocarbon wastes from diaphragm cells that use graphite anodes. Thus these wastes were designated as hazardous regardless of the amount generated or the concentration of one or more toxic materials, for example, mercury, which they may contain.

The US chlor-alkali industry, through the Chlorine Institute, felt that this action was insupportable and, indeed, contrary to the intent of the law. And so we chose to sue the government. By now the Reagan Administration was ensconced and fortunately reason prevailed. We were able to negotiate a settlement which provided for expeditious, site-specific delisting of wastes if the generator could establish that the leachate from the waste contained no more than 25-times the 2 p.p.b. limit for mercury in drinking water, that is, no more than 50 p.p.b. of mercury. We considered this approach was consistent with a degree of hazard concept that we felt the legislators had intended. We did not challenge the diaphragm cell chlorinated hydrocarbon wastes designation because they are generated in small amounts, they are indeed hazardous and are largely becoming of academic interest as metal anode technology takes a strong hold upon the industry.

Finally, Love Canal and similar inactive or orphaned disposal sites sensationalized by the press, prompted the enactment of the Comprehensive Environmental Response, Compensation and Liability Act of 1980. The Act establishes liability – generally to a maximum of $50 million – for owners or operators of facilities or vessels which release hazardous substances into the environment. In the event that liability is established, it is transferred to and assumed by a liability fund, commonly known as 'Superfund', a \$1.6 billion cleanup program. This fund is financed by a tax of \$2.13 per dry ton weight of hazardous wastes; and, as regard our industry's specific operations, an assessment of \$2.80 per ton of chlorine, and \$0.28 per ton of caustic, production.

Time does not permit exploration of other industry impacts as a result of these more recent environmental laws. Suffice it to say they will be very significant as wastes generators begin to cope with abandonment or retrofitting of pits, ponds, lagoons and other impoundments, seek alternative, approved disposal sites that the publics do not want, etc.

23.7 CONCLUSIONS

How does the US chemical industry view the environmental regulatory scene of the future? As the environmental issue once temporarily competed with the Vietnam issue, so now it is taking a back seat to priority public concerns about an economy beset at once with high unemployment, low productivity, inflation and a declining reservoir of energy-producing resources. Most of us believe we are at a unique moment in time. We have an opportunity, indeed a responsibility, to be heard by an Administration and by a national constituency who are receptive to regulatory reform. If, during this brief window in time, we are not successful in influencing sound, necessary legislation and regulation based on good science, there might well develop a backlash resulting in even more restrictive and irrational regulations.

The key to our successes is aggressive risk management, involving acceptance of 'cradle to grave' stewardship for materials we manufacture. I believe there is mounting evidence of a growing and maturing corporate philosophy of *anticipating*, and then embracing societal needs. It is impossible to have a no risk society. The realistic objective is to eliminate unacceptable risks, based on competent scientific, economic and social evaluation. This mandates a co-operative dialogue with the regulators, as contracted to the adversarial relationship of the past, and a public understanding of the nature and benefits of the chemical industry.

This, then is what I believe to be the challenge before us. In the face of continuing, albeit less emotional public pressure for a 'clean environment', inaction on our part would surely be perceived as irresponsibility. What we accomplish, or fail to accomplish, will contribute to the determination of whether or not the chemical industry will continue to flourish, both domestically and internationally, for the balance of the century. The environment probably will not continue to dominate the National agenda, but it most assuredly will be a part of it.

CHAPTER 24

Chlorine Handling and Safety in the European Situation

R. PAPP, PCUK (Products Chimiques Ugine Kuhlmann), [Chairman of the Working Group 'Storage Transport and Safety (GEST) of the Bureau International Technique Chlore (BITC)]

24.1 INTRODUCTION

Which method should be used for improving safety? During the last decade, safety in industry has been considered in Europe as a major concern. This consideration is mainly due to the increase in plant size and the bigger impact which may result from an accident. The nuclear industry has done a considerable amount of work in the analysis of risk and risk assessment. This technique has also been developed and adapted for the Chemical Industry using check-lists, logic diagrams, analysis of piping and instrumentation diagrams, fault tree analysis, etc. . . .

This method is appreciated for its logic and detailed approach. However, it must be pointed out that the result of such an approach is largely dependent on the reliability of the input information. Unknown factors can not be taken in account. For these reasons, the nuclear industry is now making a large effort in the exchange of experience through the world.

Summating the experience of all the European chlorine producers over a great number of years was one of the major intentions of the Safety Group of the BITC created in 1972, and by so doing, the BITC has made a significant contribution to improving safety. Safety in the manufacturing and handling of highly corrosive and toxic products such as chlorine, depends on several factors:

- Design intent for equipment and plants closely adapted to the product, which means the use of designs and specifications well adapted and very precise.
- Techniques for equipment manufacture in workshops and plant construction which closely adhere to the above specifications and requirements, and are subject to strict controls by inspection.
- Operation of plant and equipment in strict accordance with the specifications, which is achieved by appropriate training, by precise definition

of operating conditions, and by a high safety consciousness among personnel.

A good safety base if therefore dependent on a combination of technical and technological conditions as well as being affected by the human element.

It is obvious that a proper definition of equipment, plants and operating manuals is easier to obtain when the experience of past incidents is widely known and can be carefully analysed. Chlorine manufacturers, for many years, have promoted such exchange of experience. For many years the BITC has been kept informed of incidents and accidents involving chlorine. Each meeting of the Groupe d'Etudes Stockage-Transport-Sécurité (GEST) includes a discussion of chlorine incidents – even with no chlorine escape – which have been reported since the last meeting. GEST Recommendations, therefore, are able to take these incidents into account.

The study of these 'dangerous situations' shows that even the smallest detail is not without importance as far as chlorine is concerned, and that safe handling of chlorine needs great care and expertise, and specially in a number of small details concerning ancillary equipment.

The following examples will illustrate the above. Table 24.1 shows the analysis of accidents or incidents reports received by GEST between 1974 and 1980. Not all these accidents or incidents involved chlorine leakage. Only one had fatal consequences: an accident in Malaga (Spain) where a one-ton chlorine drum was damaged during a shipping operation. The leak was caused by the rupture of an external non-protected vent, welded onto the drum in order to facilitate venting after water testing. It should be noted that this non-protected vent does not appear on the BITC drawings and that this addition has probably been made in order to facilitate the drying of the drum and even perhaps to increase safety!

Table 24.1

Category	Percentage
1 Leakages on pipes or equipment	16
2 Loading and unloading	14
3 Maintenance operations	17
4 Cell maintenance	4
5 Over-pressure on cells	5
6 Overloading on chlorine absorption systems	14
7 Chlorine storage	9
8 Drums and cylinders	12
9 Transport in bulk	9
	100

(1) The first comment to make on Table 24.1 is about the relatively wide range of causes. However, categories 1 and 3 concerning 'maintenance problems' represents 33% of the accidents. Many human errors during maintenance operations were reported.

(2) The second point to note concerns category 2 'loading and unloading' which represents 14% of the accidents. It is obvious that the very large tonnage involved in bulk transportation by road and rail, and even by ship, represent a great number of loading and unloading operations.

The batch characteristics of these operations and the degree of flexibility required to compensate for lack of manoeuvrability between fixed installations and mobile parts, are areas of risk. The BITC has made a great effort to reduce these risks:

- by producing a European Norm for manhole design and valve position (Norm UN 14–CEFIC)
- by studying and developing a tank valve with internal sealing which is extensively used inside Europe
- by editing several recommendations in the following areas:
  tank valves
  design and operation of loading and unloading units
  flexible hoses and articulated connections

During this period, some accidents have been prevented by the remote valves recommended by the BITC on the fixed installation and on the tank wagons. Other accidents were the result of displacement of the tank wagon during loading, movement of a tank with the dome protection in the open position, etc. . . .

(3) Overloading of the chlorine absorption systems, a factor representing 14% of the accidents, is also an important category. Amongst the incidents reported, maloperation and degassing valves remaining open during unloading, so allowing liquid chlorine into the absorption unit, were the major causes of chlorine escapes.

BITC Recommendations give the safety criteria for such operations. Of course, the BITC Recommendations do not shut off any degassing valves, but in these Recommendations, liquid chlorine in the venting system is trapped into a small tank with a level or low temperature alarm. This device should prevent more chlorine going into the scrubbers (Fig. 24.1).

(4) It should be noted that no significant leak was reported on fixed or mobile chlorine storage tanks in this period. The accidents that were reported were only minor: leaking gaskets or valves, packings, etc. . . .

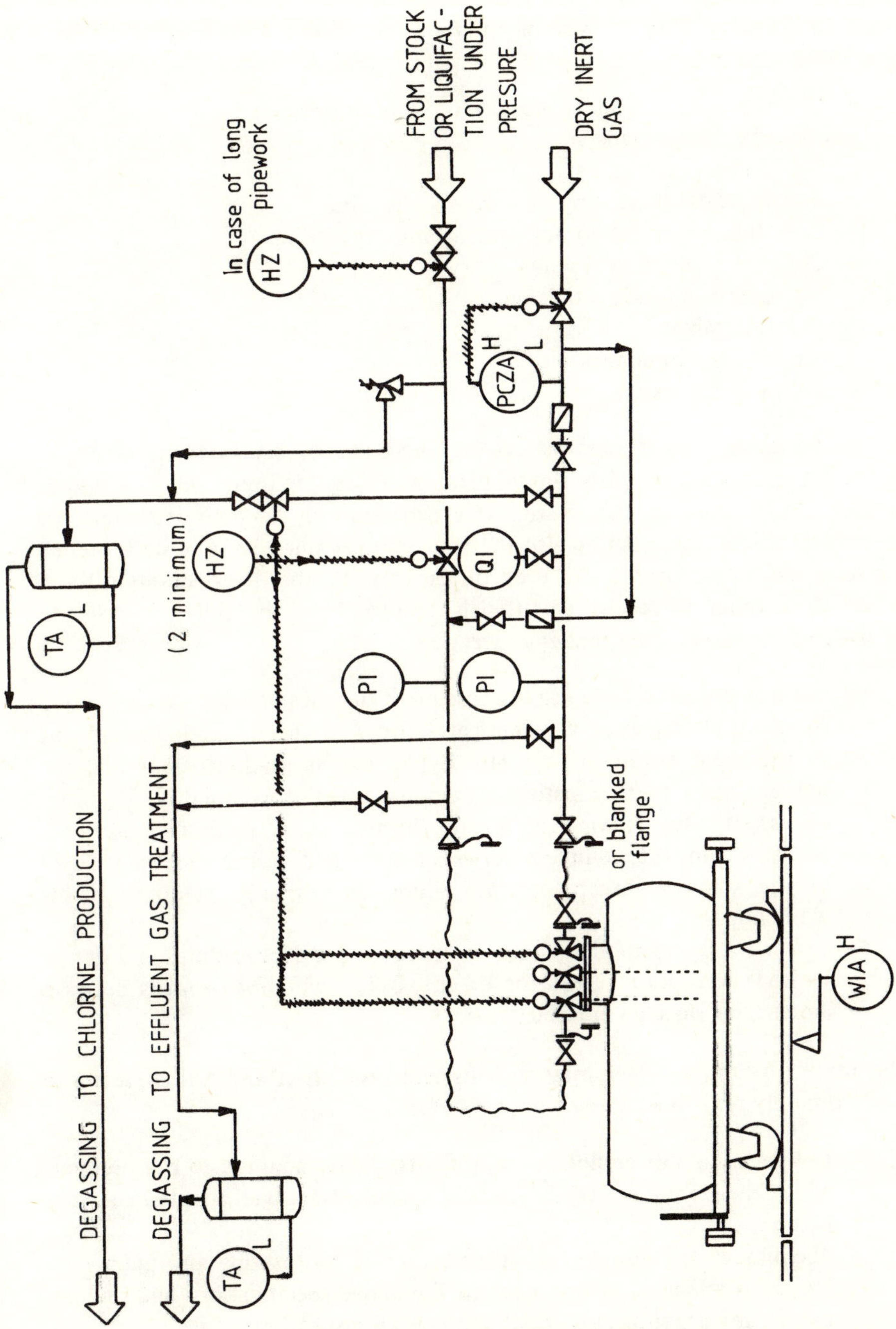

Fig. 24.1 – Scheme for the filling of rail tanks by gas pressure.

24.2 THE WORK OF BITC

Since 1972, more than 55 recommendations have been issued by GEST in a great number of areas:

- Chlorine storages, under pressure or without pressure.
- Liquid chlorine transfer.
- Chlorine vaporizers.
- Transport of chlorine by long distance pipeline.
- Bulk chlorine transport and loading and off-loading.
- Chlorine absorption systems.
- Nitrogen trichloride problems.
- Chlorine valves.
- Chlorine isocontainers.
- Chlorine dispersion.

On the particular area of chlorine valves, GEST issued very extensive codes, to define the criteria which valves should meet, when used in liquid chlorine, insofar as their design, construction, testing and inspection on dispatch or receipt, are concerned. These codes define also the inspection requirements of each model, the tests and proof tests to be used on prototypes, the documentation to be provided, in order to receive an official approval from the BITC. In order to receive such approval, the supplier undertakes:

- to meet the BITC code for all subsequent supplies of valve,
- to make all valves in an identical manner to the methods carried out in the manufacture of the prototype, insofar as dimensions and the nomenclature of the constituent components are concerned,
- to respect, when making any repairs, the criteria and methods prescribed,
- to carry out the complete series of tests and inspections laid down in the standard acceptance procedure, as normally practiced in his workshop,
- to carry out no modifications to these standard procedures and design without the prior agreement of the BITC, who may on his decision alone agree on such modifications.

The above procedure, which may be considered too strict and formal, seems to GEST the only way to achieve as we said before:

- Design intent for equipment and plants closely adapted to the product, which means the use of designs and specifications well adapted and very precise.
- Techniques for equipment manufacture in workshops and plant construction which closely adhere to the above specifications and requirements, and are subject to strict controls by inspection.

GEST has also done a great amount of work on the area of chlorine gas dispersion prediction.

In consideration of the lack of basic knowledge of the subject, the BITC discussed and prepared a programme with the Von Karman Institute (Brussels) and agreed to support technically and financially such a programme.

Fig. 24.2 – Factories providing emergency services in the event of a liquid chlorine accident.

The study into gas dispersion and water sprays was made in a series of wind tunnels of various sizes, which have indicated the effects to be expected from obstacles of various sizes on the dispersion process. Furthermore, a computer programme has been developed using a three-dimensional mathematical model of the dispersion at middle distance of a pollutant of any density. It takes into account the effects of buildings, water jets, and other obstacles in the dispersion process. The computer programme is presently operating satisfactorily in various large chlorine producer companies in Europe. Specific publication shall be made on this subject in the coming weeks.

I would also mention that in 1973, the BITC has created an International Emergency Organization to deal with transport accidents, giving precise instructions as to equipment to be carried by special emergency vehicles, and the location of such vehicles in the facilities of all the European producers. Obviously this organisation involves also the training of a special team of personnel able to provide assistance in case of spillage.

Figure 24.2 shows the organization map.

24.3 LEGISLATION – EUROPEAN DIRECTIVES

The Executive Authorities of the EEC adopted in December 1981 a Directive COM (79) 384 Final which relates to Major Accidents Hazards. This directive is also known as Directive 'Post Seveso'. This directive does not apply only to the chemical industry. But we should say that our industry is one of the most concerned. The directive requires member states to adapt their own legislation within 18 months.

What is, briefly summarized, the content of this directive? The industry shall provide information to the authorities in the following areas:

– information on the products and processes
– information on the key aspects of reliability and safety, sources of hazards and how they would occur
– information on risk assessment, analysis of risk, description of the preventive measures.
– information on emergency plans for inside and outside the plant
– information on qualified management and emergency team, able to operate the emergency plans in case of hazard.

This first safety report shall be made within a certain period of time for the plants which are recognized to be a possible source of major hazard. Thereafter, the plant management shall notify all changes which should affect safety and shall revise the report accordingly.

For chlorine, the directive shall apply in full where either more than 200 tons of chlorine is stored in isolated storage, or where there is more than 50 tons in the installation. Below this figure, the plant management shall have to prove to

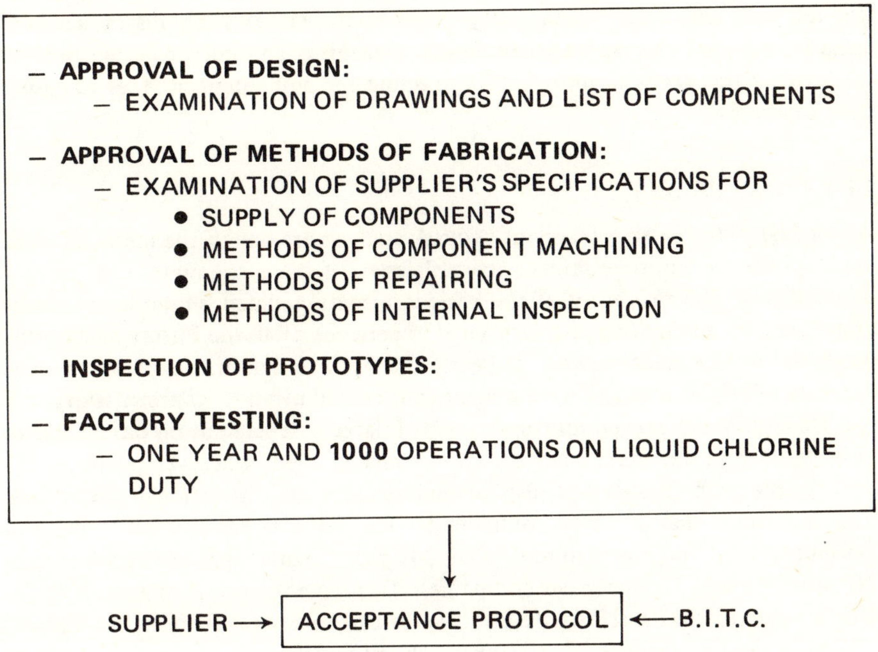

Fig. 24.3 – Inspection procedure for new valves.

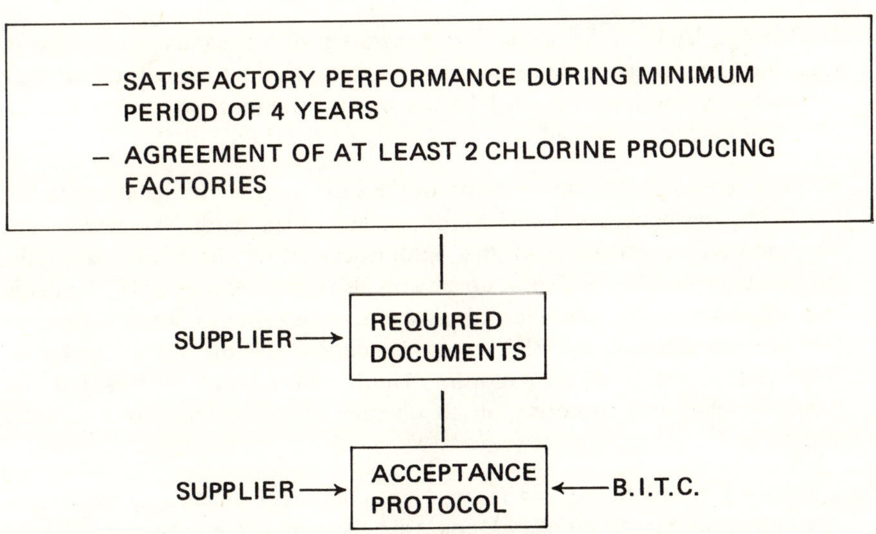

Fig. 24.4 – Direct standardisation inspection procedure.

the relevant authorities 'That they have identified existing major accident hazards, adopted the appropriate safety measures, and provided the persons working on the site with information, training and equipment in order to ensure their safety'.

24.4 CONCLUSIONS

As we said before, risk assessment and risk analysis are one of the methods to be used in order to improve safety. The efficiency of the above method is strongly dependent on the reliability of the basic information which is used in the study. The BITC, by summating the very large experience of all the European chlorine producers over a great number of years, and by including this experience in its recommendations, mean to have a significant contribution to improve safety.

Obviously the prevention of accidents is largely dependent on our efforts to have the GEST recommendations known and applied, wherever possible, not only in the plant design but also to maintenance and operational procedures. This important task is never completed. GEST held its last general meeting in December 1981 in Paris with more than 140 participants representing more than 10 countries and 50 European companies. This large interest is of great help for GEST to continue its efforts. (GEST Papers can be obtained form BIT Chlore, 250 Avenue Louise, Boite 72, 1050 Brussels, Belgium.)

24.5 DISCUSSION

1. Mr Thampi, Uhde, India, asked for clarification of the mandatory legislation regarding the storage of chlorine and its transport by road and rail and whether authority approval was needed before storage/transport systems could be commissioned and operated.

   Mr Papp replied that in each country of the EEC there is a national regulation which requires compliance with all documents of standards, etc., which have been previously provided to the authorities before the plant was built. Additionally, he felt that if these documents also complied with BITC standards then approval by the relevant authority would be easier to obtain. There are two separate things – the BITC recommendations give the best of the knowledge and safety from the chlorine manufacturers point of view but the commissioning and authorisation to operate a plant relates to the initial authorities.

2. Continuing the discussion, Mr Thampi remarked that he had heard there were conflicting rules in different EEC countries concerning the transport of liquid gases, for example, ammonia. Was the same true for chlorine?

   The problem of transport by road is strongly discussed from time to time.

In Europe some countries do not allow for that transportation, for example Germany, whilst some do, for example France, UK, Italy and Sweden. At the last meeting of the organization it was decided to allow chlorine transportation by road but that decision was obtained after long discussion by a majority of only 1 vote from the 12 or so participants.

Contributing to the discussion from the Chair, Dr Churchman emphasized the role of training in the safe transport of any dangerous chemical which was as important as any regulation.

Index

A

Alkali demand, 360
Analysers, automatic hardness, 194
Anode coatings, 122, 151, 215
  surface, 113
Anodes, dimensionally stable, chlorine evolution on, 271
  graphite, 34, 215, 222
  metal, 34, 357
    coatings, 215
Asahi Chemical's electrolyzer, 104
Asbestos, 21, 34, 38, 166, 233, 372
AZEC system, 61, 155

B

Biocides, chlorine-based, 325
Bipolar cell of metal construction, 97
Brine, chlorine content, 195
  neutralization, 101
  processing, 191
  purity, 180
  quality, 128
  secondary treatment, 194
Bureau International Technique Chlore, 376

C

Capital cost reduction, 154
Castner-Kellner process, 320, 348
Cathode, activated, 115, 166
  coated, 126, 166
  coating, 151
    nickel, 68
  surface, 115
Cathodes, low, overvoltage, 233
  nickel, 126
  oxygen, 168
Caustic soda – see Sodium hydroxide
Cell voltage, 39, 64, 130, 160, 164
Chlorinated hydrocarbons, 321, 328, 360
  insecticides, 321, 357, 367
Chlorine, 182
  compounds, toxicology, 321, 328
  consumption, 328
  drying, 199
  evolution and electrocatalysis, 246
    at high current density, 258
    on metal oxide electrodes, 271
  gas dispersion, 381
  growth, negative effects on, 353
  handling and safety, 376
  industry, Australia, 335
    Canada, 335
    future, 317
    history, 340
    Japan, 330
    Latin America, 332
    Middle East, 333
    Southeast Asia, 334
    Western Europe, 330, 340
    US, 330, 335
  leakage incidents, 377
  processing, 197
  production, 355
    new, distribution, 336
    cost, 34
  purity, 81
Chrysotile asbestos, 24
Computer simulation of chlorate process, 305
Computerization of electrochemical methods, 246
Conversion economics, 206
  of chloralkali plants to membrane cell operation, 190
Conversions, partial, 204
Cooling capacity, 198
Cost reduction, 337
Current density, 58, 67, 77, 99, 104, 228
  efficiency, 30, 47, 54, 81, 83, 160

D

Diaphragm cell, 21
        anode coatings for, 222
        plants, 190
        zero gap, 166
    characterizing parameters, 21, 24
    electrolyzer, PPG bipolar, 33
    Gore-Tex, 38
    thickness, 22
Diaphragms, asbestos, 21, 38
    microporous PTFE, 38
    synthetic, 38
Dechlorination, secondary, 195
    vacuum, 197

E

Economy, scope for, 321
Electric resistance, 83
Electricity, cost, 159
Electrocatalysis and chlorine evolution, 246
Electrocatalyst (BP), morphology, 242
    performance, 234
Electrochemical methods, computerization, 246
    ruthenium dioxide electrode, 274
Electrode area, 34
    catalytic, 170
    gap, 34, 134
        reduction, 44, 151
    kinetics, rate processes, 247
        of chloride oxidation, 251
    membrane gap, 64
    separation, 230
Electrodes, 113
    metal oxide, chlorine evolution on, 271
    oxygen reduction, 168
Electrolysis of potassium chloride, 74
Electrolyzer, Asahi Chemical, 104
    AZEC, 70
    design and operating characteristics (PPG), 36
    Glanor, 33
    PPG bipolar diaphragm, 33
Energy consumption, 77, 98
    costs, 328, 356
        and plant conversion, 208
    requirements, 142, 155, 160
    saving, 33, 132, 159
        in chlorate production, 289
Environmental Protection Agency (US), 369
Ethylene cost, 332
Exports of chlorinated hydrocarbons, 360

F

Fluorocarbons and the ozone layer, 329, 357
Fluoropolymers, 40, 76
Flemion membranes, 62, 139, 151
Fuel cell use in chlorate production, 289

G

Gas bubble erosion, 39
Gore-Tex diaphragm cell performance, 44
    manufacture, 40
    properties, 42

H

Hydrocarbons, chlorinated, 321, 328, 360
Hydrochloric acid, 101
    injection, 81
Hydrogen, 182
    evolution electrocatalyst, 233
        overvoltage, 233
    gas adhesion on membrane surface, 89
    use in fuel cell, 289
Hydrophilicity of membrane surface, 151
Hydroxides, alkaline earth, 295
Hypochlorite, effect on coated electrodes, 235

I

Insecticides, chlorinated, 321, 357, 367
Ion-exchange membrane process, 97
    polymers, transport properties, 123
Iridium anode coating, 217

L

LCD chlorate cell, 305
Leblanc process, 320, 341
Legislation, European, 382
    US, 369
Licensing, 184

M

MacMullin number, 21
    experimental determination, 22
Maintenance, 183
    chemicals, 325
Materials of construction, 183
Membrane cell, 178
        energy requirements, 134
        operation, conversion to, 190
        selection criteria, 177
        technology, 357
            Olin, 121
            Uhde, 132
    cells, anode coatings for, 228
    dimensional stability, 100
    durability, 56
    ion-exchange, 97
    life, 56
    multilayer, 82
    perfluorosulfonic acid, 88
    permeability, 99
    process characteristics, 99
    surface hydrophilicity, 151
    voltage, 166

Membranes, Flemion, 62, 139, 151
high performance, 53
ion-exchange, 61
Nafion, 53, 133, 139, 151
perfluorinated, 53
perfluorocarboxylic acid, 76
Mercury cell, early development, 348
discharges, 369
in brine, 194
plants, 190
Metal cell construction, 139
Metal anode chlorate cell, 305
oxide electrodes, chlorine evolution on, 271
MO electrode, 283

N

Nickel cathode coating, 68
molybdenum alloy electrode coating, 233
Nafion membranes, 53, 133, 139, 151

O

Olin membrane cell technology, 121
Overcapacity, 329
Overpotential, 164
oxygen, 227
Oxides as anode coatings, 217
Oxygen cathodes, 168
evolution kinetics, 224
overpotential, 227

P

Paper industry, 324
Passivation, 216
PCBs, 321
Perfluoroether monomers, 88
Perfluoropolymers, 123
carboxylated, 61
Perfluorosulfonic acid, 76
membrane, 88
Perfluorovinyl ether, 77
carboxylated, 64
Phosphoric acid electrolyte fuel cell, 290
Plastics industry, 323
Platinum anode coating, 216, 222
electrodes, chlorine formation on, 253
Pollution control, air, 372
water, 370
Pore size, effective, 44
Porosimetry, 44
Potassium chloride electrolysis, 74
Power consumption, 34, 37, 55, 92, 105, 181, 350
Pressurized operation of membrane process, 103
PTFE diaphragm, 38
PVC, 320, 349, 357

R

Raney nickel, 68
Recycling, 323
Regulatory constraints, 177, 329, 367
Resins, chelating, 194
Resistance ratio, 21
Resources Conservation and Recovery Act (US), 373
Ruthenium dioxide, 222
anode coating, 217
electrode, chemical, 276
electrochemical, 274
titanium oxide electrodes, 271, 278
chlorine formation on, 253

S

Safe Drinking Water Act (US), 373
Salt, 180
Silica content of brine, 192
Soda ash, US, 361
Sodium chlorate, LCD cell for, 305
plants, energy requirements (steam), 293
production, 289
Sodium hypochlorite, 293
Sodium hydroxide, 182, 341
energy requirement, 160
evaporation, 202
optimum concentration, 103
processing, 201
production, 166
recycle, 201
Sodium ion transport, 123
Sodium sulphate removal, 192
Solvay soda ash process, 347
Steel industry, 322
Sulphate content of brine, 192

T

Tafel slopes, 240, 271
Titanium anode coating, 215
dioxide anode coating, 217
Tetrafluoroethylene, 77
Tortuosity, 44
Toxicology of chlorine compounds, 321, 328 328
of mercury, 373
Trichloroethylene, 329

U

Uhde membrane cell technology, 132

V

Vinyl chloride monomer, 328, 332, 349, 357, 372
Voltage drop across membrane, 166
Volmer-Heyrovsky mechanism, 272
Tafel mechanism, 272